Lecture Notes in Mathematics

1962

Editors:
J.-M. Morel, Cachan
F. Takens, Groningen
B. Teissier, Paris

Robert Dalang · Davar Khoshnevisan · Carl Mueller
David Nualart · Yimin Xiao

A Minicourse on Stochastic Partial Differential Equations

Editors:

Davar Khoshnevisan
Firas Rassoul-Agha

 Springer

Authors and Editors

Robert C. Dalang
Institut de Mathématiques
Ecole Polytechnique Fédérale de Lausanne
Station 8
1015 Lausanne
Switzerland
robert.dalang@epfl.ch

Carl Mueller
Department of Mathematics
University of Rochester
Rochester, NY 14627
USA
cmlr@math.rochester.edu

Firas Rassoul-Agha
Department of Mathematics
University of Utah
Salt Lake City, UT 84112-0090
USA
firas@math.utah.edu

Davar Khoshnevisan
Department of Mathematics
University of Utah
Salt Lake City, UT 84112-0090
USA
davar@math.utah.edu

David Nualart
Department of Mathematics
University of Kansas
Lawrence, KS 66045
USA
nualart@math.ku.edu

Yimin Xiao
Department of Statistics and Probability
Michigan State University
A-413 Wells Hall
East Lansing, MI 48824
USA
xiao@stt.msu.edu

ISBN 978-3-540-85993-2   ISBN 978-3-540-85994-9 (eBook)

DOI 10.1007/978-3-540-85994-9

Lecture Notes in Mathematics ISSN print edition: 0075-8434
ISSN electronic edition: 1617-9692

Library of Congress Control Number: 2008934459

Mathematics Subject Classification (2000): 60H15, 60H20, 60H25

Cover design: SPi Publishing Services

Printed on acid-free paper

9 8 7 6 5 4 3 2 1

springer.com

Preface

From May 8 to May 19th of 2006, the Department of Mathematics at the University of Utah hosted a minicourse on some modern topics in stochastic partial differential equations [SPDEs]. The participants included graduate students and recent PhDs from across North America, as well as research mathematicians at diverse stages of their careers. Intensive courses were given by Robert C. Dalang, Davar Khoshnevisan, An Le, Carl Mueller, David Nualart, Boris Rozovsky, and Yimin Xiao. The present book is comprised of most of those lectures.

For nearly three decades, the topic of SPDEs has been an area of active research in pure and applied mathematics, fluid mechanics, geophysics, and theoretical physics. The theory of SPDEs has a similar flavor as PDEs and interacting particle systems in the sense that most of the interesting developments generally evolve in two directions: There is the general theory; and then there are specific problem-areas that arise from concrete questions in applied science. As such, it is unlikely that there ever will be a cohesive all-encompassing theory of stochastic partial differential equations. With that in mind, the present volume follows the style of the Utah minicourse in SPDEs and attempts to present a selection of interesting themes within this interesting area. The presentation, as well as the choice of the topics, were motivated primarily by our desire to bring together a combination of methods and deep ideas from SPDEs (Chapters 1, 2, and 4) and Gaussian analysis (Chapters 3 and 5), as well as potential theory and geometric measure theory (Chapter 5). Ours is a quite novel viewpoint, and we believe that the interface of the mentioned theories is fertile ground that shows excellent potential for continued future research.

We are aware of at least four books on SPDEs that have appeared since we began to collect the material for this project [4; 8; 12; 14]. Although there is little overlap between those books and the present volume, the rapidly-growing number of books on different aspects of SPDEs represents continued, as well as a growing, interest in both the theory as well as the applications of the subject. The reader is encouraged to consult the references for examples in: (i) Random media [2; 4; 18] and filtering theory [15]; (ii) applications in fluid dynamics and turbulence [1; 2; 17]; and (iii) in statistical physics of disordered media [2; 6; 7; 10]. Further references are scattered throughout the lectures that follow. The reader is invited to consult the references to this preface, together with their voluminous bibliographies, for some of the other viewpoints on this exciting topic.

The Utah Minicourse on SPDEs was funded by a generous VIGRE grant by the National Science Foundation, to whom we are grateful. We thank also the lecturers and participants of the minicourse for their efforts. Finally, we extend our wholehearted thanks to the anonymous referee; their careful reading and thoughtful remarks have led to a more effective book.

Salt Lake City, Utah   Davar Khoshnevisan
July 1, 2008   Firas Rassoul-Agha

References

[1] A. Bourlioux, A. J. Majda, and O. Volkov (2006). Conditional statistics for a passive scalar with a mean gradient and intermittency. *Phys. Fluids* **18***(10)*, 104102, 10 pp.

[2] René A. Carmona and S. A. Molchanov (1994). Parabolic Anderson problem and intermittency, *Mem. Amer. Math. Soc.* **108***(518)*

[3] *Stochastic Partial Differential Equations: Six Perspectives* (1999). Edited by Rene A. Carmona and Boris Rozovskii, American Math. Society, Providence, Rhode Island

[4] Pao-Liu Chow (2007). *Stochastic Partial Differential Equations.* Chapman & Hall/CRC, Boca Raton, Florida

[5] Gopinath Kallianpur and Jie Xiong (1995). *Stochastic Differential Equations in Infinite Dimensional Spaces.* Institute of Math. Statist. Lecture Notes— Monograph Series, Hayward, California

[6] Mehran Kardar (1987). Replica Bethe ansatz studies of two-dimensional interfaces with quenched random impurities. *Nuclear Phys.* B290 [FS20], 582–602

[7] Mehran Kardar, Giorgio Parisi, and Yi-Cheng Zhang (1986). Dynamic scaling of growing interfaces, *Phys. Rev. Lett.* **56***(9)*, 889–892

[8] Peter Kotelenez (2008). *Stochastic Ordinary and Stochastic Partial Differential Equations*, Springer, New York

[9] Nicolai V. Krylov (2006). On the foundations of the L_p-theory of stochastic partial differential equations. In: *Stochastic Partial Differential Equations and Applications—VII*, 179–191, Lecture Notes Pure Appl. Math., Chapman & Hall/CRC, Boca Raton, Florida

[10] Ernesto Medina, Terrence Hwa, and Mehran Kardar (1989). Burgers equation with correlated noise: Renormalization-group analysis and applications to directed polymers and interface growth. *Phys. Rev. A*, **38***(6)*, 3053–3075

[11] Pierre-Louis Lions and Panagiotis Souganidis (2000). Fully nonlinear stochastic pde with semilinear stochastic dependence. *C.R. Acad. Sci. Paris Sér. I Math.* **331***(8)*, 617–624

[12] S. Peszat and J. Zabczyk (2007). *Stochastic Partial Differential Equations with Lévy Noise*. Encyclopedia of Mathematics and Its Applications, Cambridge University Press, Cambridge

[13] Guisseppe Da Prato and Jerzy Zabczyk (1992). *Stochastic Equations in Infinite Dimensions*, Encyclopedia of Mathematics and Its Applications, Cambridge University Press, Cambridge

[14] Claudia Prévôt and Michael Röckner (2007). *A Concise Course on Stochastic Partial Differential Equations*. Lecture Notes in Mathematics **1905**, Springer-Verlag, Berlin, Heidelberg

[15] B. L. Rozovskiĭ (1990). *Stochastic Evolution Systems*. Kluwer Academic Publishing Group, Dordrecht (Translated from the original Russian by A. Yarkho, *Math. and Its Applicatons* (Soviet Series), 35)

[16] J. B. Walsh (1986). *An Introduction to Stochastic Partial Differential Equations*. In: Ecole d'Ete de Probabilites de Saint Flour XIV, Lecture Notes in Mathematics **1180**, 265–438

[17] Wojbor A. Woyczyński (1998). *Burgers–KPZ Turbulence*, Lecture Notes in Mathematics **1700**, Springer, Berlin

[18] Ya. B. Zeldovich, S. A. Molchanov, A. S. Ruzmaĭkin, and D. D. Solokov (1987). Intermittency in random media, *Uspekhi Fiz. Nauk* **152***(1)*, 3–32 (in Russian) [English translation in *Soviet Phys. Uspekhi* **30***(5)*, 353–369, 1987]

Contents

Some Tools and Results for Parabolic Stochastic Partial Differential Equations

Sample Path Properties of Anisotropic Gaussian Random Fields

Contributors

Davar Khoshnevisan
Department of Mathematics
The University of Utah
Salt Lake City
UT 84112–0090, USA
davar@math.utah.edu
http://www.math.utah.edu/~davar

Robert C. Dalang
Institut de Mathématiques
Ecole Polytechnique Fédérale de
Lausanne
Station 8, CH-1015
Lausanne, Switzerland
robert.dalang@epfl.ch
http://ima.epfl.ch/~rdalang/

David Nualart
Department of Mathematics
University of Kansas
Lawrence, KS 66045, USA
nualart@math.ku.edu
http://www.math.ku.edu/
~nualart/

Carl Mueller
Department of Mathematics
University of Rochester, Rochester
NY 14627, USA
cmlr@math.rochester.edu
http://www.math.rochester.edu/
~cmlr

Yimin Xiao
Department of Statistics and
Probability
A-413 Wells Hall
Michigan State University
East Lansing
MI 48824, USA
xiao@stt.msu.edu
http://www.stt.msu.edu/
~xiaoyimi

A Primer on Stochastic Partial Differential Equations

Davar Khoshnevisan

Summary. These notes form a brief introductory tutorial to elements of Gaussian noise analysis and basic stochastic partial differential equations (SPDEs) in general, and the stochastic heat equation, in particular. The chief aim here is to get to the heart of the matter quickly. We achieve this by studying a few concrete equations only. This chapter provides sufficient preparation for learning more advanced theory from the remainder of this volume.

1 What is an SPDE?

Let us consider a perfectly even, infinitesimally-thin wire of length L. We lay it down flat, so that we can identify the wire with the interval $[0, L]$. Now we apply pressure to the wire in order to make it vibrate.

Let $F(t, x)$ denote the amount of pressure per unit length applied in the direction of the y-axis at place $x \in [0, L]$: $F < 0$ means we are pressing down toward $y = -\infty$; and $F > 0$ means the opposite is true. Classical physics tells us that the position $u(t, x)$ of the wire solves the partial differential equation,

$$\frac{\partial^2 u(t, x)}{\partial t^2} = \kappa \frac{\partial^2 u(t, x)}{\partial x^2} + F(t, x) \qquad (t \geq 0, \ 0 \leq x \leq L), \qquad (1)$$

where κ is a physical constant that depends only on the linear mass density and the tension of the wire.

Equation (1) is the so-called *one-dimensional wave equation*. Its solution— via separation of variables and superposition—is a central part of the classical theory of partial differential equations.

D. Khoshnevisan and F. Rassoul-Agha (eds.) *A Minicourse on Stochastic Partial Differential Equations.*
Lecture Notes in Mathematics 1962.
© Springer-Verlag Berlin Heidelberg 2009

We are interested in addressing the question, "*What if F is random noise*"? There is an amusing interpretation, due to Walsh [30], of (1) for random noise F: If a guitar string is bombarded by particles of sand, then the induced vibrations of the string are determined by a suitable version of (1).

It turns out that in most cases of interest to us, when F is random noise, Equation (1) does not have a classical meaning. But it can be interpreted as an infinite-dimensional integral equation. These notes are a way to get you started thinking in this direction. They are based mostly on the Saint-Flour lecture notes of Walsh from 1986 [30, Chapters 1–3]. Walsh's lecture notes remain as one of the exciting introductions to this subject to date.

2 Gaussian Random Vectors

Let $g := (g_1, \ldots, g_n)$ be an n-dimensional random vector. We say that the distribution of g is *Gaussian* if $t \cdot g := \sum_{j=1}^{n} t_j g_j$ is a Gaussian random variable for all $t := (t_1, \ldots, t_n) \in \mathbf{R}^n$. It turns out that g is Gaussian if and only if there exist $\mu \in \mathbf{R}^n$ and an $n \times n$, symmetric nonnegative-definite matrix C such that

$$\mathrm{E}\left[\exp\left(it \cdot g\right)\right] = \exp\left(it \cdot \mu - \frac{1}{2}t \cdot Ct\right). \tag{2}$$

Exercise 2.1. Prove this assertion. It might help to recall that C is *nonnegative definite* if and only if $t \cdot Ct \geq 0$ for all $t \in \mathbf{R}^n$. That is, all eigenvalues of C are nonnegative.

3 Gaussian Processes

Let T be a set, and $G = \{G(t)\}_{t \in T}$ a collection of random variables indexed by T. We might refer to G as either a *random field*, or a [*stochastic*] *process indexed by T*.

We say that G is a *Gaussian process*, or a *Gaussian random field*, if $(G(t_1), \ldots, G(t_k))$ is a k-dimensional Gaussian random vector for every $t_1, \ldots, t_k \in T$. The *finite-dimensional distributions* of the process G are the collection of all probabilities obtained as follows:

$$\mu_{t_1, \ldots, t_k}(A_1, \ldots, A_k) := \mathrm{P}\left\{G(t_1) \in A_1, \ldots, G(t_k) \in A_k\right\}, \tag{3}$$

as $A_1, \ldots, A_k$ range over Borel subsets of $\mathbf{R}$ and k ranges over all positive integers. In principle, these are the only pieces of information that one has about the random process G. All properties of G are supposed to follow from properties of these distributions.

The consistency theorem of Kolmogorov [19] implies that the finite-dimensional distributions of G are uniquely determined by two functions:

1. The *mean function* $\mu(t) := \mathrm{E}[G(t)]$; and
2. the covariance function

$$C(s,t) := \mathrm{Cov}(G(s), G(t)).$$

Of course, μ is a real-valued function on T, whereas C is a real-valued function on $T \times T$.

Exercise 3.1. Prove that if G is a Gaussian process with mean function μ and covariance function C then $\{G(t) - \mu(t)\}_{t \in T}$ is a Gaussian process with mean function zero and covariance function C.

Exercise 3.2. Prove that C is *nonnegative definite*. That is, prove that for all $t_1, \ldots, t_k \in T$ and all $z_1, \ldots, z_k \in \mathbf{C}$,

$$\sum_{j=1}^{k} \sum_{l=1}^{k} C(t_j, t_l) z_j \overline{z_l} \geq 0. \tag{4}$$

Exercise 3.3. Prove that whenever $C : T \times T \to \mathbf{R}$ is nonnegative definite and symmetric,

$$|C(s,t)|^2 \leq C(s,s) \cdot C(t,t) \qquad \text{for all } s, t \in T. \tag{5}$$

This is the *Cauchy–Schwarz inequality*. In particular, $C(t,t) \geq 0$ for all $t \in T$.

Exercise 3.4. Suppose there exist $E, F \subset T$ such that $C(s,t) = 0$ for all $s \in E$ and $t \in F$. Then prove that $\{G(s)\}_{s \in E}$ and $\{G(t)\}_{t \in F}$ are *independent* Gaussian processes. That is, prove that for all $s_1, \ldots, s_n \in E$ and all $t_1, \ldots, t_m \in F$, $(G(s_1), \ldots, G(s_n))$ and $(G(t_1), \ldots, G(t_m))$ are independent Gaussian random vectors.

A classical theorem—due in various degrees of generality to Herglotz, Bochner, Minlos, etc.—states that the collection of all nonnegative definite functions f on $T \times T$ matches all covariance functions, as long as f is symmetric. [*Symmetry* means that $f(s,t) = f(t,s)$.] This, and the afore-mentioned theorem of Kolmogorov, together imply that given a function $\mu : T \to \mathbf{R}$ and a nonnegative-definite function $C : T \times T \to \mathbf{R}$ there exists a Gaussian process $\{G(t)\}_{t \in T}$ whose mean function is μ and covariance function is C.

Example 3.5 (Brownian motion). Let $T = \mathbf{R}_+ := [0, \infty)$, $\mu(t) := 0$, and $C(s, t) := \min(s, t)$ for all $s, t \in \mathbf{R}_+$. I claim that C is nonnegative definite. Indeed, for all $z_1, \ldots, z_k \in \mathbf{C}$ and $t_1, \ldots, t_k \geq 0$,

$$
\sum_{j=1}^{k} \sum_{l=1}^{k} \min(t_j, t_l) z_j \overline{z_l} = \sum_{j=1}^{k} \sum_{l=1}^{k} z_j \overline{z_l} \int_0^\infty \mathbf{1}_{[0,t_j]}(x) \mathbf{1}_{[0,t_l]}(x) \, dx
$$

$$
= \int_0^\infty \left| \sum_{j=1}^{k} \mathbf{1}_{[0,t_j]}(x) z_j \right|^2 \, dx, \tag{6}
$$

which is greater than or equal to zero. Because C is also symmetric, it must be the covariance function of *some* mean-zero Gaussian process $B := \{B(t)\}_{t \geq 0}$. That process B is called *Brownian motion*; it was first invented by Bachelier [1].

Brownian motion has the following additional property. *Let $s > 0$ be fixed. Then the process $\{B(t+s) - B(s)\}_{t \geq 0}$ is independent of $\{B(u)\}_{0 \leq u \leq s}$.* This is the socalled *Markov property* of Brownian motion, and is not hard to derive. Indeed, thanks to Exercise 3.4 it suffices to prove that for all $t \geq 0$ and $0 \leq u \leq s$,

$$
\mathrm{E}[(B(t+s) - B(s))B(u)] = 0. \tag{7}
$$

But this is easy to see because

$$
\begin{aligned}
\mathrm{E}[(B(t+s) - B(s))B(u)] &= \mathrm{Cov}(B(t+s), B(u)) - \mathrm{Cov}(B(s), B(u)) \\
&= \min(t+s, u) - \min(s, u) \\
&= u - u \\
&= 0.
\end{aligned} \tag{8}
$$

By d-dimensional Brownian motion we mean the d-dimensional Gaussian process $B := \{(B_1(t), \ldots, B_d(t))\}_{t \geq 0}$, where $B_1, \ldots, B_d$ are independent [one-dimensional] Brownian motions.

Exercise 3.6. Prove that if $s > 0$ is fixed and B is Brownian motion, then the process $\{B(t+s) - B(s)\}_{t \geq 0}$ is a *Brownian motion* independent of $\{B(u)\}_{0 \leq u \leq s}$. This and the independent-increment property of B [Example 3.5] together prove that B is a *Markov process*.

Example 3.7 (Brownian bridge). The *Brownian bridge* is a mean-zero Gaussian process $\{b(x)\}_{0 \leq x \leq 1}$ with covariance,

$$
\mathrm{Cov}(b(x), b(y)) := \min(x, y) - xy \qquad \text{for all } 0 \leq x, y \leq 1. \tag{9}
$$

The next exercise shows that the process b looks locally like a Brownian motion. Note also that $b(0) = b(1) = 0$; this follows because $\mathrm{Var}(b(0)) = \mathrm{Var}(b(1)) = 0$, and motivates the ascription "bridge." The next exercise explains why b is "brownian."

Exercise 3.8. Prove that if B is Brownian motion, then b is Brownian bridge, where

$$b(x) := B(x) - xB(1) \qquad \text{for all } 0 \le x \le 1. \tag{10}$$

Also prove that the process b is independent of $B(1)$.

Example 3.9 (OU process). Let $B := \{B(t)\}_{t \ge 0}$ denote a d-dimensional Brownian motion, and define

$$X(t) := \frac{B(e^t)}{e^{t/2}} \qquad \text{for all } t \ge 0. \tag{11}$$

The coordinate processes $X_1, \dots, X_d$ are i.i.d. Gaussian processes with mean function $\mu(t) := 0$ and covariance function

$$
\begin{aligned}
C(s,t) &:= \mathrm{E}\left[\frac{B_1(e^s) B_1(e^t)}{e^{(s+t)/2}}\right] \\
&= \exp\left(-\tfrac{1}{2}|s - t|\right).
\end{aligned}
\tag{12}
$$

Note that $C(s,t)$ depends on s and t only through $|s - t|$. Such processes are called *stationary Gaussian processes.* This particular stationary Gaussian process was predicted in the works of Dutch physicists Leonard S. Ornstein and George E. Uhlenbeck [29], and bears their name as a result. The existence of the Ornstein–Uhlenbeck process was proved rigorously in a landmark paper of Doob [10].

Example 3.10 (Brownian sheet). Let $T := \mathbf{R}_+^N := [0, \infty)^N$, $\mu(t) := 0$ for all $t \in \mathbf{R}_+^N$, and define

$$C(s,t) := \prod_{j=1}^{N} \min(s_j, t_j) \qquad \text{for all } s, t \in \mathbf{R}_+^N. \tag{13}$$

Then C is a nonnegative-definite, symmetric function on $\mathbf{R}_+^N \times \mathbf{R}_+^N$, and the resulting mean-zero Gaussian process $B = \{B(t)\}_{t \in \mathbf{R}_+^N}$ is the N-parameter *Brownian sheet.* This generalizes Brownian motion to an N-parameter random field. One can also introduce d-dimensional, N-parameter Brownian sheet as the d-dimensional process whose coordinates are independent, [one-dimensional] N-parameter Brownian sheets.

Example 3.11 (OU sheet). Let $\{B(t)\}_{t \in \mathbf{R}_+^N}$ denote N-parameter Brownian sheet, and define a new N-parameter stochastic process X as follows:

$$X(t) := \frac{B(e^{t_1}, \dots, e^{t_N})}{e^{(t_1 + \cdots + t_N)/2}} \qquad \text{for all } t := (t_1, \dots, t_N) \in \mathbf{R}_+^N. \tag{14}$$

This is called the N-paramerter *Ornstein–Uhlenbeck sheet,* and generalizes the Ornstein–Uhlenbeck process of Example 3.9.

Exercise 3.12. Prove that the Ornstein–Uhlenbeck sheet is a mean-zero, N-parameter Gaussian process and its covariance function $C(s,t)$ depends on (s,t) only through $|s-t| := \sum_{i=1}^{N} |s_i - t_i|$.

Example 3.13 (White noise). Let $T := \mathscr{B}(\mathbf{R}^N)$ denote the collection of all Borel-measurable subsets of $\mathbf{R}^N$, and $\mu(A) := 0$ for all $A \in \mathscr{B}(\mathbf{R}^N)$. Define $C(A,B) := \lambda^N(A \cap B)$, where λ^N denotes the N-dimensional Lebesgue measure. Clearly, C is symmetric. It turns out that C is also nonnegative definite (Exercise 3.14 on page 6). The resulting Gaussian process $\dot{W} := \{\dot{W}(A)\}_{A \in \mathscr{B}(\mathbf{R}^N)}$ is called *white noise* on $\mathbf{R}^N$.

Exercise 3.14. Complete the previous example by proving that the covariance of white noise is indeed a nonnegative-definite function on $\mathscr{B}(\mathbf{R}^N) \times \mathscr{B}(\mathbf{R}^N)$.

Exercise 3.15. Prove that if $A, B \in \mathscr{B}(\mathbf{R}^N)$ are disjoint then $\dot{W}(A)$ and $\dot{W}(B)$ are independent random variables. Use this to prove that if $A, B \in \mathscr{B}(\mathbf{R}^N)$ are nonrandom, then with probability one,

$$\dot{W}(A \cup B) = \dot{W}(A) + \dot{W}(B) - \dot{W}(A \cap B). \tag{15}$$

Exercise 3.16. Despite what the preceding may seem to imply, $\dot{W}$ is not a random signed measure in the obvious sense. Let $N = 1$ for simplicity. Then, prove that with probability one,

$$\lim_{n \to \infty} \sum_{j=0}^{2^n - 1} \left| \dot{W}\left(\left[\frac{j-1}{2^n}, \frac{j}{2^n} \right] \right) \right|^2 = 1. \tag{16}$$

Use this to prove that with probability one,

$$\lim_{n \to \infty} \sum_{j=0}^{2^n - 1} \left| \dot{W}\left(\left[\frac{j-1}{2^n}, \frac{j}{2^n} \right] \right) \right| = \infty. \tag{17}$$

Conclude that if $\dot{W}$ *were* a random measure then with probability one $\dot{W}$ is not sigma-finite. Nevertheless, the following example shows that one can integrate some things against $\dot{W}$.

Example 3.17 (The isonormal process). Let $\dot{W}$ denote white noise on $\mathbf{R}^N$. We wish to define $\dot{W}(h)$ where h is a nice function. First, we identify $\dot{W}(A)$ with $\dot{W}(\mathbf{1}_A)$. More generally, we define for all disjoint $A_1, \ldots, A_k \in \mathscr{B}(\mathbf{R}^N)$ and $c_1, \ldots, c_k \in \mathbf{R}$,

$$\dot{W}\left(\sum_{j=1}^{k} c_j \mathbf{1}_{A_j} \right) := \sum_{j=1}^{k} c_j \dot{W}(A_j). \tag{18}$$

The random variables $\dot{W}(A_1), \ldots, \dot{W}(A_k)$ are independent, thanks to Exercise 3.15. Therefore,

$$
\left\| \dot{W} \left(\sum_{j=1}^{k} c_j \mathbf{1}_{A_j} \right) \right\|_{L^2(\mathrm{P})}^2 = \sum_{j=1}^{k} c_j^2 |A_j|
$$

$$
= \left\| \sum_{j=1}^{k} c_j \mathbf{1}_{A_j} \right\|_{L^2(\mathbf{R}^N)}^2 . \tag{19}
$$

Classical integration theory tells us that for all $h \in L^2(\mathbf{R}^N)$ we can find h_n of the form $\sum_{j=1}^{k(n)} c_{jn} \mathbf{1}_{A_{j,n}}$ such that $A_{1,n}, \ldots, A_{k(n),n} \in \mathscr{B}(\mathbf{R}^N)$ are disjoint and $\|h - h_n\|_{L^2(\mathbf{R}^N)} \to 0$ as $n \to \infty$. This, and (19) tell us that $\{\dot{W}(h_n)\}_{n=1}^{\infty}$ is a Cauchy sequence in $L^2(\mathrm{P})$. Denote their limit by $\dot{W}(h)$. This is the *Wiener integral of* $h \in L^2(\mathbf{R}^N)$, and is sometimes written as $\int h \, dW$ [no dot!]. Its key feature is that

$$
\left\| \dot{W}(h) \right\|_{L^2(\mathrm{P})} = \|h\|_{L^2(\mathbf{R}^N)}. \tag{20}
$$

That is, $\dot{W} : L^2(\mathbf{R}^N) \to L^2(\mathrm{P})$ is an isometry; (20) is called *Wiener's isometry* [32]. [Note that we now know how to construct the stochastic integral $\int h \, dW$ only if $h \in L^2(\mathbf{R}^N)$ is *nonrandom*.] The process $\{\dot{W}(h)\}_{h \in L^2(\mathbf{R}^N)}$ is called the *isonormal process* [11]. It is a Gaussian process; its mean function is zero; and its covariance function is $C(h, g) = \int_{\mathbf{R}^N} h(x)g(x) \, dx$—the $L^2(\mathbf{R}^N)$ inner product—for all $h, g \in L^2(\mathbf{R}^N)$.

Exercise 3.18. Prove that for all [nonrandom] $h, g \in L^2(\mathbf{R}^N)$ and $a, b \in \mathbf{R}$,

$$
\int (ah + bg) \, dW = a \int h \, dW + b \int h \, dW, \tag{21}
$$

almost surely.

Exercise 3.19. Let $\{h_j\}_{j=1}^{\infty}$ be a complete orthonormal system [c.o.n.s.] in $L^2(\mathbf{R}^N)$. Then prove that $\{\dot{W}(h_j)\}_{j=1}^{\infty}$ is a complete orthonormal system in $L^2(\mathrm{P})$. In particular, for all Gaussian random variables $Z \in L^2(\mathrm{P})$ that are measurable with respect to the white noise,

$$
Z = \sum_{j=1}^{\infty} a_j \dot{W}(h_j) \quad \text{almost surely, with} \quad a_j := \mathrm{Cov}\left(Z, \dot{W}(h_j) \right), \tag{22}
$$

and the infinite sum converges in $L^2(\mathrm{P})$. This permits one possible entry into the "Malliavin calculus." For this, and much more, see the course by D. Nualart in this volume.

Exercise 3.20. Verify that (18) is legitimate. That is, prove that if $B_1, \ldots,$ $B_\ell \in \mathscr{B}(\mathbf{R}^N)$ are disjoint, then

$$\dot{W}\left(\sum_{j=1}^{k} c_j \mathbf{1}_{A_j}\right) = \dot{W}\left(\sum_{l=1}^{\ell} d_l \mathbf{1}_{B_l}\right) \quad \text{almost surely,} \tag{23}$$

provided that $d_1, \ldots, d_\ell \in \mathbf{R}$ satisfy $\sum_{j=1}^{k} c_j \mathbf{1}_{A_j} = \sum_{l=1}^{\ell} d_l \mathbf{1}_{B_l}$.

4 Regularity of Random Processes

Our construction of Gaussian processes is very general. This generality makes our construction both useful and useless. It is useful because we can make sense of fundamental mathematical objects such as Brownian motion, Brownian sheet, white noise, etc. It is useless because our "random functions," namely the Brownian motion and more generally sheet, are not yet nice random functions. This problem has to do with the structure of Kolmogorov's existence theorem. But instead of discussing this technical subject directly, let us consider a simple example first.

Let $\{B(t)\}_{t\geq 0}$ denote the Brownian motion, and suppose U is an independent positive random variable with an absolutely continuous distribution. Define

$$B'(t) := \begin{cases} B(t) & \text{if } t \neq U, \\ 5000 & \text{if } t = U. \end{cases} \tag{24}$$

Then B' and B have the same finite-dimensional distributions. Therefore, B' is also a Brownian motion. This little example shows that there is no hope of proving that a given Brownian motion is, say, a continuous random function. [Sort the logic out!] Therefore, the best one can hope to do is to produce a *modification* of Brownian motion that is continuous.

Definition 4.1. *Let X and X' be two stochastic processes indexed by some set T. We say that X' is a* modification *of X if*

$$\mathrm{P}\left\{X'(t) = X(t)\right\} = 1 \quad \text{for all } t \in T. \tag{25}$$

Exercise 4.2. Prove that any modification of a stochastic process X is a process with the same finite-dimensional distributions as X. Construct an example where X' is a modification of X, but $\mathrm{P}\{X' = X\} = 0$.

A remarkable theorem of Wiener [31] states that we can always find a continuous modification of a Brownian motion. According to the previous exercise, this modification is itself a Brownian motion. Thus, a *Wiener process* is a Brownian motion B such that the random function $t \mapsto B(t)$ is continuous; it is also some times known as *standard Brownian motion*.

4.1 A Diversion

In order to gel the ideas we consider first a simple finite-dimensional example. Let $f \in L^1(\mathbf{R})$ and denote its Fourier transform by $\mathscr{F}f$. We normalize the Fourier transform as follows:

$$(\mathscr{F}f)(z) := \int_{-\infty}^{\infty} e^{izx} f(x) \, dx \quad \text{for all } z \in \mathbf{R}. \tag{26}$$

Let $\mathscr{W}(\mathbf{R})$ denote the collection of all $f \in L^1(\mathbf{R})$ such that $\hat{f} \in L^1(\mathbf{R})$ as well. The space $\mathscr{W}$ is the so-called Wiener algebra on $\mathbf{R}$. If $f \in \mathscr{W}(\mathbf{R})$, then we can proceed, intentionally carelessly, and use the inversion formula to arrive at the following:

$$f(x) = \frac{1}{2\pi} \int_{-\infty}^{\infty} e^{-izx} (\mathscr{F}f)(z) \, dz. \tag{27}$$

It follows readily from this and the dominated convergence theorem that f is uniformly continuous. But this cannot be so! In order to see why, let us consider the function

$$g(x) = \begin{cases} f(x) & \text{if } x \neq 0, \\ f(0) + 1 & \text{if } x = 0. \end{cases} \tag{28}$$

If f were a continuous function, then g is not. But because $\mathscr{F}f = \mathscr{F}g$ the preceding argument would "show" that g is continuous too, which is a contradiction. The technical detail that we overlooked is that *a priori* (27) holds only for almost all $x \in \mathbf{R}$. Therefore,

$$x \mapsto \frac{1}{2\pi} \int_{-\infty}^{\infty} e^{-izx} (\mathscr{F}f)(z) \, dz \tag{29}$$

defines a "modification" of f which happens to be uniformly continuous. That is, we have proven that every $f \in \mathscr{W}(\mathbf{R})$ has a uniformly-continuous modification.

4.2 Kolmogorov's Continuity Theorem

Now we come to the question, "when does a stochastic process X have a continuous modification?" If X is a Gaussian process then the answer is completely known, but is very complicated [11; 12; 26; 27; 28]. When X is a fairly general process there are also complicated sufficient conditions for the existence of a continuous modification. In the special case that X is a process indexed by $\mathbf{R}^N$, however, there is a very useful theorem of Kolmogorov which gives a sufficient condition as well.

Theorem 4.3. *Suppose* $\{X(t)\}_{t \in T}$ *is a stochastic process indexed by a compact cube* $T := [a_1, b_1] \times \cdots \times [a_N, b_N] \subset \mathbf{R}^N$. *Suppose also that there exist constants* $C > 0$, $p > 0$, *and* $\gamma > N$ *such that uniformly for all* $s, t \in T$,

$$\mathrm{E}\left(|X(t) - X(s)|^p\right) \le C|t - s|^{\gamma}. \tag{30}$$

Then X *has a continuous modification* $\bar{X}$. *Moreover, if* $0 \le \theta < (\gamma - N)/p$ *then*

$$\left\| \sup_{s \ne t} \frac{|\bar{X}(s) - \bar{X}(t)|}{|s - t|^{\theta}} \right\|_{L^p(\mathrm{P})} < \infty. \tag{31}$$

Remark 4.4. Here, $|x|$ could be any of the usual Euclidean ℓ^p norms for $x \in \mathbf{R}^k$. That is,

$$|x| := \max\left(|x_1|, \ldots, |x_k|\right);$$
$$|x| := \left(|x_1|^p + \cdots + |x_k|^p\right)^{1/p} \quad \text{for } p \ge 1; \tag{32}$$
$$|x| := |x_1|^p + \cdots + |x_k|^p \quad \text{for } 0 < p < 1.$$

Proof. We prove Theorem 4.3 in the case that $N = 1$ and $T := [0, 1]$. The general case is not much more difficult to prove, but requires introducing further notation. Also, we extend the domain of the process by setting

$$X(t) := \begin{cases} X(0) & \text{if } t < 0, \\ X(1) & \text{if } t > 1. \end{cases} \tag{33}$$

First we introduce some notation: For every integer $n \ge 0$ we define $\mathscr{D}_n := \{j2^{-n} : 0 \le j < 2^n\}$ to be the collection of all dyadic points in $[0, 1)$. The totality of all dyadic points is denoted by $\mathscr{D}_\infty := \cup_{n=0}^{\infty} \mathscr{D}_n$.

Suppose $n > k \ge 1$, and consider $u, v \in \mathscr{D}_n$ that are within 2^{-k} of one another. We can find two sequences of points $u_k, \ldots, u_n$ and $v_k, \ldots, v_n$ with the following properties:

1. $u_j, v_j \in \mathscr{D}_j$ for all $j = k, \ldots, n$;
2. $|u_{j+1} - u_j| \le 2^{-j-1}$ for all $j = k, \ldots, n$;
3. $|v_{j+1} - v_j| \le 2^{-j-1}$ for all $j = k, \ldots, n$;
4. $u_n = u$, $v_n = v$, and $u_k = v_k$.

(Draw a picture.) Because $|X(u) - X(u_k)| \le \sum_{j=k}^{n-1} |X(u_{j+1}) - X(u_j)|$, this yields

$$|X(u) - X(u_k)| \le \sum_{j=k}^{\infty} \max_{s \in \mathscr{D}_{j+1}} \max_{t \in B(s, 2^{-j-1}) \cap \mathscr{D}_j} |X(s) - X(t)|, \tag{34}$$

where $B(x, r) := [x - r, x + r]$. The right-most term does not depend on u, nor on the sequences $\{u_j\}_{j=k}^n$ and $\{v_j\}_{j=k}^n$. Moreover, $|X(v) - X(v_k)| =$

$|X(v) - X(u_k)|$ is bounded above by the same quantity. Hence, by the triangle inequality,

$$|X(u) - X(v)| \leq 2 \sum_{j=k}^{\infty} \max_{s \in \mathscr{D}_{j+1}} \max_{t \in B(s, 2^{-j-1}) \cap \mathscr{D}_j} |X(s) - X(t)|, \qquad (35)$$

uniformly for all $u, v \in \mathscr{D}_n$ that are within 2^{-k} of one another. Because its right-hand side is independent of n, the preceding holds uniformly for all $u, v \in \mathscr{D}_\infty$ that are within distance 2^{-k} of one another. This and the Minkowski inequality together imply that

$$\left\| \sup_{\substack{u,v \in \mathscr{D}_\infty \\ |u-v| \leq 2^{-k}}} |X(u) - X(v)| \right\|_{L^p(P)}$$

$$\leq 2 \sum_{j=k}^{\infty} \left\| \max_{s \in \mathscr{D}_{j+1}} \max_{t \in B(s, 2^{-j-1}) \cap \mathscr{D}_j} |X(s) - X(t)| \right\|_{L^p(P)}. \qquad (36)$$

A crude bound yields

$$\mathrm{E}\left(\max_{s \in \mathscr{D}_{j+1}} \max_{t \in B(s, 2^{-j-1}) \cap \mathscr{D}_j} |X(s) - X(t)|^p \right)$$

$$\leq \sum_{s \in \mathscr{D}_{j+1}} \sum_{t \in B(s, 2^{-j-1}) \cap \mathscr{D}_j} \mathrm{E}\left(|X(s) - X(t)|^p \right) \qquad (37)$$

$$\leq C \sum_{s \in \mathscr{D}_{j+1}} \sum_{t \in B(s, 2^{-j-1}) \cap \mathscr{D}_j} |s - t|^\gamma,$$

thanks to Condition (30) of the theorem. For the range in question: $|s - t|^\gamma \leq 2^{-(j+1)\gamma}$; the sum over t then contributes a factor of 2; and the sum over s yields a factor of 2^{j+1}. Therefore,

$$\mathrm{E}\left(\max_{s \in \mathscr{D}_{j+1}} \max_{t \in B(s, 2^{-j-1}) \cap \mathscr{D}_j} |X(s) - X(t)|^p \right) \leq \frac{2^{2-\gamma}C}{2^{j(\gamma-1)}}. \qquad (38)$$

We can plug this into (36) to deduce that

$$\left\| \sup_{\substack{u,v \in \mathscr{D}_\infty \\ |u-v| \leq 2^{-k}}} |X(u) - X(v)| \right\|_{L^p(P)} \leq \frac{\tilde{C}}{2^{k\gamma/p}}, \qquad (39)$$

where

$$\tilde{C} := \frac{2^{(2-\gamma+p)/p} C^{1/p}}{1 - 2^{-(\gamma-1)/p}}. \qquad (40)$$

Now let us define

$$\bar{X}(s) := \limsup X(t), \tag{41}$$

where the lim sup is taken over all $t \in \mathscr{D}_\infty$ such that $t \to s$. Because $\bar{X}(s) = X(s)$ for all $s \in \mathscr{D}_\infty$, Equation (39) continues to hold, even if we replace X by $\bar{X}$. In that case, we can also replace the condition "$s, t \in \mathscr{D}_\infty$" with "$s, t \in [0, 1]$" at no extra cost. This proves, among other things, that $\bar{X}$ is a.s. continuous [Borel–Cantelli lemma].

It is not hard to check that $\bar{X}$ is a modification of X because: (i) X and $\bar{X}$ agree on $\mathscr{D}_\infty$; (ii) X is continuous in probability[1] by (30); and (iii) $\bar{X}$ is continuous a.s., as we just proved.

It remains to verify (31). For θ as given, (39) implies that for all integers $k \geq 1$,

$$\left\| \sup_{\substack{0 \leq s \neq t \leq 1: \\ 2^{-k} < k \leq 2^{-k+1}}} \frac{|\bar{X}(s) - \bar{X}(t)|}{|s - t|^\theta} \right\|_{L^p(\mathrm{P})} \leq \frac{\tilde{C}}{2^{k(\gamma - \theta)/p}}. \tag{42}$$

Sum both sides of this inequality from $k = 1$ to infinity to deduce (31), and hence the theorem. □

Exercise 4.5. Suppose the conditions of Theorem 4.3 are met, but we have the following in place of (30):

$$\mathrm{E}\left(|X(t) - X(s)|^p\right) \leq h\left(|t - s|\right), \tag{43}$$

where $h : [0, \infty) \to \mathbf{R}_+$ is continuous and increasing, and $h(0) = 0$. Prove that X has a continuous modification provided that

$$\int_0^\eta \frac{h(r)}{r^{1+N}} \, dr < \infty \qquad \text{for some } \eta > 0. \tag{44}$$

Definition 4.6 (Hölder continuity). *A function $f : \mathbf{R}^N \to \mathbf{R}$ is said to be globally Hölder continuous with index α if there exists a constant A such that for all $x, y \in \mathbf{R}^N$,*

$$|f(x) - f(y)| \leq A|x - y|^\alpha. \tag{45}$$

It is said to be [locally] Hölder continuous with index α if for all compact sets $K \subset \mathbf{R}^N$ there exists a constant A_K such that

$$|f(x) - f(y)| \leq A_K|x - y|^\alpha \qquad \text{for all } x, y \in K. \tag{46}$$

Exercise 4.7. Suppose $\{X(t)\}_{t \in T}$ is a process indexed by a compact set $T \subset \mathbf{R}^N$ that satisfies (30) for some $C, p > 0$ and $\gamma > N$. Choose and fix $\alpha \in (0, (\gamma - N)/p)$. Prove that with probability one, X has a modification which is Hölder continuous with index α.

---

[1] This means that $X(s)$ converges to $X(t)$ in probability as $s \to t$.

Exercise 4.8. Suppose $\{X(t)\}_{t\in\mathbf{R}^N}$ is a process indexed by $\mathbf{R}^N$. Suppose for all compact $T \subset \mathbf{R}^N$ there exist constants $C_T, p_T > 0$ and $\gamma := \gamma_T > N$ such that

$$\mathrm{E}\left(|X(s) - X(t)|^{p_T}\right) \leq C_T|s - t|^\gamma, \qquad \text{for all } s, t \in T. \tag{47}$$

Then, prove that X has a modification $\bar{X}$ which is [locally] Hölder continuous with some index ε_T. Warning: Mind your null sets!

Exercise 4.9 (Regularity of Gaussian processes). Suppose $\{X(t)\}_{t\in T}$ is a Gaussian random field, and $T \subseteq \mathbf{R}^N$ for some $N \geq 1$. Then, check that for all $p > 0$,

$$\mathrm{E}\left(|X(t) - X(s)|^p\right) = c_p\left[\mathrm{E}\left(|X(t) - X(s)|^2\right)\right]^{p/2}, \tag{48}$$

where

$$c_p := \frac{1}{(2\pi)^{1/2}} \int_{-\infty}^{\infty} |x|^p e^{-x^2/2}\,dx = \frac{2^{p/2}}{\pi^{1/2}}\Gamma\left(\frac{p+1}{2}\right). \tag{49}$$

Suppose we can find $\varepsilon > 0$ with the following property: For all compact sets $K \subset T$ there exists a positive and finite constant $A(K)$ such that

$$\mathrm{E}\left(|X(t) - X(s)|^2\right) \leq A(K)|t - s|^\varepsilon \qquad \text{for all } t, s \in K. \tag{50}$$

Then prove that X has a modification that is [locally] Hölder continuous of any given order $< \varepsilon/2$.

Example 4.10 (Brownian motion). Let $B := \{B(t)\}_{t\geq 0}$ denote a Brownian motion. Note that for all $s, t \geq 0$, $X(t) - X(s)$ is normally distributed with mean zero and variance $|t - s|$. Therefore, $\mathrm{E}(|X(t) - X(s)|^2) = |t - s|$ for all $s, t \geq 0$. It follows that X has a modification that is Hölder of any given order $\alpha < \frac{1}{2}$. This is due to Wiener [31].

Warning: This is not true for $\alpha = \frac{1}{2}$. Let B denote the modification as well. [This should not be confusing.] Then, "the law of the iterated logarithm" of [18] asserts that

$$P\left\{\limsup_{t\downarrow s} \frac{|B(t) - B(s)|}{(2(t-s)\ln|\ln(t-s)|)^{1/2}} = 1\right\} = 1 \qquad \text{for all } s > 0. \tag{51}$$

In particular, for all $s > 0$,

$$P\left\{\limsup_{t\downarrow s} \frac{|B(t) - B(s)|}{|t - s|^{1/2}} = \infty\right\} = 1. \tag{52}$$

Thus, B is not Hölder continuous of order $\frac{1}{2}$ at $s = 0$, for instance.

Exercise 4.11. Let B denote N-parameter Brownian sheet. Prove that B has a modification which is [locally] Hölder continuous with any nonrandom index $\alpha \in (0, \frac{1}{2})$. This generalized Wiener's theorem on Brownian motion.

Exercise 4.12. Let B be a continuous Brownian motion, and define $\mathcal{H}_t$ to be the smallest sigma algebra that makes the random variables $\{B(s)\}_{s \in [0,t]}$ measurable. Then prove that the event in (51), whose probability is one, is measurable with respect to $\vee_{t \geq 0} \mathcal{H}_t$. Do the same for the event in (52).

Theorem 4.3 and the subsequent exercises all deal with distances on $\mathbf{R}^N$ that are based on norms. We will need a version based on another distance as well. This we state—without proof—in the case that $N = 2$.

Choose and fix some $p \in (0, 1]$ and an integer $1 \leq k \leq 1/p$, and define for all $u, v, s, t \in [0, 1]$,

$$|(s, t) - (u, v)| := |s - u|^p + |t - v|^{kp}. \tag{53}$$

This defines a distance on $[0, 1]^2$, but it is inhomogeneous, when $k > 1$, in the sense that it scales differently in different directions. The following is essentially 1.4.1 of Kunita [23, p. 31]; see also Corollary A.3 of [6]. I omit the proof.

Theorem 4.13. *Let $\{Y(s, t)\}_{s,t \in [0,1)^2}$ be a 2-parameter stochastic process taking value in $\mathbf{R}$. Suppose that there exist $C, p > 1$ and $\gamma > (k + 1)/k$ such that for all $s, t, u, v \in [0, 1)$,*

$$\|Y(s, t) - Y(u, v)\|_{L^p(\mathrm{P})} \leq C |(s, t) - (u, v)|^\gamma. \tag{54}$$

Then, Y has a Hölder-continuous modification $\bar{Y}$ that satisfies the following for every $\theta \geq 0$ which satisfies $k\gamma - (k + 1) - k\theta > 0$:

$$\left\| \sup_{(s,t) \neq (u,v)} \frac{|Y(s, t) - Y(u, v)|}{|(s, t) - (u, v)|^\theta} \right\|_{L^p(\mathrm{P})} < \infty. \tag{55}$$

5 Martingale Measures

5.1 A White Noise Example

Let $\dot{W}$ be white noise on $\mathbf{R}^N$. We have seen already that $\dot{W}$ is *not* a signed sigma-finite measure with any positive probability. However, it is not hard to deduce that it has the following properties:

1. $\dot{W}(\varnothing) = 0$ a.s.

2. For all disjoint [nonrandom] sets $A_1, A_2, \ldots \in \mathscr{B}(\mathbf{R}^N)$,

$$\mathrm{P}\left\{\dot{W}\left(\bigcup_{i=1}^{\infty} A_i\right) = \sum_{i=1}^{\infty} \dot{W}(A_i)\right\} = 1, \tag{56}$$

where the infinite sum converges in $L^2(\mathrm{P})$.

That is,

Proposition 5.1. *White noise is an $L^2(\mathrm{P})$-valued, sigma-finite, signed measure.*

Proof. In light of Exercise 3.15 it suffices to prove two things: (a) If $A_1 \supset A_2 \supset \cdots$ are all in $\mathscr{B}(\mathbf{R}^N)$ and $\cap A_n = \varnothing$, then $\dot{W}(A_n) \to 0$ in $L^2(\mathrm{P})$ as $n \to \infty$; and (b) For all compact sets K, $\mathrm{E}[(\dot{W}(K))^2] < \infty$.

It is easy to prove (a) because $\mathrm{E}[(\dot{W}(A_n))^2]$ is just the Lebesgue measure of A_n, and $|A_n| \to 0$ because Lebesgue measure is a measure. (b) is even easier to prove because $\mathrm{E}[(\dot{W}(K))^2] = |K| < \infty$ because Lebesgue measure is sigma-finite. □

Oftentimes in SPDEs one studies the "white-noise process" $\{W_t\}_{t\geq 0}$ defined by $W_t(A) := \dot{W}([0,t] \times A)$, where $A \in \mathscr{B}(\mathbf{R}^{N-1})$. This is a proper stochastic process as t varies, but an $L^2(\mathrm{P})$-type noise in A.

Let $\mathscr{F}$ be the filtration of the process $\{W_t\}_{t\geq 0}$. By this I mean the following: For all $t \geq 0$, we define $\mathscr{F}_t$ to be the sigma-algebra generated by $\{W_s(A);\ 0 \leq s \leq t,\ A \in \mathscr{B}(\mathbf{R}^{N-1})\}$.

Exercise 5.2. Check that $\mathscr{F} := \{\mathscr{F}_t\}_{t\geq 0}$ is a *filtration* in the sense that $\mathscr{F}_s \subseteq \mathscr{F}_t$ whenever $s \leq t$.

Lemma 5.3. $\{W_t(A)\}_{t\geq 0, A\in\mathscr{B}(\mathbf{R}^{N-1})}$ *is a "martingale measure" in the sense that:*

1. *For all $A \in \mathscr{B}(\mathbf{R}^{N-1})$, $W_0(A) = 0$ a.s.;*
2. *If $t > 0$ then W_t is a sigma-finite, $L^2(\mathrm{P})$-valued signed measure; and*
3. *For all $A \in \mathscr{B}(\mathbf{R}^{N-1})$, $\{W_t(A)\}_{t\geq 0}$ is a mean-zero martingale.*

Proof. Note that $\mathrm{E}[(W_t(A))^2] = t|A|$ where $|A|$ denotes the $(N-1)$-dimensional Lebesgue measure of A. Therefore, $W_0(A) = 0$ a.s. This proves (1).

Equation (2) is proved in almost exactly the same way that Proposition 5.1 was. [Check the details!]

Finally, choose and fix $A \in \mathscr{B}(\mathbf{R}^{N-1})$. Then, whenever $t \geq s \geq u \geq 0$,

$$\mathrm{E}\big[(W_t(A) - W_s(A))\,W_u(A)\big]$$
$$= \mathrm{E}\left[\left(\dot{W}([0,t] \times A) - \dot{W}([0,s] \times A)\right)\dot{W}([0,u] \times A)\right] \tag{57}$$
$$= \min(t,u)|A| - \min(s,u)|A| = 0.$$

Therefore, $W_t(A) - W_s(A)$ is independent of $\mathscr{F}_s$ (Exercise 3.4, page 3). As a result, with probability one,

$$
\begin{aligned}
\mathrm{E}\left[W_t(A) \mid \mathscr{F}_s\right] &= \mathrm{E}\left[W_t(A) - W_s(A) \mid \mathscr{F}_s\right] + W_s(A) \\
&= \mathrm{E}\left[W_t(A) - W_s(A)\right] + W_s(A) \qquad (58) \\
&= W_s(A).
\end{aligned}
$$

This is the desired martingale property.   □

Exercise 5.4. Choose and fix $A \in \mathscr{B}(\mathbf{R}^{N-1})$ such that $1/c := |A|^{1/2} > 0$. Then prove that $\{cW_t(A)\}_{t \geq 0}$ is a Brownian motion.

Exercise 5.5 (Important). Suppose $h \in L^2(\mathbf{R}^{N-1})$. Note that $t^{-1/2}W_t$ is white noise on $\mathbf{R}^{N-1}$. Therefore, we can define $W_t(h) := \int h(x)W_t(dx)$ for all $h \in L^2(\mathbf{R}^{N-1})$. Prove that $\{W_t(h)\}_{t \geq 0}$ is a continuous martingale with quadratic variation

$$
\langle W_\bullet(h) \,,\, W_\bullet(h) \rangle_t = t \int_{\mathbf{R}^{N-1}} h^2(x)\,dx. \qquad (59)
$$

It might help to recall that if $\{Z_t\}_{t \geq 0}$ is a continuous $L^2(\mathrm{P})$-martingale, then its quadratic variation is uniquely defined as the continuous increasing process $\{\langle Z, Z \rangle_t\}_{t \geq 0}$ such that $\langle Z, Z \rangle_0 = 0$ and $t \mapsto Z_t^2 - \langle Z, Z \rangle_t$ is a continuous martingale. More generally, if Z and Y are two continuous $L^2(\mathrm{P})$-martingales then $Z_t Y_t - \langle Z, Y \rangle_t$ is a continuous $L^2(\mathrm{P})$-martingale, and $\langle Z, Y \rangle_t$ is the only such "compensator." In fact prove that for all $t \geq 0$ and $h, g \in L^2(\mathbf{R}^{N-1})$, $\langle W_\bullet(h), W_\bullet(g) \rangle_t = t \int_{\mathbf{R}^{N-1}} h(x)g(x)\,dx$.

5.2 More General Martingale Measures

Let $\mathscr{F} := \{\mathscr{F}_t\}_{t \geq 0}$ be a filtration of sigma-algebras. We assume that $\mathscr{F}$ is *right-continuous*; i.e.,

$$
\mathscr{F}_t = \bigcap_{s > t} \mathscr{F}_s \qquad \text{for all } t \geq 0. \qquad (60)
$$

[This ensures that continuous-time martingale theory works.]

Definition 5.6 (Martingale measures). *A process $\{M_t(A)\}_{t \geq 0, A \in \mathscr{B}(\mathbf{R}^n)}$ is a martingale measure [with respect to $\mathscr{F}$] if:*

1. $M_0(A) = 0$ *a.s.;*
2. *If $t > 0$ then M_t is a sigma-finite $L^2(\mathrm{P})$-valued signed measure; and*
3. *For all $A \in \mathscr{B}(\mathbf{R}^n)$, $\{M_t(A)\}_{t \geq 0}$ is a mean-zero martingale with respect to the filtration $\mathscr{F}$.*

Exercise 5.7. Double-check that you understand that if $\dot{W}$ is white noise on $\mathbf{R}^N$ then $W_t(A)$ defines a martingale measure on $\mathscr{B}(\mathbf{R}^{N-1})$.

Exercise 5.8. Let μ be a sigma-finite $L^2(\mathrm{P})$-valued signed measure on $\mathscr{B}(\mathbf{R}^n)$, and $\mathscr{F} := \{\mathscr{F}_t\}_{t \geq 0}$ a right-continuous filtration. Define $\mu_t(A) := \mathrm{E}[\mu(A) \,|\, \mathscr{F}_t]$ for all $t \geq 0$ and $A \in \mathscr{B}(\mathbf{R}^n)$. Then prove that $\{\mu_t(A)\}_{t \geq 0, A \in \mathscr{B}(\mathbf{R}^n)}$ is a martingale measure.

Exercise 5.9. Let $\{M_t(A)\}$ be a martingale measure. Prove that for all $T \geq t \geq 0$, $M_t(A) = \mathrm{E}[M_T(A) \,|\, \mathscr{F}_t]$ a.s. Thus, every martingale measure locally look like those of the preceding exercise.

It turns out that martingale measures are a good class of integrators. In order to define stochastic integrals we follow [30, Chapter 2], and proceed as one does when one constructs ordinary Itô integrals.

Definition 5.10. *A function $f : \mathbf{R}^n \times \mathbf{R}_+ \times \Omega \to \mathbf{R}$ is elementary if*

$$f(x, t, \omega) = X(\omega) \mathbf{1}_{(a,b]}(t) \mathbf{1}_A(x), \qquad (61)$$

where: (a) X is bounded and $\mathscr{F}_a$-measurable; and (b) $A \in \mathscr{B}(\mathbf{R}^n)$. Finite [nonrandom] linear combinations of elementary functions are called simple *functions. Let $\mathscr{S}$ denote the class of all simple functions.*

If M is a martingale measure and f is an elementary function of the form (61), then we define the stochastic-integral process of f as

$$(f \cdot M)_t(B)(\omega) := X(\omega) \left[M_{t \wedge b}(A \cap B) - M_{t \wedge a}(A \cap B) \right](\omega). \qquad (62)$$

Exercise 5.11 (Important). Prove that if f is an elementary function then $(f \cdot M)$ is a martingale measure. This constructs new martingale measures from old ones. For instance, if f is elementary and $\dot{W}$ is white noise then $(f \cdot W)$ is a martingale measure.

If $f \in \mathscr{S}$ then we can write f as $f = c_1 f_1 + \cdots + c_k f_k$ where $c_1, \ldots, c_k \in \mathbf{R}$ and $f_1, \ldots, f_k$ are elementary. We can then define

$$(f \cdot M)_t(B) := \sum_{j=1}^{k} c_j (f_j \cdot M)_t(B). \qquad (63)$$

Exercise 5.12. Prove that the preceding is well defined. That is, prove that the definition of $(f \cdot M)$ does not depend on a particular representation of f in terms of elementary functions.

Exercise 5.13. Prove that if $f \in \mathscr{S}$ then $(f \cdot M)$ is a martingale measure. Thus, if $\dot{W}$ is white noise and $f \in \mathscr{S}$ then $(f \cdot W)$ is a martingale measure.

The right class of integrands are functions f that are "predictable." That is, they are measurable with respect to the "predictable sigma-algebra" $\mathscr{P}$ that is defined next.

Definition 5.14. *Let* $\mathscr{P}$ *denote the sigma-algebra generated by all functions in* $\mathscr{S}$. $\mathscr{P}$ *is called the* predictable sigma-algebra.

In order to go beyond stochastic integration of $f \in \mathscr{S}$ we need a technical condition—called "worthiness"—on the martingale measure M. This requires a little background.

Definition 5.15. *Let* M *be a martingale measure. The* covariance functional *of* M *is defined as*

$$\overline{Q}_t(A, B) := \langle M_\bullet(A), M_\bullet(B) \rangle_t, \qquad \text{for all } t \geq 0, \ A, B \in \mathscr{B}(\mathbf{R}^n). \qquad (64)$$

Exercise 5.16. Prove that:

1. $\overline{Q}_t(A, B) = \overline{Q}_t(B, A)$ almost surely;
2. If $B \cap C = \emptyset$ then $\overline{Q}_t(A, B \cup C) = \overline{Q}_t(A, B) + \overline{Q}_t(A, C)$ almost surely;
3. $|\overline{Q}_t(A, B)|^2 \leq \overline{Q}_t(A, A)\overline{Q}_t(B, B)$ almost surely; and
4. $t \mapsto \overline{Q}_t(A, A)$ is almost surely non-decreasing.

Exercise 5.17. Let $\dot{W}$ be white noise on $\mathbf{R}^N$ and consider the martingale measure defined by $W_t(A) := \dot{W}((0, t] \times A)$, where $t \geq 0$ and $A \in \mathscr{B}(\mathbf{R}^{N-1})$. Verify that the quadratic functional of this martingale measure is described by $\overline{Q}_t(A, B) := t\lambda^{N-1}(A \cap B)$, where λ^k denotes the Lebesgue measure on $\mathbf{R}^k$.

Next we define a random set function Q, in steps, as follows: For all $t \geq s \geq 0$ and $A, B \in \mathscr{B}(\mathbf{R}^n)$ define

$$Q(A, B; (s, t]) := \overline{Q}_t(A, B) - \overline{Q}_s(A, B). \qquad (65)$$

If $A_i \times B_i \times (s_i, t_i]$ $(1 \leq i \leq m)$ are disjoint, then we can define

$$Q\left(\bigcup_{i=1}^n (A_i \times B_i \times (s_i, t_i])\right) := \sum_{i=1}^n Q(A_i, B_i; (s_i, t_i]). \qquad (66)$$

This extends the definition of Q to rectangles. It turns out that, in general, one cannot go beyond this; this will make it impossible to define a completely general theory of stochastic integration in this setting. However, all works fine if M is "worthy" [30]. Before we define worthy martingale measures we point out a result that shows the role of Q.

Proposition 5.18. *Suppose* $f \in \mathscr{S}$ *and* M *is a worthy martingale measure. Then,*

$$\mathrm{E}\left[((f \cdot M)_t(B))^2\right] = \mathrm{E}\left[\iiint_{B \times B \times (0, t]} f(x, t)f(y, t) Q(dx\, dy\, dt)\right]. \qquad (67)$$

Question 5.19. Although Q is not a proper measure, the triple-integral is well-defined. Why?

Proof. First we do this when f is elementary, and say has form (61). Then,

$$
\begin{aligned}
&\mathrm{E}\left[(f \cdot M)_t^2(B)\right] \\
&= \mathrm{E}\left[X^2\left(M_{t \wedge b}(A \cap B) - M_{t \wedge a}(A \cap B)\right)^2\right] \\
&= \mathrm{E}\left[X^2 M_{t \wedge b}^2(A \cap B)\right] - 2\mathrm{E}\left[X^2 M_{t \wedge b}(A \cap B) M_{t \wedge a}(A \cap B)\right] \\
&\qquad\qquad\qquad\qquad\qquad\qquad + \mathrm{E}\left[X^2 M_{t \wedge a}^2(A \cap B)\right].
\end{aligned} \tag{68}
$$

Recall that X is $\mathscr{F}_a$-measurable. Therefore, by the definition of quadratic variation,

$$
\begin{aligned}
&\mathrm{E}\left[X^2\left(M_{t \wedge b}^2(A \cap B) - \langle M(A \cap B), M(A \cap B)\rangle_{t \wedge b}\right)\right] \\
&= \mathrm{E}\left[X^2\left(M_{t \wedge a}^2(A \cap B) - \langle M(A \cap B), M(A \cap B)\rangle_{t \wedge a}\right)\right].
\end{aligned} \tag{69}
$$

Similarly,

$$
\begin{aligned}
&\mathrm{E}\left[X^2\left(M_{t \wedge b}(A \cap B) M_{t \wedge a}(A \cap B) - \langle M(A \cap B), M(A \cap B)\rangle_{t \wedge a}\right)\right] \\
&= \mathrm{E}\left[X^2\left(M_{t \wedge a}^2(A \cap B) - \langle M(A \cap B), M(A \cap B)\rangle_{t \wedge a}\right)\right].
\end{aligned} \tag{70}
$$

Combine to deduce the result in the case that f has form (61).

If $f \in \mathscr{S}$ then we can write $f = c_1 f_1 + \cdots + c_k f_k$ where $f_1, \ldots, f_k$ are elementary with disjoint support, and $c_1, \ldots, c_k$ are reals. [Why disjoint support?] Because $\mathrm{E}[(f_j \cdot M)_t] = 0$, we know that $\mathrm{E}[(f \cdot M)_t^2(B)] = \sum_{j=1}^k c_j^2 \mathrm{E}[(f_j \cdot M)_t^2(B)]$. The first part of the proof finishes the derivation. $\square$

Definition 5.20. *A martingale measure M is* worthy *if there exists a random sigma-finite measure $K(A \times B \times C, \omega)$ —where $A, B \in \mathscr{B}(\mathbf{R}^n)$, $C \in \mathscr{B}(\mathbf{R}_+)$, and $\omega \in \Omega$ – such that:*

1. $A \times B \mapsto K(A \times B \times C, \omega)$ *is nonnegative definite and symmetric;*
2. $\{K(A \times B \times (0, t])\}_{t \geq 0}$ *is a predictable process (i.e., $\mathscr{P}$-measurable) for all $A, B \in \mathscr{B}(\mathbf{R}^n)$;*
3. *For all compact sets $A, B \in \mathscr{B}(\mathbf{R}^n)$ and $t > 0$,*

$$
\mathrm{E}[K(A \times B \times (0, t])] < \infty;
$$

4. *For all $A, B \in \mathscr{B}(\mathbf{R}^n)$ and $t > 0$,*

$$
|Q(A \times B \times (0, t])| \leq K(A \times B \times (0, t]) \qquad a.s.
$$

[As usual, we drop the dependence on ω.] If and when such a K exists then it is called a dominating measure *for M.*

Remark 5.21. If M is worthy then Q_M can be extended to a measure on $\mathscr{B}(\mathbf{R}^n) \times \mathscr{B}(\mathbf{R}^n) \times \mathscr{B}(\mathbf{R}_+)$. This follows, basically, from the dominated convergence theorem.

Exercise 5.22 (Important). Suppose $\dot{W}$ denotes white noise on $\mathbf{R}^N$, and consider the martingale measure on $\mathscr{B}(\mathbf{R}^{N-1})$ defined by $W_t(A) = W((0,t] \times A)$. Prove that it is worthy. Hint: Try the dominating measure $K(A \times B \times C) := \lambda^{N-1}(A \cap B)\lambda^1(C)$, where λ^k denotes the Lebesgue measure on $\mathbf{R}^k$. Is this different than Q?

Proposition 5.23. *If M is a worthy martingale measure and $f \in \mathscr{S}$, then $(f \cdot M)$ is a worthy martingale measure. If Q_N and K_N respectively define the covariance functional and dominating measure of a worthy martingale measure N, then*

$$Q_{f\cdot M}(dx\,dy\,dt) = f(x,t)f(y,t)\,Q_M(dx\,dy\,dt),$$
$$K_{f\cdot M}(dx\,dy\,dt) = |f(x,t)f(y,t)|\,K_M(dx\,dy\,dt). \tag{71}$$

Proof. We will do this for elementary functions f; the extension to simple functions is routine. In light of Exercise 5.11 it suffices to compute $Q_{f\cdot M}$. The formula for $K_{f\cdot M}$ follows from this immediately as well.

Now, suppose f has the form (61), and note that for all $t \geq 0$ and $B, C \in \mathscr{B}(\mathbf{R}^n)$,

$$
\begin{aligned}
(f \cdot M)_t&(B)(f \cdot M)_t(C) \\
&= X^2\left[M_{t\wedge b}(A \cap B) - M_{t\wedge a}(A \cap B)\right] \\
&\qquad\qquad \times \left[M_{t\wedge b}(A \cap C) - M_{t\wedge a}(A \cap C)\right] \\
&= \text{martingale} + X^2\langle M(A \cap B), M(A \cap C)\rangle_{t\wedge b} \\
&\qquad\qquad\qquad - X^2\langle M(A \cap B), M(A \cap C)\rangle_{t\wedge a} \\
&= \text{martingale} + X^2 Q_M((A \cap B) \times (A \cap B) \times (s,t]) \\
&= \text{martingale} + \iiint_{B\times C\times(0,t]} f(x,s)f(y,s)\,Q_M(dx\,dy\,ds).
\end{aligned}
\tag{72}
$$

This does the job. □

From now on we will be interested only in the case where the time variable t is in some finite interval $(0,T]$.

If K_M is the dominating measure for a worthy martingale measure M, then we define $\|f\|_M$, for all predictable function f, via

$$\|f\|_M^2 := \mathrm{E}\left[\iiint_{\mathbf{R}^n \times \mathbf{R}^n \times (0,T]} |f(x,t)f(y,t)|\,K_M(dx\,dy\,dt)\right]. \tag{73}$$

Let $\mathscr{P}_M$ denote the collection of all predictable functions f such that $\mathrm{E}(\|f\|_M)$ is finite.

Exercise 5.24. $\|\cdot\|_M$ is a norm on $\mathscr{P}$, and $\mathscr{P}_M$ is complete [hence a Banach space] in this norm.

I will not prove the following technical result. For a proof see [30, p. 293, Proposition 2.3].

Theorem 5.25. $\mathscr{S}$ *is dense in* $\mathscr{P}_M$.

Note from Proposition 5.18 that

$$\mathrm{E}\left[(f\cdot M)_t^2(B)\right] \leq \|f\|_M^2 \quad \text{for all } t \in (0,T], \ f \in \mathscr{S}, \ B \in \mathscr{B}(\mathbf{R}^n). \tag{74}$$

Consequently, if $\{f_m\}_{m=1}^\infty$ is a Cauchy sequence in $(\mathscr{S},\|\cdot\|_M)$ then the sequence $\{(f_m \cdot M)_t(B)\}_{m=1}^\infty$ is Cauchy in $L^2(\mathrm{P})$. If $f_m \to f$ in $\|\cdot\|_M$ then write the $L^2(\mathrm{P})$-limit of $(f_m \cdot M)_t(B)$ as $(f \cdot M)_t(B)$. A few more lines imply the following.

Theorem 5.26. *Let* M *be a worthy martingale measure. Then for all* $f \in \mathscr{P}_M$, $(f \cdot M)$ *is a worthy martingale measure that satisfies (71). Moreover, for all* $t \in (0,T]$ *and* $A, B \in \mathscr{B}(\mathbf{R}^n)$,

$$\langle (f\cdot M)(A),(f\cdot M)(B)\rangle_t = \iiint\limits_{A\times B\times(0,t]} f(x,s)f(y,s)\,Q_M(dx\,dy\,ds),$$
$$\mathrm{E}\left[(f\cdot M)_t^2(B)\right] \leq \|f\|_M^2. \tag{75}$$

The above $L^2(\mathrm{P})$ bound has an L^p version as well.

Theorem 5.27 (Burkholder's inequality [3]). *For all* $p \geq 2$ *there exists* $c_p \in (0,\infty)$ *such that for all predictable* f *and all* $t > 0$,

$$\mathrm{E}\left[|(f\cdot M)_t(B)|^p\right]$$
$$\leq c_p \mathrm{E}\left[\left(\iiint\limits_{\mathbf{R}^n\times\mathbf{R}^n\times(0,T]} |f(x,t)f(y,t)|\,K_M(dx\,dy\,dt)\right)^{p/2}\right]. \tag{76}$$

Proof (Special Case). It is enough to prove that if $\{N_t\}_{t\geq 0}$ is a martingale with $N_0 := 0$ and quadratic variation $\langle N,N\rangle_t$ at time t, then

$$\|N_t\|_{L^p(\mathrm{P})}^p \leq c_p \|\langle N,N\rangle_t\|_{L^{p/2}(\mathrm{P})}^{p/2}, \tag{77}$$

but this is precisely the celebrated Burkholder inequality [3]. Here is why it is true in the case that N is a *bounded* and continuous martingale. Recall Itô's formula [15; 16; 17]: For all f that is C^2 a.e.,

$$f(N_t) = f(0) + \int_0^t f'(N_s)\,dN_s + \frac{1}{2}\int_0^t f''(N_s)\,d\langle N, N\rangle_s. \qquad (78)$$

Apply this with $f(x) := |x|^p$ for $p > 2$ $[f''(x) = p(p-1)|x|^{p-2}$ a.e.$]$ to find that

$$|N_t|^p = \frac{p(p-1)}{2}\int_0^t |N_s|^{p-2}\,d\langle N, N\rangle_s + \text{mean-zero martingale.} \qquad (79)$$

Take expectations to find that

$$\mathrm{E}\left(|N_t|^p\right) \le \frac{p(p-1)}{2}\mathrm{E}\left(\sup_{0\le u\le t}|N_u|^{p-2}\langle N, N\rangle_t\right). \qquad (80)$$

Because $|N_t|^p$ is a submartingale, Doob's maximal inequality asserts that

$$\mathrm{E}\left(\sup_{0\le u\le t}|N_u|^p\right) \le \left(\frac{p}{p-1}\right)^p \mathrm{E}\left(|N_t|^p\right). \qquad (81)$$

Therefore, $\phi_p(t) := \mathrm{E}(\sup_{0\le u\le t}|N_u|^p)$ satisfies

$$
\begin{aligned}
\phi_p(t) &\le \frac{p(p-1)}{2}\left(\frac{p}{p-1}\right)^p \mathrm{E}\left(\sup_{0\le u\le t}|N_u|^{p-2}\langle N, N\rangle_t\right) \\
&:= a_p\mathrm{E}\left(\sup_{0\le u\le t}|N_u|^{p-2}\langle N, N\rangle_t\right).
\end{aligned}
\qquad (82)
$$

Apply Hölder's inequality to find that

$$\phi_p(t) \le a_p\left(\phi_p(t)\right)^{(p-2)/p}\left(\mathrm{E}\left[\langle N, N\rangle_t^{p/2}\right]\right)^{2/p}. \qquad (83)$$

We can solve this inequality for $\phi_p(t)$ to finish. $\qquad\qquad\square$

Exercise 5.28. In the context of the preceding prove that for all $p \ge 2$ there exists $c_p \in (0, \infty)$ such that for all bounded stopping times T,

$$\mathrm{E}\left(\sup_{0\le u\le T}|N_u|^p\right) \le c_p\mathrm{E}\left(\langle N, N\rangle_T^{p/2}\right). \qquad (84)$$

In addition, prove that we do not need N to be a bounded martingale in order for the preceding to hold. [Hint: Localize.]

Exercise 5.29 (Harder). In the context of the preceding prove that for all $p \ge 2$ there exists $c_p' \in (0, \infty)$ such that for all bounded stopping times T,

$$\mathrm{E}\left(\langle N, N\rangle_T^{p/2}\right) \le c_p'\mathrm{E}\left(\sup_{0\le u\le T}|N_u|^p\right). \qquad (85)$$

Hint: Start with $\langle N, N\rangle_t = N_t^2 - \int_0^t N_s\,dN_s \le N_t^2 + |\int_0^t N_s\,dN_s|$.

From now on we adopt a more standard stochastic-integral notation:

$$(f \cdot M)_t(A) := \iint\limits_{A \times (0,t]} f \, dM := \iint\limits_{A \times (0,t]} f(x,s) \, M(dx \, ds). \qquad (86)$$

[N.B.: The last $f(x,s)$ is actually $f(x,s,\omega)$, but we have dropped the ω as usual.] These martingale integrals have the Fubini–Tonelli property:

Theorem 5.30. *Suppose M is a worthy martingale measure with dominating measure K. Let $(A, \mathscr{A}, \mu)$ be a measure space and $f : \mathbf{R}^n \times \mathbf{R}_+ \times \Omega \times A \to \mathbf{R}$ measurable such that the following expectation is finite:*

$$\int \cdots \int\limits_{\Omega \times \mathbf{R}^n \times \mathbf{R}^n \times [0,T] \times A} |f(x,t,\omega,u)f(y,t,\omega,u)| \, K(dx \, dy \, dt) \, \mu(du) \, \mathrm{P}(d\omega). \qquad (87)$$

Then almost surely,

$$\int_A \left(\iint\limits_{\mathbf{R}^n \times [0,t]} f(x,s,\bullet,u) \, M(dx \, ds) \right) \mu(du)$$
$$= \iint\limits_{\mathbf{R}^n \times [0,t]} \left(\int_A f(x,s,\bullet,u) \, \mu(du) \right) M(dx \, ds). \qquad (88)$$

It suffices to prove this for elementary functions of the form (61). You can do this yourself, or consult the lecture notes of Walsh [30, p. 297].

6 A Nonlinear Heat Equation

We are ready to try and study a class of nonlinear elliptic SPDEs that is an example of the equations studied by Baklan [2], Daleckiĭ [7], Dawson [8; 9], Pardoux [24; 25], Krylov and Rozovski [20; 21; 22], and Funaki [13; 14]. It is possible to adapt the arguments to study hyperbolic SPDEs as well. For an introductory example see the paper by Cabaña [4]. The second chapter, by R. C. Dalang, of this volume contains more advanced recent results on hyperbolic SPDEs.

Let $L > 0$ be fixed, and consider

$$\left| \begin{array}{ll} \dfrac{\partial u}{\partial t} = \dfrac{\partial^2 u}{\partial x^2} + f(u)\dot{W}, & t > 0, \ x \in [0,L], \\[2mm] \dfrac{\partial u}{\partial x}(0,t) = \dfrac{\partial u}{\partial x}u(L,t) = 0, & t > 0, \\[2mm] u(x,0) = u_0(x), & x \in [0,L], \end{array} \right. \qquad (89)$$

where $\dot{W}$ is white noise with respect to some given filtration $\{\mathscr{F}_t\}_{t \geq 0}$, and $u_0 : [0,L] \to \mathbf{R}$ is a nonrandom, measurable, and bounded function. As regards the function $f : \mathbf{R} \to \mathbf{R}$, we assume that

$$K := \sup_{0 \leq x \neq y \leq L} \frac{|f(x) - f(y)|}{|y - x|} + \sup_{0 \leq x \leq L} |f(x)| < \infty. \qquad (90)$$

In other words, we assume that f is globally Lipschitz, as well as bounded.

Exercise 6.1. Recall that $f : \mathbf{R} \to \mathbf{R}$ is *globally Lipschitz* if there exists a constant A such that $|f(x) - f(y)| \leq A|x - y|$ for all $x, y \in \mathbf{R}$. Verify that any globally Lipschitz function $f : \mathbf{R} \to \mathbf{R}$ satisfies $|f(x)| = O(|x|)$ as $|x| \to \infty$. That is, prove that f has at most linear growth.

Now we multiply (89) by $\phi(x)$ and integrate $[dt\,dx]$ to find (formally, again) that for all $\phi \in C^\infty([0,L])$ with $\phi'(0) = \phi'(L) = 0$,

$$\begin{aligned}
&\int_0^L u(x,t)\phi(x)\,dx - \int_0^L u_0(x)\phi(x)\,dx \\
&= \int_0^t \int_0^L \frac{\partial^2 u}{\partial x^2}(x,s)\phi(x)\,dx\,ds + \int_0^t \int_0^L f(u(x,s))\,\phi(x)W(dx\,ds).
\end{aligned} \qquad (91)$$

Certainly we understand the stochastic integral now. But $\partial_{xx}u$ is not well defined. Therefore, we try and integrate by parts (again formally!): Because $\phi'(0) = \phi'(L) = 0$, the boundary-values of $\partial_x u$ [formally speaking] imply that

$$\int_0^t \int_0^L \frac{\partial^2 u}{\partial x^2}u(x,s)\phi(x)\,dx\,ds = \int_0^t \int_0^L u(x,s)\phi''(x)\,dx\,ds. \qquad (92)$$

And now we have ourselves a proper stochastic-integral equation: Find u such that for all $\phi \in C^\infty([0,L])$ with $\phi'(0) = \phi'(L) = 0$,

$$\begin{aligned}
&\int_0^L u(x,t)\phi(x)\,dx - \int_0^L u_0(x)\phi(x)\,dx \\
&= \int_0^t \int_0^L u(x,s)\phi''(x)\,dx\,ds + \int_0^t \int_0^L f(u(x,s))\,\phi(x)W(dx\,ds).
\end{aligned} \qquad (93)$$

Exercise 6.2 (Important). Argue that if u solves (93), then for all C^∞ functions $\psi(x,t)$ with $\partial_x \psi(0,t) = \partial_x \psi(L,t) = 0$,

$$\begin{aligned}
\int_0^L u(x,t)\psi(x,t)\,dx - \int_0^L u_0(x)\psi(x,0)\,dx \\
= \int_0^t \int_0^L u(x,s)\left[\frac{\partial^2 u}{\partial x^2}\psi(x,s) + \frac{\partial \psi}{\partial s}(x,s)\right] dx\,ds \\
+ \int_0^t \int_0^L f(u(x,s))\,\psi(x,s)W(dx\,ds).
\end{aligned} \qquad (94)$$

This is formal, but important.

Let $G_t(x, y)$ denote the Green's function for the linear heat equation. [The subscript t is *not* a derivative, but a variable.] Then it follows from the method of images that

$$G_t(x, y) = \sum_{n=-\infty}^{\infty} \left[\Gamma(t; x - y - 2nL) + \Gamma(t; x + y - 2nL) \right], \qquad (95)$$

where Γ is the fundamental solution to the linear heat equation (89); i.e.,

$$\Gamma(t; a) = \frac{1}{(4\pi t)^{1/2}} \exp\left(-\frac{a^2}{4t} \right). \qquad (96)$$

Define for all smooth $\phi : [0, L] \to \mathbf{R}$,

$$G_t(\phi, y) := \int_0^L G_t(x, y)\phi(x) \, dx, \qquad (97)$$

if $t > 0$, and $G_0(\phi, y) := \phi(y)$. We can integrate (89)—with $f(u, t) \equiv 0$—by parts for all C^∞ functions $\phi : [0, L] \to \mathbf{R}$ such that $\phi'(0) = \phi'(L) = 0$, and obtain the following:

$$G_t(\phi, y) = \phi(y) + \int_0^t G_s(\phi'' - \phi, y) \, ds. \qquad (98)$$

Fix $t > 0$ and define $\psi(x, s) := G_{t-s}(\phi, x)$ to find that ψ solves

$$\frac{\partial^2 \psi}{\partial x^2}(x, s) + \frac{\partial \psi}{\partial s}(x, s) = 0, \quad \psi(x, t) = \phi(x), \ \psi(x, 0) = G_t(\phi, x). \qquad (99)$$

Use this ψ in Exercise 6.2 to find that any solution to (89) must satisfy

$$\int_0^L u(x, t)\phi(x) \, dx - \int_0^L u_0(y)G_t(\phi, y) \, dy$$
$$= \int_0^t \int_0^L f(u(y, s)) \, G_{t-s}(\phi, y)W(dy \, ds). \qquad (100)$$

This must hold for all smooth ϕ with $\phi'(0) = \phi'(L) = 0$. Therefore, we would expect that for Lebesgue-almost all (x, t),

$$u(x, t) - \int_0^L u_0(y)G_t(x, y) \, dy$$
$$= \int_0^t \int_0^L f(u(y, s)) \, G_{t-s}(x, y)W(dy \, ds). \qquad (101)$$

If $\dot{W}$ were smooth then this reasoning would be rigorous and honest. As things are, it is still merely a formality. However, we are naturally led to a place where we have an honest stochastic-integral equation.

Definition 6.3. *By a "solution" to the formal stochastic heat equation* (89) *we mean a solution u to* (101) *that is adapted. Sometimes this is called a* mild solution.

With this nomenclature in mind, let us finally prove something.

Theorem 6.4. *The stochastic heat equation* (93) *subject to* (90) *has an a.s.-unique solution u that satisfies the following for all $T > 0$:*

$$\sup_{0 \le x \le L} \sup_{0 \le t \le T} \mathrm{E}\left(|u(x,t)|^2\right) < \infty. \tag{102}$$

For its proof we will need the following well-known result.

Lemma 6.5 (Gronwall's lemma). *Suppose $\phi_1, \phi_2, \ldots : [0,T] \to \mathbf{R}_+$ are measurable and non-decreasing. Suppose also that there exist a constant A such that for all integers $n \ge 1$, and all $t \in [0,T]$,*

$$\phi_{n+1}(t) \le A \int_0^t \phi_n(s)\, ds. \tag{103}$$

Then,

$$\phi_n(t) \le \phi_1(T) \frac{(At)^{n-1}}{(n-1)!} \quad \text{for all } n \ge 1 \text{ and } t \in [0,T]. \tag{104}$$

The preceding is proved by applying induction. I omit the details.

Remark 6.6. As a consequence of Gronwall's lemma, any positive power of $\phi_n(t)$ is summable in n. Also, if ϕ_n does not depend on n then it follows that $\phi_n \equiv 0$.

Proof (Theorem 6.4: Uniqueness). Suppose u and v both solve (101), and both satisfy the integrability condition (102). We wish to prove that u and v are modifications of one another. Let $d(x,t) := u(x,t) - v(x,t)$. Then,

$$d(x,t) = \int_0^t \int_0^L \left[f(u(y,s)) - f(v(y,s)) \right] G_{t-s}(x,y)\, W(dy\, ds). \tag{105}$$

According to Theorem 5.26 (p. 21) and (90),

$$\mathrm{E}\left(|d(x,t)|^2\right) \le K^2 \int_0^t \int_0^L \mathrm{E}\left(|d(y,s)|^2\right) G_{t-s}^2(x,s)\, dy\, ds. \tag{106}$$

Let $H(t) := \sup_{0 \le x \le L} \sup_{0 \le s \le t} \mathrm{E}[d^2(x,s)]$. The preceding implies that

$$H(t) \le K^2 \int_0^t H(s) \left(\int_0^L G_{t-s}^2(x,y)\, dy \right) ds. \tag{107}$$

Now from (95) and the semigroup properties of Γ it follows that

$$\int_0^L G_t(x,y)G_s(y,z)\,dy = G_{t+s}(x,z), \quad \text{and} \quad G_t(x,y) = G_t(y,x). \tag{108}$$

Consequently, $\int_0^L G_t^2(x,y)\,dy = G_{2t}(x,x) = Ct^{-1/2}$. Hence,

$$H(t) \le CK^2 \int_0^t \frac{H(s)}{|t-s|^{1/2}}\,ds. \tag{109}$$

Now choose and fix some $p \in (1,2)$, let q be the conjugate to p [i.e., $p^{-1} + q^{-1} = 1$], and apply Hölder's inequality to find that there exists $A = A_T$ such that uniformly for all $t \in [0,T]$,

$$H(t) \le A \left(\int_0^t H^q(s)\,ds \right)^{1/q}. \tag{110}$$

We can apply Gronwall's Lemma 6.5 with $\phi_1 = \phi_2 = \phi_3 = \cdots = H^q$ to find that $H(t) \equiv 0$. □

Proof (Theorem 6.4: Existence). Note from (95) that $\int_0^L G_t(x,y)\,dy$ is a number in $[0,1]$. Because u_0 is assumed to be bounded $\int_0^L u_0(y)G_t(x,y)\,dy$ is bounded; this is the first term in (101). Now we proceed with a Picard-type iteration scheme. Let $u_0(x,t) := u_0(x)$, and then iteratively define

$$u_{n+1}(x,t)$$
$$= \int_0^L u_0(y)G_t(x,y)\,dy + \int_0^t \int_0^L f(u_n(y,s))\,G_{t-s}(x,y)W(dy\,ds). \tag{111}$$

Define $d_n(x,t) := u_{n+1}(x,t) - u_n(x,t)$ to find that

$$d_n(x,t)$$
$$= \int_0^t \int_0^L [f(u_{n+1}(y,s)) - f(u_n(y,s))]\,G_{t-s}(x,y)\,W(dy\,ds). \tag{112}$$

Consequently, by (90),

$$\mathrm{E}\left(|d_n(x,t)|^2\right) \le K^2 \int_0^t \int_0^L \mathrm{E}\left(|d_{n-1}(y,s)|^2\right) G_{t-s}^2(x,y)\,dy\,ds. \tag{113}$$

Let $H_n^2(t) := \sup_{0 \le x \le L} \sup_{0 \le s \le t} \mathrm{E}(|d_n(x,s)|^2)$ to find that

$$H_n^2(t) \le CK^2 \int_0^t \frac{H_{n-1}^2(s)}{|t-s|^{1/2}}\,ds. \tag{114}$$

Choose and fix $p \in (0,2)$, and let q denote its conjugate so that $q^{-1} + p^{-1} = 1$. Apply Hölder's inequality to find that there exists $A = A_T$ such that uniformly for all $t \in [0,T]$,

$$H_n^2(t) \leq A \left(\int_0^t H_{n-1}^{2q}(s)\,ds \right)^{1/q}. \tag{115}$$

Apply Gronwall's Lemma 6.5 with $\phi_n := H_n^{2q}$ to find that $\sum_{n=1}^{\infty} H_n(t) < \infty$. Therefore, $u_n(t,x)$ converges in $L^2(P)$ to some $u(t,x)$ for each t and x. This proves also that

$$\lim_{n \to \infty} \int_0^t \int_0^L f(u_n(y,s))G_{t-s}(x,y)\,W(dy\,ds)$$
$$= \int_0^t \int_0^L f(u(y,s))G_{t-s}(x,y)\,W(dy\,ds), \tag{116}$$

where the convergence holds in $L^2(P)$. This proves that u is a solution to (101). □

We are finally ready to complete the picture by proving that the solution to (89) is continuous [up to a modification, of course].

Theorem 6.7. *There exists a continuous modification $u(x,t)$ of (89).*

Remark 6.8. In Exercise 6.9, on page 31 below, you will be asked to improve this to the statement that there exists a Hölder-continuous modification.

Proof (Sketch). We need the following easy-to-check fact about the Green's function G:

$$G_t(x,y) = \Gamma(t;x-y) + H_t(x,y), \tag{117}$$

where $H_t(x,y)$ is smooth in $(t,x,y) \in \mathbf{R}_+ \times \mathbf{R} \times \mathbf{R}$, and Γ is the "heat kernel" defined in (96). Define

$$U(x,t) := \int_0^t \int_0^L f(u(y,s))\Gamma(t-s;x-y)\,W(dy\,ds). \tag{118}$$

The critical step is to prove that U has a continuous modification. Because u_0 is bounded it is then not too hard to complete the proof based on this, and the fact that the difference between Γ and G is smooth and bounded. From here on I prove things honestly.

Let $0 \leq t \leq t'$ and note that

$$U(x,t') - U(x,t)$$
$$= \int_0^t \int_0^L f(u(y,s)) \left[\Gamma(t'-s;x-y) - \Gamma(t-s;x-y) \right] W(dy\,ds)$$
$$+ \int_t^{t'} \int_0^L f(u(y,s))\Gamma(t'-s;x-y)\,W(dy\,ds). \tag{119}$$

By Burkholder's inequality (Theorem 5.27, page 21) and the elementary inequality $|a + b|^p \le 2^p|a|^p + 2^p|b|^p$,

$$\mathrm{E}\left(|U(x,t) - U(x,t')|^p\right)$$

$$\le 2^p c_p \mathrm{E}\left[\left(\int_0^t \int_0^L f^2(u(y,s))\Lambda(s,t,t';x,y)\,dy\,ds\right)^{p/2}\right] \tag{120}$$

$$+ 2^p c_p \mathrm{E}\left[\left(\int_t^{t'} \int_0^L f^2(u(y,s))\Gamma^2(t - s;x - y)\,dy\,ds\right)^{p/2}\right].$$

where

$$\Lambda(s,t,t';x,y) := [\Gamma(t' - s;x - y) - \Gamma(t - s;x - y)]^2. \tag{121}$$

Because of (90), $\sup|f| \le K$. Therefore,

$$\mathrm{E}\left(|U(x,t) - U(x,t')|^p\right)$$

$$\le (2K)^p c_p \left(\int_0^t \int_{-\infty}^\infty \Lambda(s,t,t';x,y)\,dy\,ds\right)^{p/2} \tag{122}$$

$$+ (2K)^p c_p \left(\int_t^{t'} \int_{-\infty}^\infty \Gamma^2(t - s;x - y)\,dy\,ds\right)^{p/2}.$$

[Notice the change from $\int_0^L$ to $\int_{-\infty}^\infty$.] Because $\int_{-\infty}^\infty \Gamma^2(t - s;a)\,da$ is a constant multiple of $|t - s|^{-1/2}$,

$$\left(\int_t^{t'} \int_{-\infty}^\infty \Gamma^2(t - s;x - y)\,dy\,ds\right)^{p/2} = C_p|t' - t|^{p/4}. \tag{123}$$

For the other integral we use a method that is motivated by the ideas in [5]. Recall Plancherel's theorem: For all $g \in L^1(\mathbf{R}) \cap L^2(\mathbf{R})$,

$$\|g\|_{L^2(\mathbf{R})}^2 = \frac{1}{2\pi}\|\mathscr{F}g\|_{L^2(\mathbf{R})}^2, \tag{124}$$

where $(\mathscr{F}g)(z) := \int_{-\infty}^\infty g(x)e^{ixz}\,dx$ denotes the Fourier transform in the space variable. Because $(\mathscr{F}\Gamma)(t;\xi) = \exp(-t\xi^2)$,

$$\int_{-\infty}^\infty [\Gamma(t' - s;x - y) - \Gamma(t - s;x - y)]^2\,dy$$

$$= \frac{1}{2\pi}\int_{-\infty}^\infty \left[e^{-(t'-s)\xi^2} - e^{-(t-s)\xi^2}\right]^2\,d\xi \tag{125}$$

$$= \frac{1}{2\pi}\int_{-\infty}^\infty e^{-2(t-s)\xi^2}\left[1 - e^{-(t'-t)\xi^2}\right]^2\,d\xi.$$

Therefore,

$$\int_0^t \int_{-\infty}^\infty \left[\Gamma(t'-s\,;x-y) - \Gamma(t-s\,;x-y)\right]^2 dy\, ds$$

$$= \frac{1}{2\pi} \int_{-\infty}^\infty \left(\int_0^t e^{-2(t-s)\xi^2}\, ds\right) \left[1 - e^{-(t'-t)\xi^2}\right]^2 d\xi \quad (126)$$

$$= \frac{1}{4\pi} \int_{-\infty}^\infty \frac{1 - e^{-2t\xi^2}}{\xi^2} \left[1 - e^{-(t'-t)\xi^2}\right]^2 d\xi.$$

A little thought shows that $(1 - e^{-2t\xi^2})/\xi^2 \le C_T/(1+\xi^2)$, uniformly for all $0 \le t \le T$. Also, $[1 - e^{-(t'-t)\xi^2}]^2 \le 2\min[(t'-t)\xi^2\,,1]$. Therefore,

$$\int_0^t \int_{-\infty}^\infty \left[\Gamma(t'-s\,;x-y) - \Gamma(t-s\,;x-y)\right]^2 dy\, ds$$

$$\le \frac{C_T}{\pi} \int_0^\infty \frac{\min[(t'-t)\xi^2\,,1]}{1+\xi^2}\, d\xi \quad (127)$$

$$\le \frac{C_T}{\pi} \left(\int_{|t'-t|^{-1/2}}^\infty \frac{d\xi}{\xi^2} + \int_0^{|t'-t|^{-1/2}} \frac{(t'-t)\xi^2}{1+\xi^2}\, d\xi\right).$$

The first term is equal to $A|t'-t|^{1/2}$, and the second term is also bounded above by $|t'-t|^{1/2}$ because $\xi^2/(1+\xi^2) \le 1$. This, (122) and (123) together prove that

$$\mathrm{E}\left(|U(x\,,t) - U(x\,,t')|^p\right) \le C_p |t'-t|^{p/4}. \quad (128)$$

Similarly, we can prove that for all $x, x' \in [0\,,L]$,

$$\mathrm{E}\left(|U(x\,,t) - U(x'\,,t)|^p\right)$$

$$\le c_p K^p \left(\int_0^t \int_{-\infty}^\infty \left|\Gamma(t-s\,;y) - \Gamma(t-s\,;x'-x-y)\right|^2 dy\, ds\right)^{p/2}. \quad (129)$$

By Plancherel's theorem, and because the Fourier transform of $x \mapsto g(x+a)$ is $e^{-i\xi a}(\mathscr{F}g)(\xi)$,

$$\int_{-\infty}^\infty \left|\Gamma(t-s\,;y) - \Gamma(t-s\,;x'-x-y)\right|^2 dy$$

$$= \frac{1}{2\pi} \int_{-\infty}^\infty e^{-2(t-s)\xi^2} \left|1 - e^{i\xi(x'-x)}\right|^2 d\xi. \quad (130)$$

Consequently, we can apply Tonelli's theorem to find that

$$\int_0^t \int_{-\infty}^\infty \left|\Gamma(t-s\,;y) - \Gamma(t-s\,;x'-x-y)\right|^2 dy\, ds$$

$$= \frac{1}{2\pi} \int_{-\infty}^\infty \frac{1 - e^{-2t\xi^2}}{2\xi^2} \left|1 - e^{i\xi(x'-x)}\right|^2 d\xi \quad (131)$$

$$= \frac{1}{4\pi} \int_0^\infty \frac{1 - e^{-2t\xi^2}}{\xi^2} \left|1 - \cos(\xi(x'-x))\right| d\xi.$$

We use the elementary bounds $1 - \exp(-|\theta|) \leq 1$, and $1 - \cos\theta \leq \min(1, \theta^2)$—valid for all $\theta \in \mathbf{R}$—in order to bound the preceding, and obtain

$$\int_0^t \int_{-\infty}^\infty \left| \Gamma(t - s\,;y) - \Gamma(t - s\,;x' - x - y) \right|^2 dy\, ds$$
$$\leq \frac{1}{4\pi} \int_0^\infty \frac{\xi^2 (x' - x)^2 \wedge 1}{\xi^2}\, d\xi. \tag{132}$$

We split the domain of integration into two domains: Where $\xi < |x' - x|^{-1}$; and where $\xi \geq |x' - x|^{-1}$. Each of the two resulting integrals is easy enough to compute explicitly, and we obtain

$$\int_0^t \int_{-\infty}^\infty \left| \Gamma(t - s\,;y) - \Gamma(t - s\,;x' - x - y) \right|^2 dy\, ds \leq \frac{|x' - x|}{2\pi} \tag{133}$$

as a result. Hence, it follows that

$$\sup_{t \geq 0} \mathrm{E}\left(|U(x, t) - U(x', t)|^p \right) \leq a_p |x' - x|^{p/2}. \tag{134}$$

For all $(x, t) \in \mathbf{R}^2$ define $|(x, t)| := |x|^{1/2} + |t|^{1/4}$. This defines a norm on $\mathbf{R}^2$, and is equivalent to the usual Euclidean norm $(x^2 + t^2)^{1/2}$ in the sense that both generate the same topology. Moreover, we have by (128) and (134): For all $t, t' \in [0, T]$ and $x, x' \in [0, L]$,

$$\mathrm{E}\left(|U(x, t) - U(x', t')|^p \right) \leq A\, |(x, t) - (x', t')|^p. \tag{135}$$

This and Kolmogorov's continuity theorem (Theorem 4.13, page 14) together prove that U has a modification which is continuous, in our inhomogeneous norm on (x, t), of any order < 1. Because our norm is equivalent to the usual Euclidean norm, this proves continuity in the ordinary sense.   □

Exercise 6.9. Complete the proof. Be certain that you understand why we have derived Hölder continuity. For example, prove that there is a modification of our solution which is Hölder continuous in x of any given order $< \frac{1}{2}$; and it is Hölder continuous in t of any given order $< \frac{1}{4}$.

Exercise 6.10. Consider the constant-coefficient, free-space stochastic heat equation in two space variables. For instance, here is one formulation: Let $\dot{W}(x, t)$ denote white noise on $(x, t) \in \mathbf{R}^2 \times \mathbf{R}_+$, and consider

$$\left| \begin{array}{ll} \dfrac{\partial u}{\partial t} = \left(\dfrac{\partial^2 u}{\partial x_1^2} + \dfrac{\partial^2 u}{\partial x_2^2} \right) + \dot{W} & t > 0,\ x \in \mathbf{R}^2, \\[3mm] u(x, 0) = 0 & x \in \mathbf{R}^2. \end{array} \right. \tag{136}$$

Interpret the adapted solution to the preceding as

$$u(x,t) = \int_0^t \int_{\mathbf{R}^2} \Gamma(t-s\,;x-y)\,W(dy\,ds), \tag{137}$$

subject to $(t,x) \mapsto \mathrm{E}[u^2(t,x)]$ being continuous (say!). Here, Γ is the heat kernel on $\mathbf{R}^2$; that is, $\Gamma(t,x) := (4\pi t)^{-1}\exp(-\|x\|^2/(4t))$. Prove that $\mathrm{E}[u^2(x,t)] = \infty$ for all $x \in \mathbf{R}^2$ and $t > 0$. Prove also that if $u(x,t)$ were a proper stochastic process then it would have to be a Gaussian process, but this cannot be because Gaussian processes have finite moments. Therefore, in general, one cannot hope to find function-valued solutions to the stochastic heat equation in spatial dimensions ≥ 2.

7 From Chaos to Order

Finally, I mention an example of SPDEs that produce smooth solutions for all times $t > 0$, and yet the solution is white noise at time $t = 0$. In this way, one can think of the solution to the forthcoming SPDE as a smooth deformation of white noise, where the deformation is due to the action of the heat operator.

Now consider the heat equation on $[0,1]$, but with random initial data instead of random forcing terms. More specifically, we consider the stochastic process $\{u(x,t)\}_{0 \leq x \leq 1, t \geq 0}$ that is formally defined by

$$\left| \begin{aligned} &\frac{\partial u}{\partial t}(x,t) = \frac{\partial^2 u}{\partial x^2}(x,t) && 0 < x < 1,\ t \geq 0 \\ &u(0,t) = u(1,t) = 0 && t > 0 \\ &u(x,0) = \dot{W}(x) && 0 < x < 1, \end{aligned} \right. \tag{138}$$

where $\dot{W}$ denotes white noise.

A classical interpretation of (138) follows: Consider an infinitesimally-thin wire of length one that has even density and width. Interpret this wire as the interval $[0,1]$, and apply totally random heat to the wire, the heat amount at x being $\dot{W}(x)$ units. The endpoints of the wire are perfectly cooled. If we watch the wire cool as time passes, then the amount of heat retained at position x at time $t > 0$ is $u(x,t)$.

If $\dot{W}$ were replaced by a square-integrable function then the solution is classical, and is given by

$$u(x,t) = \sqrt{2}\sum_{n=1}^{\infty} \xi_n \sin(n\pi x)\exp\left(-n^2\pi^2 t\right), \tag{139}$$

where

$$\xi_n := \sqrt{2}\int_0^1 \dot{W}(x)\sin(n\pi x)\,dx, \tag{140}$$

and the infinite sum in (139) converges in $L^2(dx)$ for each $t > 0$, for example. Although $\dot{W}$ is not a square-integrable function, one can first consider "weak solutions," and then proceed to integrate by parts, and thus arrive at the *mild solution* to (138). That is described by (139), but with (140) replaced by the Wiener stochastic integrals

$$\xi_n := \sqrt{2} \int_0^1 \sin(n\pi x) W(dx), \qquad n = 1, 2, \dots . \tag{141}$$

It follows from our construction of Wiener integrals that $\{\xi_n\}_{n=1}^\infty$ is a mean-zero Gaussian process. Thanks to the Wiener isometry (20), we also can compute its covariance structure to find that

$$\mathrm{Cov}(\xi_n, \xi_m) = 2 \int_0^1 \sin(n\pi x) \sin(m\pi x) \, dx = \begin{cases} 1 & \text{if } m = n, \\ 0 & \text{if } m \neq n. \end{cases} \tag{142}$$

Consequently, $\{\xi_n\}_{n=1}^\infty$ is an i.i.d. sequence of standard-normal variates. The following lemma controls the rate of growth of the ξ_n's.

Lemma 7.1. *With probability one,*

$$|\xi_n| = O\left(\sqrt{\ln n}\right) \qquad \text{as } n \to \infty. \tag{143}$$

Proof. We can apply Chebyshev's inequality to find that for all $a, \lambda > 0$,

$$P\{\xi_n \geq a\} \leq e^{-\lambda a} E \exp(\lambda \xi_1) = \exp\left(-\lambda a + \frac{\lambda^2}{2}\right). \tag{144}$$

The optimal choice of λ is $a/2$, and this yields the following well-known bound: $P\{\xi_n \geq a\} \leq \exp(-a^2/2)$, valid for all $a > 0$. By symmetry,

$$P\{|\xi_n| \geq a\} \leq 2\exp(-a^2/2) \qquad \text{for all } a > 0. \tag{145}$$

We plug in $a := 2\sqrt{\ln n}$ and deduce (143) from

$$\sum_{n \geq 100} P\left\{|\xi_n| \geq 2\sqrt{\ln n}\right\} \leq \sum_{n \geq 100} \frac{2}{n^2} < \infty \tag{146}$$

and the Borel–Cantelli lemma. $\qquad \square$

Exercise 7.2. Improve Lemma 7.1 to the statement that

$$\limsup_{n \to \infty} \frac{\xi_n}{\sqrt{2\ln n}} = -\liminf_{n \to \infty} \frac{\xi_n}{\sqrt{2\ln n}} = 1 \qquad \text{a.s.} \tag{147}$$

An immediate consequence of Lemma 7.1 is that for all fixed $0 < r < R$, the infinite series in (139) converges a.s., uniformly for $(x, t) \in [0, 1] \times [r, R]$.

Among other things, this proves that u is infinitely differentiable in both variables, away from time zero.

Thus, the random function u is smooth except near time zero, where its behavior is chaotic. In words, the heat operator takes the pure-noise initial condition "$u(x,0) = \dot{W}(x)$" and immediately smooths it to generate nice random functions $u(x,t)$, one for every $t > 0$. Thus, it is interesting to investigate the transition from "chaos" $[t = 0]$ to "order" $[t > 0]$ in greater depth.

Here we study the mentioned blowup problem for average x-values, and plan to prove that there is a sense in which the following holds for all "typical values of x":

$$u(x,t) \approx t^{-1/4} \qquad \text{when } t \approx 0. \tag{148}$$

Define

$$\mathscr{E}(t) := \left(\int_0^1 |u(x,t)|^2 \, dx \right)^{1/2}. \tag{149}$$

A classical interpretation of $\mathscr{E}(t)$ is the average heat—in the sense of $L^2(dx)$—in the wire at time t, where the wire at time 0 is subjected to heat amount $\dot{W}(x)$ at position $x \in [0,1]$. The following rigorous interpretation of (148) is a rather simple result that describes roughly the nature of the blowup of the solution near time zero.

Theorem 7.3. *With probability one,*

$$\lim_{t \searrow 0} t^{1/4} \mathscr{E}(t) = \frac{1}{(2\pi)^{3/4}}. \tag{150}$$

The proof of Theorem 7.3 relies on a lemma from calculus.

Lemma 7.4. *The following holds:*

$$\lim_{\lambda \searrow 0} \lambda^{1/2} \sum_{n=1}^{\infty} e^{-n^2 \lambda} = \frac{1}{2\sqrt{\pi}}. \tag{151}$$

Proof. Because $\int_0^\infty \exp(-x^2 \lambda) \, dx = 1/(2\sqrt{\pi\lambda})$,

$$\int_8^\infty e^{-x^2 \lambda} \, dx = O(1) + \frac{1}{2\sqrt{\pi\lambda}} \qquad \text{as } \lambda \searrow 0. \tag{152}$$

Because $\sum_{k=1}^8 \exp(-n^2 \lambda) = O(1)$ as $\lambda \searrow 0$, it therefore suffices to prove that

$$T := \left| \sum_{n=9}^{\infty} e^{-n^2 \lambda} - \int_8^\infty e^{-x^2 \lambda} \, dx \right| = o\left(\frac{1}{\sqrt{\lambda}} \right) \qquad \text{as } \lambda \searrow 0. \tag{153}$$

To prove this we first write T as

$$T = \sum_{n=9}^{\infty} \int_{n-1}^{n} e^{-x^2\lambda} \left(1 - e^{-(n^2-x^2)\lambda}\right) dx. \tag{154}$$

Because $1 - \exp(-\theta) \leq 1 \wedge \theta$ for all $\theta \geq 0$, and since $n^2 - x^2 \leq 4x$ for all $x \in [n-1, n]$ and $n \geq 1$,

$$\begin{aligned} T &\leq 4 \int_8^{\infty} e^{-x^2\lambda} \left(1 \wedge x\lambda\right) dx \\ &\leq \frac{4}{\sqrt{\lambda}} \int_0^{\infty} e^{-y^2} \left(1 \wedge y\sqrt{\lambda}\right) dy, \end{aligned} \tag{155}$$

and this is $o(1/\sqrt{\lambda})$ by the dominated convergence theorem. This proves (153), and hence the lemma.

$\square$

Next we prove Theorem 7.3.

Proof (Theorem 7.3). Equation (142) and the uniform convergence of the series in (139) together imply that for all $t > 0$,

$$\mathscr{E}^2(t) = \sum_{n=1}^{\infty} \xi_n^2 e^{-2n^2\pi^2 t} \qquad \text{a.s.} \tag{156}$$

Consequently, Lemma 7.4 implies that

$$E\left(|\mathscr{E}(t)|^2\right) = \sum_{n=1}^{\infty} e^{-2n^2\pi^2 t} = \frac{1 + o(1)}{(2\pi)^{3/2}\sqrt{t}} \qquad \text{as } t \searrow 0. \tag{157}$$

Because the ξ_n's are independent, a second application of Lemma 7.4 yields

$$\begin{aligned} \text{Var}\left(|\mathscr{E}(t)|^2\right) &= \text{Var}(\xi_1^2) \sum_{n=1}^{\infty} e^{-4n^2\pi^2 t} \\ &= O\left(E\left(|\mathscr{E}(t)|^2\right)\right) \qquad \text{as } t \searrow 0. \end{aligned} \tag{158}$$

These remarks, together with the Chebyshev inequality, yield two constants $C, \varepsilon > 0$ such that for all $t \in (0, \varepsilon)$ and $\delta > 0$,

$$P\left\{\left|\frac{\mathscr{E}^2(t)}{E\left(|\mathscr{E}(t)|^2\right)} - 1\right| > \delta\right\} \leq C\sqrt{t}. \tag{159}$$

We can replace t by k^{-4}, sum both sides from $k = 1$ to $k = \infty$, apply the Borel–Cantelli lemma, and then finally deduce that

$$\lim_{k \to \infty} \frac{\mathscr{E}^2\left(k^{-4}\right)}{E\left(|\mathscr{E}\left(k^{-4}\right)|^2\right)} = 1 \qquad \text{a.s.} \tag{160}$$

Because $\mathscr{E}^2$ is non-increasing, (157) and a monotonicity argument together finish the proof.

$\square$

Exercise 7.5 (Rapid cooling). Prove that with probability one,

$$\lim_{t \nearrow \infty} \exp\left(\pi^2 t\right) \mathscr{E}(t) = 1. \tag{161}$$

That is, the wire cools rapidly as time goes by, as it does for classical initial heat profiles. Thus, the only new phenomenon occurs near time zero.

Exercise 7.6. Define the *average heat flux* in the wire as

$$\mathscr{F}(t) := \left(\int_0^1 \left| \frac{\partial u}{\partial x}(x,t) \right|^2 dx \right)^{1/2}. \tag{162}$$

Describe the blowup rate of $\mathscr{F}(t)$ as t tends down to zero.

For a greater challenge try the following.

Exercise 7.7. Prove that as $t \searrow 0$, and after suitable centering and normalization, $\mathscr{E}(t)$ converges in distribution to a non-degenerate law. Describe that law.

Exercise 7.8. Prove that $\{b(x)\}_{0 \leq x \leq 1}$ is a Brownian bridge, where

$$b(x) := \frac{1}{\sqrt{\pi}} \int_0^\infty \frac{u(x,t)}{\sqrt{t}} dt \qquad \text{for all } x \in [0,1]. \tag{163}$$

References

[1] Louis Bachelier (1900). Théorie de la Spéculation, *Ann. Sci. École Norm. Sup.* **17**, 21–86. [See also the 1995 reprint. Sceaux: Gauthier–Villars.]

[2] V. V. Baklan (1965). The existence of solutions for a class of equations involving variational derivatives, *Dopovidi Akad. Nauk Ukraïn. RSR* **1965**, 554–556 (Ukranian. Russian, English summary)

[3] D. L. Burkholder (1971). Martingale inequalities, In: *Lecture Notes in Math.* **190**, 1–8 Springer-Verlag, Berlin

[4] E. M. Cabaña (1970). The vibrating string forced by white noise, *Z. Wahrscheinlichkeitstheorie Verw. Gebiete* **15**, 111–130

[5] Robert C. Dalang (1999). Extending the martingale measure stochastic integral with applications to spatially homogeneous s.p.d.e.'s, *Electron. J. Probab.* **4**, no. 6, 29 pages (electronic)

[6] Robert C. Dalang, Davar Khoshnevisan, and Eulalia Nualart (2007). Hitting probabilities for systems of non-linear stochastic heat equations with additive noise, *Latin American J. Probab. and Math. Statist.* (or *Alea*; http://alea.impa.br/english), Vol. III, 231–371

[7] Ju. L. Daleckiĭ(1967). Infinite-dimensional elliptic operators and the corresponding parabolic equations, *Uspehi Mat. Nauk* **22**(4) (136), 3–54 (In Russian) [English translation in: *Russian Math. Surveys* **22**(4), 1–53, 1967]

[8] D. A. Dawson (1975). Stochastic evolution equations and related measure processes, *J. Multivariate Anal.* **5**, 1–52

[9] D. A. Dawson (1972). Stochastic evolution equations, *Math. Biosci.* **15**, 287–316

[10] J. L. Doob (1942). The Brownian movement and stochastic equations, *Ann. of Math.* **43**(2), 351–369

[11] R. M. Dudley (1967). The sizes of compact subsets of Hilbert space and continuity of Gaussian processes, *J. Functional Analysis*, **1**, 290–330

[12] X. Fernique (1975). Regularité des trajectoires des fonctions aléatoires gaussiennes, In: *Lecture Notes in Math.* **480**, 1–96 Springer-Verlag, Berlin (in French)

[13] Tadahisa Funaki (1984). Random motion of strings and stochastic differential equations on the space $C([0,1], \mathbf{R}^d)$, In: *Stochastic Analysis (Katata/ Kyoto, 1982)*, North-Holland Math. Library, **32**, 121–133, North-Holland, Amsterdam

[14] Tadahisa Funaki (1983). Random motion of strings and related stochastic evolution equations, *Nagoya Math. J.* **89**, 129–193

[15] Kiyosi Itô (1944). Stochastic integral, *Proc. Imp. Acad. Tokyo* **20**, 519–524

[16] Kiyosi Itô (1950). Stochastic differential equations in a differentiable manifold, *Nagoya Math. J.* **1**, 35–47

[17] Kiyosi Itô (1951). On a formula concerning stochastic differentials, *Nagoya Math. J.* **3**, 55–65

[18] A. Ya. Khintchine (1933). *Asymptotische Gesetz der Wahrscheinlichkeitsrechnung*, Springer, Berlin

[19] A. N. Kolmogorov (1933). *Grundbegriffe der Wahrscheinlichkeitsrechnung*, Springer, Berlin

[20] N. V. Krylov and B. L. Rozovskiĭ (1979a). Itô equations in Banach spaces and strongly parabolic stochastic partial differential equations, *Dokl. Akad. Nauk SSSR* **249**(2), 285–289 (in Russian)

[21] N. V. Krylov and B. L. Rozovskiĭ (1979b). Stochastic evolution equations, In: *Current Problems in Mathematics, Vol. 14* Akad. Nauk SSSR, Vsesoyuz. Inst. Nauchn. i Tekhn. Informatsii, Moscow, 71–147, 256 (in Russian)

[22] N. V. Krylov, N. V. and B. L. Rozovskiĭ (1977). The Cauchy problem for linear stochastic partial differential equations, *Izv. Akad. Nauk SSSR Ser. Mat.* **41**(6), 1329–1347, 1448 (in Russian)

[23] H. Kunita (1991). *Stochastic Flows and Stochastic Differential Equations*, Cambridge University Press, Cambridge

[24] Étienne Pardoux (1975). *Equations aux dérivées partielles stochastiques non linéaires monotones—Étude de solutions fortes de type Itô*, Thése d'État, Univ. Paris XI, Orsay

[25] Étienne Pardoux (1972). Sur des équations aux dérivées partielles stochastiques monotones, *C. R. Acad. Sci. Paris Sér. A–B* **275**, A101–A103

[26] Christopher Preston (1972). Continuity properties of some Gaussian processes, *Ann. Math. Statist.* **43**, 285–292

[27] Michel Talagrand (1985). Régularité des processus gaussiens, C. R. Acad. Sci. Paris Sér. I Math. **301**(7), 379–381 (French, with English summary)

[28] Michel Talagrand (1987). Regularity of Gaussian processes, *Acta Math.* **159***(1–2)*, 99–149

[29] G. E. Uhlenbeck and L. S. Ornstein (1930). On the theory of Brownian Motion, *Phys. Rev.* **36**, 823–841

[30] John B. Walsh (1986). *An Introduction to Stochastic Partial Differential Equations*, In: Lecture Notes in Math. **1180**, 265–439, Springer, Berlin

[31] N. Wiener (1923). Differential space, *J. Math. Phys.* **2**, 131–174

[32] Norbert Wiener (1938). The Homogeneous Chaos, *Amer. J. Math.* **60***(4)*, 897–936

The Stochastic Wave Equation

Robert C. Dalang

Summary. These notes give an overview of recent results concerning the non-linear stochastic wave equation in spatial dimensions $d \geq 1$, in the case where the driving noise is Gaussian, spatially homogeneous and white in time. We mainly address issues of existence, uniqueness and Hölder–Sobolev regularity. We also present an extension of Walsh's theory of stochastic integration with respect to martingale measures that is useful for spatial dimensions $d \geq 3$.

1 Introduction

The stochastic wave equation is one of the fundamental stochastic partial differential equations (SPDEs) of hyperbolic type. The behavior of its solutions is significantly different from those of solutions to other SPDEs, such as the stochastic heat equation. In this introductory section, we present two real-world examples that can motivate the study of this equation, even though in neither case is the mathematical technology sufficiently developed to answer the main questions of interest. It is however pleasant to have such examples in order to motivate the development of rigorous mathematics.

Example 1.1 (The motion of a strand of DNA). A DNA molecule can be viewed as a long elastic string, whose diameter is essentially infinitely small compared to its length. We can describe the position of the string by using a parameterization defined on $\mathbf{R}_+ \times [0, 1]$ with values in $\mathbf{R}^3$:

$$\mathbf{u}(t, x) = \begin{pmatrix} u_1(t, x) \\ u_2(t, x) \\ u_3(t, x) \end{pmatrix}. \tag{1}$$

Here, $\mathbf{u}(t, x)$ is the position at time t of the point labelled x on the string, where $x \in [0, 1]$ represents the distance from this point to one extremity of

D. Khoshnevisan and F. Rassoul-Agha (eds.) *A Minicourse on Stochastic Partial Differential Equations.*
Lecture Notes in Mathematics 1962.

the string if the string were straightened out. The unit of length is chosen so that the entire string has length 1.

A DNA molecule typically "floats" in a fluid, so it is constantly in motion, just as a particle of pollen floating in a fluid moves according to Brownian motion. The motion of the string can be described by Newton's law of motion, which equates the sum of forces acting on the string with the product of the mass and the acceleration. Let $\mu = 1$ be the mass of the string per unit length. The acceleration at position x along the string, at time t, is

$$\frac{\partial^2 \mathbf{u}}{\partial t^2}(t, x), \tag{2}$$

and the forces acting on the string are mainly of three kinds: elastic forces $\mathbf{F_1}$, which include torsion forces, friction due to viscosity of the fluid $\mathbf{F_2}$, and random impulses $\mathbf{F_3}$ due the the impacts on the string of the fluid's molecules. Newton's equation of motion can therefore be written

$$1 \cdot \frac{\partial^2 \mathbf{u}}{\partial t^2} = \mathbf{F_1} - \mathbf{F_2} + \mathbf{F_3}. \tag{3}$$

This is a rather complicated system of three stochastic partial differential equations, and it is not even clear how to write down the torsion forces or the friction term. Elastic forces are generally related to the second derivative in the spatial variable, and the molecular forces are reasonably modelled by a stochastic noise term.

The simplest 1-dimensional equation related to this problem, in which one only considers vertical displacement and forgets about torsion, is the following one, in which $u(t, x)$ is now scalar valued:

$$\frac{\partial^2 u}{\partial t^2}(t, x) = \frac{\partial^2 u}{\partial x^2}(t, x) - \int_0^1 k(x, y)\, u(t, y)\, dy + \dot{F}(t, x), \tag{4}$$

where the first term on the right-hand side represents the elastic forces, the second term is a (non-local) friction term, and the third term $\dot{F}(t, y)$ is a Gaussian noise, with spatial correlation $k(\cdot, \cdot)$, that is,

$$\mathrm{E}(\dot{F}(t, x)\, \dot{F}(s, y)) = \delta_0(t - s)\, k(x, y), \tag{5}$$

where δ_0 denotes the Dirac delta function. The function $k(\cdot, \cdot)$ is the same in the friction term and in the correlation.

Why is the motion of a DNA strand of biological interest? When a DNA strand moves around and two normally distant parts of the string get close enough together, it can happen that a biological event occurs: for instance, an enzyme may be released. Therefore, some biological events are related to the motion of the DNA string. Some mathematical results for equation (4) can be found in [20]. Some of the biological motivation for the specific form of equation (4) can be found in [8].

Example 1.2 (The internal structure of the sun). The study of the internal structure of the sun is an active area of research. One important international project is known as Project SOHO (Solar and Heliospheric Observatory) [9]. Its objective was to use measurements of the motion of the sun's surface to obtain information about the internal structure of the sun. Indeed, the sun's surface moves in a rather complex manner: at any given time, any point on the surface is typically moving towards or away from the center. There are also waves going around the surface, as well as shock waves propagating through the sun itself, which cause the surface to pulsate.

A question of interest to solar geophysicists is to determine the origin of these shock waves. One school of thought is that they are due to turbulence, but the location and intensities of the shocks are unknown, so a probabilistic model can be considered.

A model that was proposed by P. Stark of U.C. Berkeley is that the main source of shocks is located in a spherical zone inside the sun, which is assumed to be a ball of radius R. Assuming that the shocks are randomly located on this sphere, the equation (known as the Navier equation) for the dilatation (see [6, Section 8.3]) throughout the sun would be

$$\frac{\partial^2 u}{\partial t^2}(t\,,x) = c^2(x)\,\rho_0(x)\left(\boldsymbol{\nabla}\cdot\left(\frac{1}{\rho_0(x)}\,\boldsymbol{\nabla}u\right) + \boldsymbol{\nabla}\cdot\mathbf{F}(t\,,x)\right), \qquad (6)$$

where $x \in B(0\,,R)$, the ball centered at the origin with radius R, $c^2(x)$ is the speed of wave propagation at position x, $\rho_0(x)$ is the density at position x and the vector $\mathbf{F}(t\,,x)$ models the shock that originates at time t and position x.

A model for $\mathbf{F}$ that corresponds to the description of the situation would be 3-dimensional Gaussian noise concentrated on the sphere $\partial B(0\,,r)$, where $0 < r < R$. A possible choice of the spatial correlation for the components of $\mathbf{F}$ would be

$$\delta(t - s)\,f(x\cdot y), \qquad (7)$$

where $x\cdot y$ denotes the Euclidean inner product. A problem of interest is to estimate r from the available observations of the sun's surface. Some mathematical results relevant to this problem are developed in [3].

2 The Stochastic Wave Equation

Equation (6) is a wave equation for a medium with non-constant density. The (simpler) constant coefficient stochastic wave equation with real-valued noise that we will be studying in these notes reads as follows: For all $(t\,,x) \in [0\,,T] \times \mathbf{R}^d$,

$$
\begin{cases}
\left(\dfrac{\partial^2 u}{\partial t^2} - \Delta u \right)(t,x) = \sigma(t,x,u(t,x))\,\dot{F}(t,x) + b(t,x,u(t,x)), \\[2mm]
u(0,x) = v_0(x), \\[2mm]
\dfrac{\partial u}{\partial t}(0,x) = \tilde{v}_0(x),
\end{cases}
\tag{8}
$$

where $\dot{F}(t,x)$ is a (real-valued) Gaussian noise, which we take to be space-time white noise for the moment, and $\sigma, b : \mathbf{R}_+ \times \mathbf{R}^d \times \mathbf{R} \to \mathbf{R}$ are functions that satisfy standard properties, such as being Lipschitz in the third variable. The term Δu denotes the Laplacian of u in the x-variables.

Mild Solutions of the Stochastic Wave Equation

It is necessary to specify the notion of solution to (8) that we are considering. We will mainly be interested in the notion of *mild solution*, which is the following integral form of (8):

$$
\begin{aligned}
u(t,x) &= \int_{[0,t] \times \mathbf{R}^d} G(t-s,x-y)\left[\sigma(s,y,u(s,y))\,\dot{F}(s,y) + b(s,y,u(s,y))\right] ds\,dy \\
&\quad + \left(\frac{d}{dt} G(t) * v_0 \right)(x) + (G(t) * \tilde{v}_0)(x).
\end{aligned}
\tag{9}
$$

In this equation, $G(t-s,x-y)$ is Green's function of (8), which we discuss next, and $*$ denotes convolution in the x-variables. For the term involving $\dot{F}(s,y)$, a notion of stochastic integral is needed, that we will discuss later on.

Green's Function of a PDE

We consider first the case of an equation with constant coefficients. Let L be a partial differential operator with constant coefficients. A basic example is the wave operator

$$
Lf = \frac{\partial^2 f}{\partial t^2} - \Delta f.
\tag{10}
$$

Then there is a (Schwartz) distribution $G \in \mathscr{S}'(\mathbf{R}_+ \times \mathbf{R}^d)$ such that the solution of the PDE

$$
L u = \varphi, \qquad \varphi \in \mathscr{S}(\mathbf{R}^d),
\tag{11}
$$

is

$$
u = G \underset{(t,x)}{*} \varphi
\tag{12}
$$

where $\underset{(t,x)}{*}$ denotes convolution in the (t,x)-variables. We recall that $\mathscr{S}(\mathbf{R}^d)$ denotes the space of smooth test functions with rapid decrease, and $\mathscr{S}'(\mathbf{R}_+ \times \mathbf{R}^d)$ denotes the space of tempered distributions [15].

When G is a function, this convolution can be written

$$u(t,x) = \int_{\mathbf{R}_+ \times \mathbf{R}^d} G(t-s,x-y)\,\varphi(s,y)\,ds\,dy. \tag{13}$$

We note that this is the solution with vanishing initial conditions.

In the case of an operator with non-constant coefficients, such as

$$Lf = \frac{\partial^2 f}{\partial t^2} + 2c(t,x)\,\frac{\partial f}{\partial t} + \frac{\partial^2 f}{\partial x^2} \qquad (d=1), \tag{14}$$

Green's function has the form $G(t,x\,;\,s,y)$ and the solution of

$$Lu = \varphi \tag{15}$$

is given by the expression

$$u(t,x) = \int_{\mathbf{R}_+ \times \mathbf{R}^d} G(t,x\,;\,s,y)\,\varphi(s,y)\,ds\,dy. \tag{16}$$

Example 2.1 (The heat equation). The partial differential operator L is

$$Lu = \frac{\partial u}{\partial t} - \Delta u, \qquad d \geq 1, \tag{17}$$

and Green's function is

$$G(t,x) = (2\pi t)^{-d/2}\,\exp\left(-\frac{|x|^2}{2t}\right). \tag{18}$$

This function is smooth except for a singularity at $(0,0)$.

Example 2.2 (The wave equation). The partial differential operator L is

$$Lu = \frac{\partial^2 u}{\partial t^2} - \Delta u. \tag{19}$$

The form of Green's function depends on the dimension d. We refer to [18] for $d \in \{1,2,3\}$ and to [7] for $d > 3$. For $d = 1$, it is

$$G(t,x) = \frac{1}{2}\,1_{\{|x|<t\}}, \tag{20}$$

which is a bounded but discontinuous function. For $d = 2$, it is

$$G(t,x) = \frac{1}{\sqrt{2\pi}}\,\frac{1}{\sqrt{t^2-|x|^2}}\,1_{\{|x|<t\}}. \tag{21}$$

This function is unbounded and discontinuous. For $d = 3$, the "Green's function" is

$$G(t, dx) = \frac{1}{4\pi} \frac{\sigma_t(dx)}{t},$$ (22)

where σ_t is uniform measure on $\partial B(0, t)$, with total mass $4\pi t^2$. In particular, $G(t, \mathbf{R}^3) = t$. This Green's function is in fact *not* a function, but a measure. Its convolution with a test function φ is given by

$$
\begin{aligned}
(G * \varphi)(t, x) &= \frac{1}{4\pi} \int_0^t ds \int_{\partial B(0,s)} \varphi(t - s, x - y) \frac{\sigma_s(dy)}{s} \\
&= \frac{1}{4\pi} \int_0^t ds\, s \int_{\partial B(0,1)} \varphi(t - s, x - sy)\, \sigma_1(dy).
\end{aligned}
$$ (23)

Of course, the meaning of an expression such as

$$\int_{[0,t] \times \mathbf{R}^d} G(t - s, x - y) F(ds, dy)$$ (24)

where G is a measure and F is a Gaussian noise, is now unclear: it is certainly outside of Walsh's theory of stochastic integration [10].

In dimensions greater than 3, Green's function of the wave equation becomes even more irregular. For $d \geq 4$, set

$$
N(d) = \begin{cases} \dfrac{d - 3}{2} & \text{if } d \text{ is odd,} \\[2ex] \dfrac{d - 2}{2} & \text{if } d \text{ is even.} \end{cases}
$$ (25)

For d even, set

$$\sigma_t^d(dx) = \frac{1}{\sqrt{t^2 - |x|^2}} 1_{\{|x| < t\}}\, dx,$$ (26)

and for d odd, let $\sigma_t^d(dx)$ be the uniform surface measure on $\partial B(0, t)$ with total mass t^{d-1}. Then for d odd, $G(t, x)$ can formally be written

$$G(t, x) = c_d \left(\frac{1}{s} \frac{\partial}{\partial s}\right)^{N(d)} \left(\frac{\sigma_s^d}{s}\right) ds,$$ (27)

that is, for d odd,

$$
(G * \varphi)(t, x)
$$
$$
= c_d \int_0^t ds \left(\frac{1}{r} \frac{\partial}{\partial r}\right)^{N(d)} \left(\int_{\mathbf{R}^d} \varphi(t - s, x - y) \frac{\sigma_r^d(dy)}{r}\right)\bigg|_{r=s},
$$ (28)

while for d even,

$$
(G * \varphi)(t, x)
$$
$$
= c_d \int_0^t ds \left(\frac{1}{r} \frac{\partial}{\partial r}\right)^{N(d)} \left(\int_{B(0,r)} \varphi(t - s, x - y) \frac{dy}{\sqrt{r^2 - |y|^2}}\right)\bigg|_{r=s}.
$$ (29)

The meaning of $\int_{[0,t] \times \mathbf{R}^d} G(t - s, x - y) \, F(ds, dy)$ is even less clear in these cases!

The Case of Spatial Dimension One

Existence and uniqueness of the solution to the stochastic wave equation in spatial dimension 1 is covered in [19, Exercise 3.7 p. 323]. It is a good exercise that we leave to the reader.

Exercise 2.3. Establish existence and uniqueness of the solution to the non-linear wave equation on $[0, T] \times \mathbf{R}$, driven by space-time white noise:

$$\frac{\partial^2 u}{\partial t^2} - \frac{\partial^2 u}{\partial x^2} = \sigma(u(t, x)) \, \dot{W}(t, x), \tag{30}$$

with initial conditions

$$u(0, \cdot) = \frac{\partial u}{\partial t}(0, \cdot) \equiv 0. \tag{31}$$

The solution uses the following standard steps, which also appear in the study of the semilinear stochastic heat equation (see [19] and [10]):

- define the Picard iteration scheme;
- establish L^2-convergence using Gronwall's lemma;
- show existence of higher moments of the solution, using Burkholder's inequality

$$\mathrm{E}(|M_t|^p) \leq c_p \, \mathrm{E}\left(\langle M \rangle_t^{p/2} \right); \tag{32}$$

- establish ρ-Hölder continuity of the solution, for $\rho \in \left]0, \frac{1}{2}\right[$.

  It is also a good exercise to do the following calculation.

Exercise 2.4. Let G be Green's function of the wave equation, as defined in Example 2.2. For $d = 1$ and $d = 2$, check that for $\varphi \in C^2([0, \infty[\times \mathbf{R}^d)$,

$$u(t, x) = \int_0^t ds \int_{\mathbf{R}^d} dy \, G(t - s, x - y) \, \varphi(s, y) \tag{33}$$

satisfies

$$\frac{\partial^2 u}{\partial t^2}(t, x) - \Delta u(t, x) = \varphi(t, x). \tag{34}$$

Space-Time White Noise in Dimension $d = 2$

Having solved the non-linear stochastic wave equation driven by space-time white noise in dimension $d - 1$, it is tempting to attempt the same thing in dimension $d = 2$. We are going to show that there is a fundamental obstacle to doing this.

To this end, consider the *linear case*, that is, $\sigma \equiv 1$ and $b \equiv 0$. The mild solution given in (9) is not an equation in this case, but a formula:

$$
\begin{aligned}
u(t,x) &= \int_{[0,t] \times \mathbf{R}^2} G(t - s, x - y) \, W(ds, dy) \\
&= \int_{[0,t] \times \mathbf{R}^2} \frac{1}{\sqrt{2\pi}} \frac{1}{\sqrt{(t-s)^2 - |y - x|^2}} 1_{\{|y-x| < t-s\}} \, W(ds, dy),
\end{aligned}
\tag{35}
$$

where $W(ds, dy)$ is space-time white noise.

The first issue is whether this stochastic integral well-defined. For this, we would need (see [10, Exercise 5.5]) to have

$$
\int_0^t ds \int_{\mathbf{R}^2} dy \, G^2(t - s, x - y) < +\infty.
\tag{36}
$$

The integral is equal to

$$
\begin{aligned}
\int_0^t ds \int_{|y-x|<t-s} \frac{dy}{(t-s)^2 + |y - x|^2} &= \int_0^t dr \int_{|z|<r} \frac{dz}{r^2 - |z|^2} \\
&= \int_0^t dr \int_0^r d\rho \, \frac{2\pi\rho}{r^2 - \rho^2} \\
&= \pi \int_0^t dr \, \ln(r^2 - \rho^2)\big|_r^0 \\
&= +\infty.
\end{aligned}
\tag{37}
$$

In particular, when $d = 2$, there is *no* mild solution to the wave equation (9) driven by space-time white noise.

There have been some attempts at overcoming this problem (see [12], for instance), but as yet, there is no satisfactory approach to studying non-linear forms of the stochastic wave or heat equations driven by space-time white noise in dimensions $d \geq 2$.

A different tack is to consider spatially homogeneous noise with some conditions on the spatial covariance. We introduce these notions in the next section.

3 Spatially Homogeneous Gaussian Noise

Let Γ be a non-negative and non-negative definite tempered measure on $\mathbf{R}^d$, so that $\Gamma(dx) \geq 0$,

$$\int_{\mathbf{R}^d} \Gamma(dx) \, (\varphi * \tilde{\varphi})(x) \geq 0, \qquad \text{for all } \varphi \in \mathscr{S}(\mathbf{R}^d), \tag{38}$$

where $\tilde{\varphi}(x) \stackrel{\text{def}}{=} \varphi(-x)$, and there exists $r > 0$ such that

$$\int_{\mathbf{R}^d} \Gamma(dx) \, \frac{1}{(1 + |x|^2)^r} < \infty. \tag{39}$$

According to the Bochner–Schwartz theorem [15], there is a nonnegative measure μ on $\mathbf{R}^d$ whose Fourier transform is Γ: we write $\Gamma = \mathscr{F}\mu$. By definition, this means that for all $\varphi \in \mathscr{S}(\mathbf{R}^d)$,

$$\int_{\mathbf{R}^d} \Gamma(dx) \, \varphi(x) = \int_{\mathbf{R}^d} \mu(d\eta) \, \mathscr{F}\varphi(\eta). \tag{40}$$

We recall that the Fourier transform of $\varphi \in \mathscr{S}(\mathbf{R}^d)$ is

$$\mathscr{F}\varphi(\eta) = \int_{\mathbf{R}^d} \exp(-i\,\eta \cdot x) \, \varphi(x) \, dx, \tag{41}$$

where $\eta \cdot x$ denotes the Euclidean inner product. The measure μ is called the *spectral measure*.

Definition 3.1. *A spatially homogeneous Gaussian noise that is white in time is an $L^2(\Omega, \mathscr{F}, \mathrm{P})$–valued mean zero Gaussian process*

$$\left(F(\varphi), \ \varphi \in C_0^\infty(\mathbf{R}^{1+d}) \right), \tag{42}$$

such that

$$\mathrm{E}(F(\varphi) \, F(\psi)) = J(\varphi, \psi), \tag{43}$$

where

$$J(\varphi, \psi) \stackrel{\text{def}}{=} \int_{\mathbf{R}_+} ds \int_{\mathbf{R}^d} \Gamma(dx) \, (\varphi(s, \cdot) * \tilde{\psi}(s, \cdot))(x). \tag{44}$$

In the case where the covariance measure Γ has a density, so that $\Gamma(dx) = f(x) \, dx$, then it is immediate to check that $J(\varphi, \psi)$ can be written as follows:

$$J(\varphi, \psi) = \int_{\mathbf{R}_+} ds \int_{\mathbf{R}^d} dx \int_{\mathbf{R}^d} dy \, \varphi(s, x) \, f(x - y) \, \psi(s, y). \tag{45}$$

Using the fact that the Fourier transform of a convolution is the product of the Fourier transforms, this can also be written

$$J(\varphi,\psi) = \int_{\mathbf{R}_+} ds \int_{\mathbf{R}^d} \mu(d\eta)\,\mathscr{F}\varphi(s)(\eta)\,\overline{\mathscr{F}\psi(s)(\eta)}. \tag{46}$$

Informally, one often writes

$$\mathrm{E}\left(\dot{F}(t,x)\dot{F}(s,y)\right) = \delta_0(t-s)\,f(x-y), \tag{47}$$

as though $F(\varphi)$ were equal to $\int_{\mathbf{R}_+ \times \mathbf{R}^d} \varphi(s,x)\dot{F}(s,x)\,dsdx$.

Example 3.2. (a) If $\Gamma(dx) = \delta_0(x)$, where δ_0 denotes the Dirac delta function, then the associated spatially homogeneous Gaussian noise is simply space-time white noise.

(b) Fix $0 < \beta < d$ and let

$$\Gamma_\beta(dx) = \frac{dx}{|x|^\beta}. \tag{48}$$

One can check [17, Chapter 5] that $\Gamma_\beta = \mathscr{F}\mu_\beta$, with

$$\mu_\beta(d\eta) = c_{d,\beta}\,\frac{d\eta}{|\eta|^{d-\beta}}. \tag{49}$$

Exercise 3.3. Show that if $\beta \uparrow d$, then the spatially homogeneous Gaussian noise F_β with the covariance measure Γ_β converges weakly to space-time white noise. (*Hint.* Find the weak limit of the spectral measure μ_β and notice that $\mathscr{F}(d\eta) = \delta_0$.)

Extension of $F(\varphi)$ to a Worthy Martingale Measure

From the spatially homogenenous Gaussian noise, we are going to construct a worthy martingale measure $M = (M_t(A),\, t \geq 0,\, A \in \mathscr{B}_b(\mathbf{R}^d))$, where $\mathscr{B}_b(\mathbf{R}^d)$ denotes the family of bounded Borel subsets of $\mathbf{R}^d$. For this, if $A \in \mathscr{B}_b(\mathbf{R}^d)$, we set

$$M_t(A) \stackrel{\text{def}}{=} \lim_{n\to\infty} F(\varphi_n), \tag{50}$$

where the limit is in $L^2(\Omega,\mathscr{F},\mathrm{P})$, $\varphi_n \in C_0^\infty(\mathbf{R}^{d+1})$ and $\varphi_n \downarrow 1_{[0,t]\times A}$.

Exercise 3.4. ([2]) Show that $(M_t(A),\, t \geq 0,\, A \in \mathscr{B}_b(\mathbf{R}^d))$ is a worthy martingale measure in the sense of Walsh; its covariation measure Q is given by

$$Q(A \times B \times]s,t]) = (t-s) \int_{\mathbf{R}^d} dx \int_{\mathbf{R}^d} dy\, 1_A(x)\, f(x-y)\, 1_B(y), \tag{51}$$

and its dominating measure is $K \equiv Q$.

The key relationship between F and M is that

$$F(\varphi) = \int_{\mathbf{R}_+ \times \mathbf{R}^d} \varphi(t\,,x)\,M(dt\,,dx), \tag{52}$$

where the stochastic integral on the right-hand side is Walsh's martingale measure stochastic integral.

The underlying filtration $(\mathscr{F}_t\,,\,t \geq 0)$ associated with this martingale measure is given by

$$\mathscr{F}_t = \sigma\left(M_s(A),\ s \leq t,\ A \in \mathscr{B}_b(\mathbf{R}^d)\right) \vee \mathscr{N}, \qquad t \geq 0, \tag{53}$$

where $\mathscr{N}$ is the σ-field generated by all P-null sets.

4 The Wave Equation in Spatial Dimension 2

We shall consider the following form of the stochastic wave equation in spatial dimension $d = 2$:

$$\left(\frac{\partial^2 u}{\partial t^2} - \Delta u\right)(t\,,x) = \sigma(u(t\,,x))\,\dot{F}(t\,,x), \qquad (t\,,x) \in\,]0\,,T] \times \mathbf{R}^2, \tag{54}$$

with vanishing initial conditions. By a solution to (54), we mean a jointly measurable adapted process $(u(t,x))$ that satisfies the associated integral equation

$$u(t\,,x) = \int_{[0,t] \times \mathbf{R}^2} G(t - s\,,x - y)\,\sigma(u(s\,,y))\,M(ds\,,dy), \tag{55}$$

where M is the worthy martingale measure associated with $\dot{F}$.

The Linear Equation

A first step is to examine the linear equation, which corresponds to the case where $\sigma \equiv 1$:

$$\left(\frac{\partial^2 u}{\partial t^2} - \Delta u\right)(t\,,x) = \dot{F}(t\,,x), \tag{56}$$

with vanishing initial conditions. The mild solution should be

$$u(t\,,x) = \int_{[0,t] \times \mathbf{R}^2} G(t - s\,,x - y)\,M(ds\,,dy). \tag{57}$$

We know that the stochastic integral on the right-hand side is not defined for space-time white noise, so let us determine for which spatially homogeneous Gaussian noises it is well defined. This is the case if

$$\int_0^t ds \int_{\mathbf{R}^2} dy \int_{\mathbf{R}^2} dz\,G(t - s\,,x - y)\,f(y - z)\,G(t - s\,,x - z) < +\infty, \tag{58}$$

or, equivalently, if

$$\int_0^t ds \int_{\mathbf{R}^2} \mu(d\eta)\,|\mathscr{F}G(s)(\eta)|^2 < +\infty. \tag{59}$$

Calculation of $\mathscr{F}G$

In principle, Green's function of a PDE solves the same PDE with $\delta_{(0,0)}(t,x) = \delta_0(t)\,\delta_0(x)$ as right-hand side:

$$\frac{\partial^2 G}{\partial t^2} - \Delta G = \delta_0(t)\,\delta_0(x). \tag{60}$$

For fixed $t \neq 0$, the right-hand side vanishes. We shall take the Fourier transform in x on both sides of this equation, but first, we observe that since

$$\mathscr{F}G(t)(\xi) = \hat{G}(t)(\xi) = \int_{\mathbf{R}^2} e^{i\,\xi\cdot x}\, G(t,x)\,dx, \tag{61}$$

it is clear that

$$\mathscr{F}\left(\frac{\partial^2 G(t)}{\partial t}\right)(\xi) = \frac{\partial^2 \hat{G}(t)}{\partial t^2}(\xi), \tag{62}$$

and, using integration by parts, that

$$\begin{aligned}
\mathscr{F}(\Delta G(t))(\xi) &= \int_{\mathbf{R}^2} e^{i\,\xi\cdot x}\, \Delta G(t,x)\,dx \\
&= \int_{\mathbf{R}^2} \Delta(e^{i\,\xi\cdot x})\, G(t,x)\,dx \\
&= -|\xi|^2\, \mathscr{F}G(t)\,(\xi).
\end{aligned} \tag{63}$$

Therefore, we deduce from (60) that for $t > 0$,

$$\frac{\partial^2 \hat{G}(t)}{\partial t^2}(\xi) + |\xi|^2\, \hat{G}(t)\,(\xi) = \delta_0(t). \tag{64}$$

For fixed ξ, the solution to the associated homogeneous ordinary differential equation in t is

$$\hat{G}(t)(\xi) = a(\xi)\,\frac{\sin(t|\xi|)}{|\xi|} + b(\xi)\,\frac{\cos(t|\xi|)}{|\xi|}. \tag{65}$$

The solution that we seek (see [18, Chapter I, Section 4] for an explanation) is the one such that $\hat{G}(0)(\xi) = 0$ and $\frac{d\hat{G}(0)}{dt}(\xi) = 1$, so we conclude that for $t \geq 0$ and $\xi \in \mathbf{R}^2$,

$$\mathscr{F}G(t)\,(\xi) = \frac{\sin(t|\xi|)}{|\xi|}. \tag{66}$$

This formula is in fact valid in all dimensions $d \geq 1$.

Condition on the Spectral Measure

Condition (59) for existence of a mild solution on $[0,T]$ to the linear wave equation (56) becomes

$$\int_0^T ds \int_{\mathbf{R}^2} \mu(d\eta) \, \frac{\sin^2(s|\eta|)}{|\eta|^2} < +\infty. \tag{67}$$

Using Fubini's theorem, one can evaluate the ds-integral explicitly, or simply check that

$$\frac{c_1}{1+|\eta|^2} \leq \int_0^T ds \, \frac{\sin^2(s|\eta|)}{|\eta|^2} \leq \frac{c_2}{1+|\eta|^2}, \tag{68}$$

so condition (59) on the spectral measure becomes

$$\int_{\mathbf{R}^2} \mu(d\eta) \, \frac{1}{1+|\eta|^2} < +\infty. \tag{69}$$

Exercise 4.1. Let $d \geq 1$. Consider the case where $f(x) = |x|^{-\beta}$, $0 < \beta < d$. In this case, $\mu(d\eta) = c_{d,\beta}|\eta|^{\beta-d} d\eta$ (see Example 3.2). Check that condition (69) holds (even when $\mathbf{R}^2$ is replaced by $\mathbf{R}^d$) if and only if $\beta < 2$. In particular, the spatially homogeneous Gaussian noise with the covariance function f is defined for $0 < \beta < d$, but a mild solution of the linear stochastic wave equation (56) exists if and only if $0 < \beta < 2$.

Reformulating (69) in Terms of the Covariance Measure

Condition (69) on the spectral measure can be reformulated as a condition on the covariance measure Γ.

Exercise 4.2. ([11]) Show that in dimension $d = 2$, (69) is equivalent to

$$\int_{|x| \leq 1} \Gamma(dx) \, \ln\left(\frac{1}{|x|}\right) < +\infty, \tag{70}$$

while in dimensions $d \geq 3$, (69) is equivalent to

$$\int_{|x| \leq 1} \Gamma(dx) \, \frac{1}{|x|^{d-2}} < +\infty. \tag{71}$$

In dimension $d = 1$, condition (69) is satisfied for any non-negative measure μ such that $\Gamma = \mathscr{F}\mu$ is also a non-negative measure.

The Non-Linear Wave Equation in Dimension $d = 2$

We consider equation (54). The following theorem is the main result on existence and uniqueness.

Theorem 4.3. *Assume $d = 2$. Suppose that σ is a Lipschitz continuous function and that condition (69) holds. Then there exists a unique solution $(u(t,x),\ t \geq 0,\ x \in \mathbf{R}^2)$ of (54) and for all $p \geq 1$, this solution satisfies*

$$\sup_{0 \leq t \leq T} \sup_{x \in \mathbf{R}^d} \mathrm{E}\left(|u(t,x)|^p\right) < \infty. \tag{72}$$

Proof. This proof follows a classical Picard iteration scheme. We set $u_0(t,x) = 0$, and, by induction, for $n \geq 0$,

$$u_{n+1}(t,x) = \int_{[0,t] \times \mathbf{R}^2} G(t-s, x-y)\, \sigma(u_n(s,y))\, M(ds, dy). \tag{73}$$

Before establishing convergence of this scheme, we first check that for $p \geq 2$,

$$\sup_{n \geq 0} \sup_{0 \leq s \leq T} \sup_{x \in \mathbf{R}^2} \mathrm{E}\left(|u_n(s,x)|^p\right) < +\infty. \tag{74}$$

We apply Burkholder's inequality (32) and use the explicit form of the quadratic variation of the stochastic integral [10, Theorem 5.26] to see that

$$\mathrm{E}\left(|u_{n+1}(t,x)|^p\right) \leq c\mathrm{E}\left[\left(\int_0^t ds \int_{\mathbf{R}^2} dy \int_{\mathbf{R}^2} dz\, G(t-s, x-y)\, \sigma(u_n(s,y))\right.\right.$$
$$\left.\left. \times f(y-z)\, G(t-s, x-z)\, \sigma(u_n(s,z))\right)^{p/2}\right]. \tag{75}$$

Since $G \geq 0$ and $f \geq 0$, we apply Hölder's inequality in the form

$$\left|\int f\, d\mu\right|^p \leq \left(\int 1\, d\mu\right)^{p/q} \left(\int |f|^p\, d\mu\right), \qquad \text{where } \frac{p}{q} = p - 1 \tag{76}$$

and μ is a non-negative measure, to see that $\mathrm{E}\left(|u_{n+1}(t,x)|^p\right)$ is bounded above by

$$c\left(\int_0^t ds \int_{\mathbf{R}^2} dy \int_{\mathbf{R}^2} dz\, G(t-s, x-y)\, f(y-z)\, G(t-s, x-z)\right)^{\frac{p}{2}-1}$$
$$\times \int_0^t ds \int_{\mathbf{R}^2} dy \int_{\mathbf{R}^2} dz\, G(t-s, x-y)\, f(y-z)\, G(t-s, x-z)$$
$$\times \mathrm{E}\left(|\sigma(u_n(s,y))\, \sigma(u_n(s,z))|^{\frac{p}{2}}\right). \tag{77}$$

We apply the Cauchy–Schwarz inequality to the expectation and use the Lipschitz property of σ to bound this by

$$C \left(\int_0^t ds \int_{\mathbf{R}^2} \mu(d\eta) \, |\mathscr{F}G(t-s)(\eta)|^2 \right)^{\frac{p}{2}-1}$$

$$\times \int_0^t ds \int_{\mathbf{R}^2} dy \int_{\mathbf{R}^2} dz \, G(t-s, x-y) \, f(y-z) \, G(t-s, x-z) \tag{78}$$

$$\times \left(\mathrm{E} \left(1 + |u_n(s,y)|^p \right) \right)^{1/2} \left(\mathrm{E} \left(1 + |u_n(s,z)|^p \right) \right)^{1/2}.$$

Let

$$J(t) = \int_0^t ds \int_{\mathbf{R}^2} \mu(d\eta) \, |\mathscr{F}G(t-s)(\eta)|^2 \le C \int_{\mathbf{R}^2} \mu(d\eta) \frac{1}{1+|\eta|^2}. \tag{79}$$

Then

$$\mathrm{E} \left(|u_{n+1}(t,x)|^p \right)$$

$$\le C \left(J(t) \right)^{\frac{p}{2}-1} \int_0^t ds \left(1 + \sup_{y \in \mathbf{R}^2} \mathrm{E} \left(|u_n(s,y)|^p \right) \right) \times \int_{\mathbf{R}^2} \mu(d\eta) \, |\mathscr{F}G(t-s)(\eta)|^2$$

$$\le \tilde{C} \int_0^t ds \left(1 + \sup_{y \in \mathbf{R}^2} \mathrm{E}(|u_n(s,y)|^p) \right). \tag{80}$$

Therefore, if we set

$$M_n(t) = \sup_{x \in \mathbf{R}^2} \mathrm{E} \left(|u_n(t,x)|^p \right), \tag{81}$$

then

$$M_{n+1}(t) \le \tilde{C} \int_0^t ds \, (1 + M_n(s)). \tag{82}$$

Using Gronwall's lemma, we conclude that

$$\sup_{n \in \mathbf{N}} \sup_{0 \le t \le T} M_n(t) < +\infty. \tag{83}$$

We now check L^2-convergence of the Picard iteration scheme. By the same reasoning as above, we show that

$$\sup_{x \in \mathbf{R}^2} \mathrm{E} \left(|u_{n+1}(t,x) - u_n(t,x)|^p \right)$$

$$\le C \int_0^t ds \sup_{y \in \mathbf{R}^2} \mathrm{E} \left(|u_n(s,y) - u_{n-1}(s,y)|^p \right). \tag{84}$$

Gronwall's lemma shows that $(u_n(t,x), \, n \ge 1)$ converges in $L^2(\Omega, \mathscr{F}, \mathrm{P})$, uniformly in $x \in \mathbf{R}^2$.

Uniqueness of the solution follows in a standard way: see [10, Proof of Theorem 6.4]. $\qquad\square$

Hölder-Continuity ($d = 2$)

In order to establish Hölder continuity of the solution to the stochastic wave equation in spatial dimension 2, we first recall the *Kolmogorov continuity theorem*. It is a good idea to compare this statement with the equivalent one in [10, Theorem 4.3].

Theorem 4.4 (The Kolmogorov Continuity Theorem). *Suppose that there is $q > 0$, $\rho \in]\frac{d}{q}, 1[$ and $C > 0$ such that for all $x, y \in \mathbf{R}^d$,*

$$\mathrm{E}\left(|u(t, x) - u(t, y)|^q\right) \leq C |x - y|^{\rho q}. \tag{85}$$

Then $x \mapsto u(t, x)$ has a $\tilde{\rho}$-Hölder continuous version, for any $\tilde{\rho} \in]0, \rho - \frac{d}{q}[$.

In order to use the statement of this theorem to establish $(\rho - \varepsilon)$-Hölder continuity, for any $\varepsilon > 0$, it is necessary to obtain estimates on arbitrarily high moments of increments, that is, to establish (85) for arbitrarily large q.

L^q-Moments of Increments

From the integral equation (55), we see that

$$
\begin{aligned}
u(t, x) &- u(s, y) \\
&= \iint (G(t - r, x - z) - G(s - r, y - z))\, \sigma(u(r, z))\, M(dr, dz),
\end{aligned} \tag{86}
$$

and so, by Burkholder's inequality (32),

$$
\begin{aligned}
\mathrm{E}\left(|u(t, x) - u(s, y)|^p\right) & \\
\leq C\mathrm{E}\Bigg(\Bigg|\int_0^t dr &\int_{\mathbf{R}^2} dz \int_{\mathbf{R}^2} dv\, (G(t - r, x - z) - G(s - r, y - z))\, f(z - v) \\
&\times (G(t - r, x - v) - G(s - r, y - v))\, \sigma(u(r, z))\, \sigma(u(r, v))\Bigg|^{p/2}\Bigg) \\
\leq C\left(\int dr \int dz \int dv\, |G(\ ) - G(\ )|\, f(\ )\, |G(\ ) - G(\ )|\right)^{\frac{p}{2} - 1} & \tag{87} \\
\times \int dr \int dz \int dv\, |G(\ ) - G(\ )|\, f(\ )\, |G(\ ) - G(\ )| & \\
\times \mathrm{E}\left(|\sigma(u(r, z))|^{p/2}\, |\sigma(u(r, v))|^{p/2}\right), &
\end{aligned}
$$

where the omitted variables are easily filled in. The Lipschitz property of σ implies a bound of the type "linear growth," and so, using also the Cauchy–Schwarz inequality, we see that the expectation is bounded by

$$C \sup_{r \leq T, z \in \mathbf{R}^2} (1 + \mathrm{E}(|u(r, z)|^p)). \tag{88}$$

Define

$$J(t,x;s,y)$$
$$= \int_0^t dr \int_{\mathbf{R}^2} dz \int_{\mathbf{R}^2} dv \, |G(t-r,x-z) - G(s-r,y-z)| \, f(z-v) \qquad (89)$$
$$\times |G(t-r,x-v) - G(s-r,y-v)|.$$

We have shown that

$$E\left(|u(t,x) - u(s,y)|^p\right) \le (J(t,x;s,y))^{p/2}. \qquad (90)$$

Therefore, we will get Hölder-continuity provided that we can establish an estimate of the following type for some $\gamma > 0$ and $\rho > 0$:

$$J(t,x;s,y) \le c(|t-s|^\gamma + |x-y|^\rho). \qquad (91)$$

Indeed, this will establish $\frac{\gamma_1}{2}$-Hölder continuity in time, and $\frac{\rho_1}{2}$-Hölder continuity in space, for all $\gamma_1 \in]0,\gamma[$ and $\rho_1 \in]0,\rho[$.

Analysis of $J(t,x;s,y)$

If there were no absolute values around the increments of G, then we could use the Fourier transform to rewrite $J(t,x;s,y)$, in the case $x = y$ and $s > t$, for instance, as

$$J(t,x;s,x) = \int_0^s dr \int_{\mathbf{R}^2} \mu(d\eta) \, |\mathscr{F}G(t-r)(\eta) - \mathscr{F}G(s-r)(\eta)|^2$$
$$+ \int_s^t dr \int_{\mathbf{R}^2} \mu(d\eta) \, |\mathscr{F}G(t-r)(\eta)|^2. \qquad (92)$$

We could then analyse this using the specific form of $\mathscr{F}G$ in (66). However, the presence of the absolute values makes this approach inoperable. By a direct analysis of $J(t,x;s,x)$, Sanz-Solé and Sarrá [14] have established the following results. If

$$\int_{\mathbf{R}^2} \mu(d\eta) \, \frac{1}{(1+|\eta|^2)^a} < \infty, \qquad \text{for some } a \in]0,1[, \qquad (93)$$

then $t \mapsto u(t,x)$ is γ_1-Hölder continuous, for

$$\gamma_1 \in \left]0, \frac{1}{2} \wedge (1-a)\right[, \qquad (94)$$

and $x \mapsto u(t,x)$ is γ_2-Hölder continuous, for $\gamma_2 \in]0, 1-a[$.
When $\mu(d\eta) = |\eta|^{-\beta} d\eta$, these intervals become

$$\gamma_1 \in \left]0, \frac{1}{2} \wedge \frac{2-\beta}{2}\right[\qquad \text{and} \qquad \gamma_2 \in \left]0, \frac{2-\beta}{2}\right[. \qquad (95)$$

The best possible interval for γ_1 is in fact $]0, \frac{2-\beta}{2}[$; see [5, Chapter 5].

5 A Function-Valued Stochastic Integral

Because Green's function in spatial dimension 3 is a measure and not a function, the study of the wave equation in this dimension requires different methods than those used in dimensions 1 and 2. In particular, we will use a function-valued stochastic integral, developed in [4].

Our first objective is to define a stochastic integral of the form

$$\int_{[0,t]\times\mathbf{R}^d} G(s,x-y)\,Z(s,y)\,M(ds,dy), \tag{96}$$

where $G(s,\cdot)$ is Green's function of the wave equation (see Example 2.2) and $Z(s,y)$ is a random field that plays the role of $\sigma(u(s,y))$.

We shall assume for the moment that $d \geq 1$ and that the following conditions are satisfied.

Hypotheses

(H1) For $0 \leq s \leq T$, $Z(s,\cdot) \in L^2(\mathbf{R}^d)$ a.s., $Z(s,\cdot)$ is $\mathscr{F}_s$−measurable, and
$s \mapsto Z(s,\cdot)$ from $\mathbf{R}_+ \to L^2(\mathbf{R}^d)$ is continuous.

(H2) For all $s \geq 0$,

$$\int_0^T ds \sup_{\xi\in\mathbf{R}^d} \int_{\mathbf{R}^d} \mu(d\eta)\,|\mathscr{F}G(s)(\xi-\eta)|^2 < +\infty.$$

We note that $\mathscr{F}G(s)(\xi-\eta)$ is given in (66), so that (H2) is a condition on the spectral measure μ, while (H1) is a condition on Z. In general, condition (H2) is stronger than (59): see [13].

Fix $\psi \in C_0^\infty(\mathbf{R}^d)$ such that $\psi \geq 0$, supp $\psi \subset B(0,1)$ and

$$\int_{\mathbf{R}^d} \psi(x)\,dx = 1. \tag{97}$$

For $n \geq 1$, set

$$\psi_n(x) = n^d\,\psi(nx). \tag{98}$$

In particular, $\psi_n \to \delta_0$ in $\mathscr{S}'(\mathbf{R}^d)$, and $\mathscr{F}\psi_n(\xi) = \mathscr{F}\psi(\xi/n)$, so that $|\mathscr{F}\psi_n(\xi)| \leq 1$. Define

$$G_n(s,\cdot) = G(s) * \psi_n, \tag{99}$$

so that G_n is a C_0^∞-function. Then,

$$v_{G_n,Z}(t,x) \stackrel{\text{def}}{=} \int_{[0,t]\times\mathbf{R}^d} G_n(s,x-y)\,Z(s,y)\,M(ds,dy) \tag{100}$$

is well-defined as a Walsh-stochastic integral, and

$$\mathrm{E}\left(\|v_{G_n,Z}(t\,,\cdot)\|_{L^2(\mathbf{R}^d)}^2\right) = I_{G_n,Z},\tag{101}$$

where

$$
\begin{aligned}
I_{G_n,Z} &= \int_{\mathbf{R}^d} dx\, \mathrm{E}\left((v_{G_n,Z}(t\,,x))^2\right) \\
&= \int_{\mathbf{R}^d} dx \int_0^t ds \int_{\mathbf{R}^d} dy \int_{\mathbf{R}^d} dz\, G_n(s\,,x-y)Z(s\,,y)\, f(y-z) \\
&\qquad\qquad \times\; G_n(s\,,x-z)\, Z(s\,,z).
\end{aligned}\tag{102}
$$

Using the fact that the Fourier transform of a convolution (respectively product) is the product (resp. convolution) of the Fourier transforms, one easily checks that

$$
\begin{aligned}
&I_{G_n,Z} \\
&= \int_0^t ds \int_{\mathbf{R}^d} d\xi\, \mathrm{E}\left(|\mathscr{F}Z(s\,,\cdot)(\xi)|^2\right) \int_{\mathbf{R}^d} \mu(d\eta)\, |\mathscr{F}G_n(s\,,\cdot)\,(\xi-\eta)|^2.
\end{aligned}\tag{103}
$$

We note that:

(a) the following inequality holds:

$$I_{G_n,Z} \le \tilde{I}_{G_n,Z},\tag{104}$$

where

$$
\begin{aligned}
&\tilde{I}_{G_n,Z} \\
&\stackrel{\mathrm{def}}{=} \int_0^t ds\, \mathrm{E}\left(\|Z(s\,,\cdot)\|_{L^2(\mathbf{R}^d)}^2\right) \sup_{\xi\in\mathbf{R}^d} \int_{\mathbf{R}^d} \mu(d\eta)\, |\mathscr{F}G_n(s\,,\cdot)(\xi-\eta)|^2;
\end{aligned}\tag{105}
$$

(b) the equality (101) plays the role of an isometry property;
(c) by elementary properties of convolution and Fourier transform,

$$\tilde{I}_{G_n,Z} \le \tilde{I}_{G,Z} < +\infty,\tag{106}$$

by (H2) and (H1).

In addition, one checks that the stochastic integral

$$v_{G,Z}(t) \stackrel{\mathrm{def}}{=} \lim_{n\to\infty} v_{G_n,Z}\tag{107}$$

exists, in the sense that

$$\mathrm{E}\left(\|v_{G,Z}(t) - v_{G_n,Z}(t\,,\cdot)\|_{L^2(\mathbf{R}^d)}^2\right) \longrightarrow 0,\tag{108}$$

and

$$\mathrm{E}\left(\|v_{G,Z}(t)\|_{L^2(\mathbf{R}^d)}^2\right) = I_{G,Z} \le \tilde{I}_{G,Z}. \tag{109}$$

We use the following notation for the stochastic integral that we have just defined:

$$v_{G,Z}(t) = \int_{[0,t]\times\mathbf{R}^d} G(s,\cdot - y)\, Z(s,y)\, M(ds,dy). \tag{110}$$

For t fixed, $v_{G,Z}(t) \in L^2(\mathbf{R}^d)$ is a square-integrable function that is defined almost-everywhere.

The definition of the stochastic integral requires in particular that hypothesis (H2) be satisfied. In the case where

$$\Gamma(dx) = k_\beta(x)\, dx, \qquad \text{with } k_\beta(x) = |x|^{-\beta}, \quad \beta > 0, \tag{111}$$

this condition becomes

$$\int_0^T ds \sup_{\xi\in\mathbf{R}^d} \int_{\mathbf{R}^d} d\eta\, |\eta|^{\beta-d}\, \frac{\sin^2(s|\xi-\eta|)}{|\xi-\eta|^2} < +\infty. \tag{112}$$

Exercise 5.1. ([4]) Show that (112) holds if and only if $0 < \beta < 2$.

6 The Wave Equation in Spatial Dimension $d \ge 1$

We consider the following stochastic wave equation in spatial dimension $d \ge 1$, driven by spatially homogeneous Gaussian noise $\dot{F}(t,x)$ as defined in Section 3:

$$\begin{cases} \left(\dfrac{\partial^2 u}{\partial t^2} - \Delta u\right)(t,x) = \sigma(x,u(t,x))\, \dot{F}(t,x), \qquad t \in\,]0,T],\ x\in\mathbf{R}^d, \\[2mm] u(0,x) = v_0(x), \qquad \dfrac{\partial u}{\partial t}(0,x) = \tilde{v}_0(x), \end{cases} \tag{113}$$

where $v_0 \in L^2(\mathbf{R}^d)$ and $\tilde{v}_0 \in H^{-1}(\mathbf{R}^d)$. By definition, $H^{-1}(\mathbf{R}^d)$ is the set of square-integrable functions $\tilde{v}_0$ such that

$$\|\tilde{v}_0\|_{H^{-1}(\mathbf{R}^d)}^2 \overset{\text{def}}{=} \int_{\mathbf{R}^d} d\xi\, \frac{1}{1+|\xi|^2}\, |\mathscr{F}\tilde{v}_0(\xi)|^2 < +\infty. \tag{114}$$

We shall restrict ourselves, though this is not really necessary (see [4]) to the case where $\Gamma(dx)$ is as in (111). In this case, Exercise 4.1 shows that the further restriction $0 < \beta < 2$ is needed.

The Past-Light Cone Property

Consider a bounded domain $D \subset \mathbf{R}^d$. A fundamental property of the wave equation (see [18, Theorem 14.1]) is that $u(T, x)$, $x \in D$, only depends on $v_0|_{K^D}$ and $\tilde{v}_0|_{K^D}$, where

$$K^D = \{y \in \mathbf{R}^d : d(y, D) \leq T\} \tag{115}$$

and $d(y, D)$ denotes the distance from y to the set D, and on the noise $\dot{F}(s, y)$ for $y \in K^D(s)$, $0 \leq s \leq T$, where

$$K^D(s) = \{y \in \mathbf{R}^d : d(y, D) \leq T - s\}. \tag{116}$$

Therefore, the solution $u(t, x)$ in D is unchanged if we take the SPDE

$$\left(\frac{\partial^2 u}{\partial t^2} - \Delta u\right)(t, x) = 1_{K^D(t)}(x) \, \sigma(x, u(t, x)) \, \dot{F}(t, x). \tag{117}$$

We shall make the following linear growth and Lipschitz continuity assumptions on the function σ.

Assumptions.

(a) $|\sigma(x, u)| \leq c(1 + |u|) 1_{K^D(T)}(x)$, for all $x \in \mathbf{R}^d$ and $u \in \mathbf{R}$;
(b) $|\sigma(x, u) - \sigma(x, v)| \leq c|u - v|$, for all $x \in \mathbf{R}^d$ and $u, v \in \mathbf{R}$.

Definition 6.1. *An adapted and mean-square continuous $L^2(\mathbf{R}^d)$-valued process $(u(t), 0 \leq t \leq T)$ is a solution of (113) in D if for all $t \in]0, T]$,*

$$u(t) 1_{K^D(t)} = 1_{K^D(t)} \cdot \left(\frac{d}{dt} G(t) * v_0 + G(t) * \tilde{v}_0 \right. \tag{118}$$

$$\left. + \int_{[0,t] \times \mathbf{R}^d} G(t - s, \cdot - y) \, \sigma(y, u(s, y)) \, M(ds, dy)\right).$$

Theorem 6.2. *Let $d \geq 1$. Suppose $0 < \beta < 2 \wedge d$ and that the assumptions above on σ are satisfied. Then (113) has a unique solution $(u(t), 0 \leq t \leq T)$ in D.*

Proof. We use a Picard iteration scheme. Set

$$u_0(t, x) = \frac{d}{dt} G(t) * v_0 + G(t) * \tilde{v}_0. \tag{119}$$

We first check that $u_0(t) \in L^2(\mathbf{R}^d)$. Indeed,

$$\left\| \frac{d}{dt} G(t) * v_0 \right\|_{L^2(\mathbf{R}^d)} = \left\| \mathscr{F}\left(\frac{d}{dt} G(t) \right) \cdot \mathscr{F} v_0 \right\|_{L^2(\mathbf{R}^d)}$$

$$= \int_{\mathbf{R}^d} d\xi \left| |\xi| \frac{\cos(t|\xi|)}{|\xi|} \cdot \mathscr{F} v_0(\xi) \right|^2 \qquad (120)$$

$$\leq \|v_0\|_{L^2(\mathbf{R}^d)},$$

and, similarly,

$$\|G(t) * \tilde{v}_0\|_{L^2(\mathbf{R}^d)} \leq \|\tilde{v}_0\|_{H^{-1}(\mathbf{R}^d)} . \qquad (121)$$

One checks in a similar way that $t \mapsto u_0(t)$ from $[0, T]$ into $L^2(\mathbf{R}^d)$ is continuous.

We now define the Picard iteration scheme. For $n \geq 0$, assume that $(u_n(t), 0 \leq t \leq T)$ has been defined, and satisfies (H1). Set

$$u_{n+1}(t) = 1_{K^D(t)} \cdot (u_0(t) + v_{n+1}(t)), \qquad (122)$$

where

$$v_{n+1}(t) = \int_{[0,t] \times \mathbf{R}^d} G(t - s, \cdot - y)\, \sigma(y, u_n(s, y))\, M(ds, dy). \qquad (123)$$

By induction, $Z_n(s, y) = \sigma(y, u_n(s, y))$ satisfies (H1). Indeed, this process is adapted, and since

$$\|\sigma(\cdot, u_n(s, \cdot)) - \sigma(\cdot, u_n(t, \cdot))\|_{L^2(\mathbf{R}^d)} \leq C\|u_n(s, \cdot) - u_n(t, \cdot)\|_{L^2(\mathbf{R}^d)}, \qquad (124)$$

it follows that $s \mapsto u_n(s, \cdot)$ is mean-square continuous. One checks that u_{n+1} also satisfies (H1): this uses assumption (a).

Therefore, the stochastic integral (123) is well-defined. Let

$$M_n(r) = \sup_{0 \leq t \leq r} \mathrm{E}\left(\|u_{n+1}(t) - u_n(t)\|_{L^2(K^D(t))}^2 \right) \qquad (125)$$

$$= \sup_{0 \leq t \leq r} \mathrm{E}\left(\|v_{n+1}(t) - v_n(t)\|_{L^2(K^D(t))}^2 \right)$$

$$= \sup_{0 \leq t \leq r} \mathrm{E}\Bigg(\left\| \int_{[0,t] \times \mathbf{R}^d} G(t - s, \cdot - y) \right.$$

$$\times (\sigma(y, u_n(s, y)) - \sigma(y, u_{n-1}(s, y)))\, M(ds, dy) \Bigg\|_{L^2(K^D(t))}^2 \Bigg)$$

$$\leq \sup_{0 \leq t \leq r} \int_0^t ds\, \mathrm{E}\left(\|\sigma(\cdot, u_n(s, \cdot)) - \sigma(\cdot, u_{n-1}(s, \cdot))\|_{L^2(K^D(t))}^2 \right) J(t - s),$$

where

$$J(s) = \sup_{\xi \in \mathbf{R}^d} \int_{\mathbf{R}^d} d\eta\, |\eta|^{\beta - d} \frac{\sin^2(s|\xi - \eta|)}{|\xi - \eta|^2} . \qquad (126)$$

A direct calculation shows that

$$\sup_{0 \le s \le T} J(s) < +\infty, \tag{127}$$

since $0 < \beta < 2$, so

$$M_n(r) \le C \sup_{0 \le t \le r} \int_0^t ds\, \mathrm{E}\left(\|u_n(s,\cdot) - u_{n-1}(s,\cdot)\|^2_{L^2(K^D(t))} \right), \tag{128}$$

that is,

$$M_n(r) \le C \int_0^r M_{n-1}(s)\, ds. \tag{129}$$

Because $M_0(T) < +\infty$, Gronwall's lemma implies that

$$\sum_{n=0}^{+\infty} (M_n(r))^{1/2} < +\infty. \tag{130}$$

Therefore, $(u_n(t,\cdot),\, n \in \mathbf{N})$ converges in $L^2(\Omega \times \mathbf{R}^d, d\mathrm{P} \times dx)$, uniformly in $t \in [0,T]$, to a limit $u(t,\cdot)$. Since u_n satisfies (H1) and u_n converges uniformly in t to $u(t,\cdot)$, it follows that $u(t,\cdot)$ is a solution to (113): indeed, it suffices to pass to the limit in (122) and (123).

Uniqueness of the solution follows by a standard argument. $\qquad\square$

7 Spatial Regularity of the Stochastic Integral ($d = 3$)

We aim now to analyze spatial regularity of the solution to the 3-dimensional stochastic wave equation (113) driven by spatially homogeneous Gaussian noise, with covariance given by a Riesz kernel $f(x) = |x|^{-\beta}$, where $0 < \beta < 2$. For this, we shall first examine the regularity in the x-variable of the function-valued stochastic integral defined in Section 5 when $d = 3$.

We recall that studying regularity properties requires information on higher moments. With these, one can use the Kolmogorov continuity theorem (Theorem 4.4) or the Sobolev embedding theorem, which we now recall.

Theorem 7.1 (The Sobolev Embedding Theorem). *Let $\mathscr{O}$ be an open subset of $\mathbf{R}^d$. Suppose that $g \in W^{\rho,q}(\mathscr{O})$. Then $x \mapsto g(x)$ is $\tilde{\rho}$-Hölder continuous, for all $\tilde{\rho} \in]0, \rho - \frac{d}{q}[$.*

We recall [16] that the norm in the space $W^{\rho,q}(\mathscr{O})$ is defined by

$$\|g\|^q_{W^{\rho,q}(\mathscr{O})} = \|g\|^q_{L^q(\mathscr{O})} + \|g\|^q_{\rho,q,\mathscr{O}}, \tag{131}$$

where

$$\|g\|^q_{L^q(\mathscr{O})} = \int_{\mathscr{O}} |g(x)|^q\, dx$$

$$\|g\|^q_{\rho,q,\mathscr{O}} = \int_{\mathscr{O}} dx \int_{\mathscr{O}} dy\, \frac{|g(x) - g(y)|^q}{|x - y|^{d + \rho q}}. \tag{132}$$

Our first objective is to determine conditions that ensure that

$$E\left(\|v_{G,Z}\|^q_{L^q(\mathscr{O})}\right) < +\infty. \tag{133}$$

For $\varepsilon > 0$, we let

$$\mathscr{O}^\varepsilon = \left\{x \in \mathbf{R}^3 : \exists z \in \mathscr{O} \text{ with } |x - z| < \varepsilon\right\} \tag{134}$$

denote the ε-enlargement of $\mathscr{O}$, and use the notation

$$v^t_{G,Z} = \int_{[0,t]\times\mathbf{R}^3} G(t - s, \cdot - y)\, Z(s, y)\, M(ds, dy). \tag{135}$$

An Estimate in L^p-Norm

Theorem 7.2. *Suppose $0 < \beta < 2$. Fix $T > 0$, $q \in [2, +\infty[$ and let $\mathscr{O} \subset \mathbf{R}^3$ be a bounded domain. There is a constant $C < \infty$ with the following property. Suppose that*

$$\int_0^t ds\, E\left(\|Z(s)\|^q_{L^q(\mathscr{O}^{t-s})}\right) < +\infty. \tag{136}$$

Then

$$E\left(\|v^t_{G,Z}\|^q_{L^q(\mathscr{O})}\right) \le C \int_0^t ds\, E\left(\|Z(s)\|^q_{L^q(\mathscr{O}^{t-s})}\right). \tag{137}$$

Proof. We present the main ideas, omitting some technical issues that are handled in [5, Proposition 3.4]. First, we check inequality (137) with G replaced by G_n:

$$E\left(\|v^t_{G_n,Z}\|^q_{L^q(\mathscr{O})}\right)$$

$$= \int_{\mathscr{O}} dx\, E\left(\left|\int_{[0,t]\times\mathbf{R}^3} G_n(t - s, x - y)\, Z(s, y)\, M(ds, dy)\right|^q\right)$$

$$\le \int_{\mathscr{O}} dx\, E\left(\left|\int_0^t ds \int_{\mathbf{R}^3} dy \int_{\mathbf{R}^3} dz\, G_n(t - s, x - y)\, Z(s, y)\, f(y - z)\right.\right.$$

$$\left.\left. \times G_n(t - s, x - z)\, Z(s, z)\right|^{q/2}\right). \tag{138}$$

Let

$$\mu_n(t, x)$$

$$= \int_0^t ds \int_{\mathbf{R}^3} dy \int_{\mathbf{R}^3} dz\, G_n(t - s, x - y)\, f(y - z)\, G_n(t - s, x - z). \tag{139}$$

Assume that

$$\sup_{n,x,\,t\leq T} \mu_n(t,x) < +\infty. \tag{140}$$

By Hölder's inequality, written in the form (76), we see, since $G_n \geq 0$, that

$$
\begin{aligned}
&\mathrm{E}\left(\|v_{G_n,Z}^t\|_{L^q(\mathcal{O})}^q \right) \\
&\leq \int_{\mathcal{O}} dx\, (\mu_n(t,x))^{\frac{q}{2}-1} \mathrm{E}\left(\int_0^t ds \int_{\mathbf{R}^3} dy \int_{\mathbf{R}^3} dz\, G_n(t-s,x-y) \right. \\
&\qquad\qquad \left. \times f(y-z)\, G_n(t-s,x-z)\, |Z(s,y)|^{q/2}\, |Z(s,z)|^{q/2} \right) \\
&= I_{G_n,|Z\,1_{\mathcal{O}^{t-s+1/n}}|^{q/2}}.
\end{aligned}
\tag{141}
$$

We apply (104), then (105), to bound this by

$$
\begin{aligned}
\tilde{I}_{G_n,|Z\,1_{\mathcal{O}^{t-s+1/n}}|^{q/2}} &= \int_0^t ds\, \mathrm{E}\left(\left\| |Z(s)|^{q/2}\, 1_{\mathcal{O}^{t-s+1/n}} \right\|_{L^2(\mathbf{R}^3)}^2 \right) \\
&\qquad \times \sup_{\xi\in\mathbf{R}^3} \int_{\mathbf{R}^3} \mu(d\eta)\, |\mathscr{F} G_n(s,\cdot)(\xi-\eta)|^2.
\end{aligned}
\tag{142}
$$

Since $0 < \beta < 2$, the supremum over ξ is finite, therefore

$$\mathrm{E}\left(\|v_{G_n,Z}^t\|_{L^q(\mathcal{O})}^q \right) \leq C \int_0^t ds\, \mathrm{E}\left(\|Z(s)\|_{L^q(\mathcal{O}^{t-s+1/n})}^q \right). \tag{143}$$

By Fatou's lemma,

$$
\begin{aligned}
\mathrm{E}\left(\|v_{G,Z}^t\|_{L^q(\mathcal{O})}^q \right) &\leq \liminf_{k\to\infty} \mathrm{E}\left(\|v_{G_{n_k},Z}^t\|_{L^q(\mathcal{O})}^q \right) \\
&\leq \liminf_{k\to\infty} \int_0^t ds\, \mathrm{E}\left(\|Z(s)\|_{L^q(\mathcal{O}^{t-s+1/n_k})}^q \right) \\
&= \int_0^t ds\, \mathrm{E}\left(\|Z(s)\|_{L^q(\mathcal{O}^{t-s})}^q \right).
\end{aligned}
\tag{144}
$$

It remains to check that (140) holds. Since

$$|\mathscr{F} G_n(t-s)(\eta)|^2 \leq |\mathscr{F} G(t-s)(\eta)|^2 = \frac{\sin^2((t-s)|\eta|)}{|\eta|^2}, \tag{145}$$

it follows that for $t \in [0,T]$ and $x \in \mathbf{R}^3$,

$$\mu_n(t,x) \leq \int_0^t ds \int_{\mathbf{R}^3} d\eta\, |\eta|^{\beta-3}\, \frac{\sin^2((t-s)|\eta|)}{|\eta|^2} \leq C(T), \tag{146}$$

since $0 < \beta < 2$. This completes the proof. $\square$

An Estimate in Sobolev Norm

We consider here a spatially homogeneous Gaussian noise $\dot{F}(t,x)$, with covariance given by $f(x) = |x|^{-\beta}$, where $0 < \beta < 2$. We seek an estimate of the Sobolev norm of the stochastic integral $v^t_{G,Z}$. We recall that the Sobolev norm is defined in (131).

Theorem 7.3. *Suppose $0 < \beta < 2$. Fix $T > 0$, $q \in {]}3, +\infty{[}$, and let $\mathcal{O} \subset \mathbf{R}^3$ be a bounded domain. Fix $\gamma \in {]}0,1{[}$, and suppose that*

$$\int_0^t ds\, \mathrm{E}\left(\|Z(s)\|^q_{W^{\gamma,q}(\mathcal{O}^{t-s})}\right) < +\infty. \tag{147}$$

Consider

$$\rho \in \left]0, \gamma \wedge \left(\frac{2-\beta}{2} - \frac{3}{q}\right)\right[. \tag{148}$$

Then there exists $C < +\infty$—depending on ρ but not on Z—such that

$$\mathrm{E}\left(\|v^t_{G,Z}\|^q_{\rho,q,\mathcal{O}}\right) \le C \int_0^t ds\, \mathrm{E}\left(\|Z(s)\|^q_{W^{\rho,q}(\mathcal{O}^{t-s})}\right). \tag{149}$$

Remark 7.4. In the case of the heat equation, spatial regularity of the stochastic integral process, that is, of $x \mapsto v^t_{G,Z}(x)$, occurs because of regularity of the heat kernel G, even if Z is merely integrable. Here, the spatial regularity of $v^t_{G,Z}$ is due to the regularity of Z.

Proof (Theorem 7.3). The key quantity that we need to estimate is

$$\mathrm{E}\left(\int_{\mathcal{O}} dx \int_{\mathcal{O}} dy\, \frac{|v^t_{G,Z}(x) - v^t_{G,Z}(y)|^q}{|x-y|^{3+\rho q}}\right). \tag{150}$$

Let $\bar{\rho} = \rho + \frac{3}{q}$, so that $3 + \rho q = \bar{\rho} q$. If we replace G by G_n, then the numerator above is equal to

$$\left|\int_0^t ds \int_{\mathbf{R}^3} (G_n(t-s, x-u) - G_n(t-s, y-u))\, Z(s,u)\, M(ds, du)\right|^q,$$

so by Burkholder's inequality (32),

$$\mathrm{E}\left(|v^t_{G_n,Z}(x) - v^t_{G_n,Z}(y)|^q\right)$$

$$\le C\, \mathrm{E}\left(\left|\int_0^t ds \int_{\mathbf{R}^3} du \int_{\mathbf{R}^3} dv\, Z(s,u)\, f(u-v)\, Z(s,v)\right.\right.$$

$$\times (G_n(t-s, x-u) - G_n(t-s, y-u)) \tag{151}$$

$$\left.\left.\times (G_n(t-s, x-v) - G_n(t-s, y-v))\right|^{q/2}\right).$$

If we had G instead of G_n, and *if* G were ρ-Hölder continuous with exponent ρ, for instance, then we would get a bound involving $|x - y|^{\rho q}$, even if Z were merely integrable.

Here we use a different idea: we shall pass the increments on the G_n over to the factors $Z f Z$ by changing variables. For instance, if there were only one factor involving increments of G_n, we could use the following calculation, where G_n is generically denoted g and $Z f Z$ is denoted ψ:

$$
\begin{aligned}
\int_{\mathbf{R}^3} du \, &(g(x - u) - g(y - u)) \, \psi(u) \\
&= \int_{\mathbf{R}^3} du \, g(x - u) \, \psi(u) - \int_{\mathbf{R}^3} du \, g(y - u) \, \psi(u) \\
&= \int_{\mathbf{R}^3} d\tilde{u} \, g(\tilde{u}) \, \psi(x - \tilde{u}) - \int_{\mathbf{R}^3} d\tilde{u} \, g(\tilde{u}) \, \psi(y - \tilde{u}) \Bigg) \\
&= \int_{\mathbf{R}^3} d\tilde{u} \, g(\tilde{u}) \, (\psi(x - \tilde{u}) - \psi(y - \tilde{u})).
\end{aligned}
\tag{152}
$$

Using this idea, it turns out that the integral on the right-hand side of (151) is equal to

$$
\sum_{i=1}^{4} J_{i,n}^t (x, y),
\tag{153}
$$

where

$$
\begin{aligned}
J_{i,n}^t &(x, y) \\
&= \int_0^t ds \int_{\mathbf{R}^3} du \int_{\mathbf{R}^3} dv \, G_n(s, u) \, G_n(s, v) \, h_i(t, s, x, y, u, v),
\end{aligned}
\tag{154}
$$

and

$$
\begin{aligned}
h_1(t, s, x, y, u, v) &= f(y - x + v - u) \, (Z(t - s, x - u) - Z(t - s, y - u)) \\
&\quad \times (Z(t - s, x - v) - Z(t - s, y - v)), \\[6pt]
h_2(t, s, x, y, u, v) &= Df(v - u, x - y)Z(t - s, x - u) \\
&\quad \times (Z(t - s, x - v) - Z(t - s, y - v)), \\[6pt]
h_3(t, s, x, y, u, v) &= Df(v - u, y - x)Z(t - s, y - v) \\
&\quad \times (Z(t - s, x - u) - Z(t - s, y - u)),
\end{aligned}
\tag{155}
$$

$$
h_4(t, s, x, y, u, v) = -D^2 f(v - u, x - y) \, Z(t - s, x - u) \, Z(t - s, x - u),
$$

and we use the notation

$$
\begin{aligned}
Df(u, x) &= f(u + x) - f(u), \\
D^2 f(u, x) &= f(u - x) - 2 f(u) + f(u + x).
\end{aligned}
\tag{156}
$$

We can now estimate separately each of the four terms

$$T_n^i(t, \mathscr{O}) = \int_{\mathscr{O}} dx \int_{\mathscr{O}} dy \, \frac{\mathrm{E}(|J_{i,n}^t(x,y)|^{q/2})}{|x-y|^{\bar{\rho} q}}, \qquad i = 1, \dots, 4. \tag{157}$$

The term $T_n^1(t, \mathscr{O})$. Set

$$\begin{aligned}
\mu_n(x,y) &= \sup_{s \in [0,T]} \int_{\mathbf{R}^3} du \int_{\mathbf{R}^3} dv \, G_n(s,u) \, G_n(s,v) \, f(y - x + v - u) \\
&= \sup_{s \in [0,T]} \int_{\mathbf{R}^3} \mu(d\eta) \, e^{i \eta \cdot (x-y)} |\mathscr{F} G_n(s)(\eta)|^2,
\end{aligned} \tag{158}$$

so that

$$\sup_{n,x,y} \mu_n(x,y) < +\infty, \tag{159}$$

since $\beta < 2$. Therefore, since $G_n(s,u) \geq 0$, by Hölder's inequality,

$$\begin{aligned}
\mathrm{E}&\left(|J_{1,n}^t(x,y)|^{q/2}\right) \\
&\leq (T \mu_n(x,y))^{\frac{q}{2}-1} \\
&\quad \times \mathrm{E}\bigg(\int_0^t ds \int_{\mathbf{R}^3} du \int_{\mathbf{R}^3} dv \, G_n(s,u) \, G_n(s,v) \, f(y - x + v - u) \\
&\qquad\qquad \times |Z(t-s,x-u) - Z(t-s,y-u)|^{q/2} \\
&\qquad\qquad \times |Z(t-s,x-v) - Z(t-s,y-v)|^{q/2} \bigg). \tag{160}
\end{aligned}$$

Apply the Cauchy–Schwarz inequality with respect to the measure

$$d\mathrm{P} \, dx \, dy \, ds \, du \, dv \, G_n(s,u) \, G_n(s,v) \, f(y - x + v - u) \tag{161}$$

to see that

$$T_n^1(t, \mathscr{O}) \leq \left(T_n^{1,1}(t, \mathscr{O}) \, T_n^{1,2}(t, \mathscr{O})\right)^{1/2}, \tag{162}$$

where

$$\begin{aligned}
T_n^{1,1}&(t, \mathscr{O}) \\
&= \int_0^t ds \int_{\mathscr{O}} dx \int_{\mathscr{O}} dy \int_{\mathbf{R}^3} du \int_{\mathbf{R}^3} dv \, G_n(s,u) \, G_n(s,v) \, f(y - x + v - u) \\
&\quad \times \frac{\mathrm{E}\left(|Z(t-s,x-u) - Z(t-s,y-u)|^q\right)}{|x-y|^{\bar{\rho} q}}, \tag{163}
\end{aligned}$$

and there is an analogous expression for $T_n^{1,2}(t, \mathscr{O})$. We note that for $x \in \mathscr{O}$, when $G_n(s,u) > 0$ (resp. for $y \in \mathscr{O}$, when $G_n(s,v) > 0$), $x - u \in \mathscr{O}^{s(1+1/n)}$ (resp. $y - u \in \mathscr{O}^{s(1+1/n)}$), so

$$T_n^{1,1}(t, \mathcal{O}) \leq \int_0^t ds \, \mathrm{E} \left(\|Z(t-s)\|_{\rho, q, \mathcal{O}^{s(1+1/n)}}^q \right) \sup_{n, x, y} \mu_n(x, y). \qquad (164)$$

The same bound arises for the term $T_n^{1,2}(t, \mathcal{O})$, so this gives the desired estimate for this term.

We shall not discuss the terms $T_n^2(t, \mathcal{O})$ and $T_n^3(t, \mathcal{O})$ here: the interested reader may consult [5, Chapter 3], but we consider the term $T_n^4(t, \mathcal{O})$.

The term $T_n^4(t, \mathcal{O})$. In order to bound $T_n^4(t, \mathcal{O})$, we aim to bring the exponent $q/2$ inside the $ds \, du \, dv$ integral, in such a way that it only affects the Z factors but not f.

Set

$$\mu_n(x, y)$$
$$= \sup_{s \in [0, T]} \int_{\mathbf{R}^3} du \int_{\mathbf{R}^3} dv \, G_n(s, u) \, G_n(s, v) \frac{|D^2 f(v - u, x - y)|}{|x - y|^{2\bar{\rho}}}. \qquad (165)$$

We will show below that

$$\sup_{n \geq 1, \, x, y \in \mathcal{O}} \mu_n(x, y) \leq C < +\infty, \qquad (166)$$

which will turn out to require a quite interesting calculation. Assuming this for the moment, let $p = q/2$. Then, by Hölder's inequality,

$$\frac{\mathrm{E} \left(|J_{4,n}^t(x, y)|^p \right)}{|x - y|^{2p\bar{\rho}}}$$
$$\leq \sup_{n, x, y} (\mu_n(x, y))^{p-1} \int_0^t ds \int_{\mathbf{R}^3} du \int_{\mathbf{R}^3} dv \, G_n(s, u) \, G_n(s, v) \qquad (167)$$
$$\times \frac{|D^2 f(v - u, x - y)|}{|x - y|^{2\bar{\rho}}} \mathrm{E} \left(|Z(t - s, x - u)|^p |Z(t - s, x - v)|^p \right).$$

This quantity must be integrated over $\mathcal{O} \times \mathcal{O}$. We apply the Cauchy–Schwarz inequality to the measure $ds \, du \, dv (\cdots) d\mathrm{P}$, and this leads to

$$T_n^4(t, \mathcal{O}) \leq \sup_{n, x, y} (\mu_n(x, y))^p \int_0^t ds \, \mathrm{E} \left(\|Z(s)\|_{L^q(\mathcal{O}^{(t-s)(1+1/n)})}^q \right). \qquad (168)$$

This is the desired bound for this term.

It remains to check (166). The main difficulty is to bound the second-order difference $|D^2 f(v - u, x - y)|$. We explain the main issues below.

Bounding Symmetric Differences

Let $g : \mathbf{R} \to \mathbf{R}$. Suppose that we seek a bound on

$$D^2 g(x,h) = g(x-h) - 2g(x) + g(x+h). \tag{169}$$

Notice that if g is differentiable only once ($g \in C^1$), then the best that we can do is essentially to write

$$|D^2 g(x,h)| \le |g(x-h) - g(x)| + |g(x+h) - g(x)| \atop \le c\,h. \tag{170}$$

On the other hand, if g is twice differentiable ($g \in C^2$), then we can do better:

$$|D^2 g(x,h)| \le c\,h^2. \tag{171}$$

In the case of a Riesz kernel $f(x) = |x|^{-\beta}$, $x \in \mathbf{R}^3$, we can write

$$\begin{aligned} |D^2 f(u,x)| &= \big|\,|u-x|^{-\beta} - 2|u|^{-\beta} + |u+x|^{-\beta}\,\big| \\ &\le C\,|f''(u)|\,|x|^2 \\ &= C\,|u|^{-\beta-2}\,|x|^2\,. \end{aligned} \tag{172}$$

Taking into account the definition of $\mu_n(x,y)$ in (165), this inequality leads to the bound

$$\mu_n(x,y)$$
$$\le \sup_{s\in[0,T]} \left(\int_{\mathbf{R}^3} du \int_{\mathbf{R}^3} dv\, G_n(s,u)\, G_n(s,v)\, |u-v|^{-(\beta+2)} \right) \frac{|x-y|^2}{|x-y|^{2\bar\rho}}\,. \tag{173}$$

However, the double integral converges to $+\infty$ as $n \to \infty$, since $\beta + 2 > 2$.

Since differentiating once does not necessarily give the best bound possible and differentiating twice gives a better exponent but with an infinite constant, it is natural to want to differentiate a fractional number of times, namely just under $2 - \beta$ times. If we "differentiate α times" and all goes well, then this should give a bound of the form $\mu_n(x,y) \le C|x-y|^\alpha$, for $\alpha \in\,]0, 2-\beta[$. We shall make this precise below.

Riesz Potentials, their Fractional Integrals and Laplacians

Let $\alpha \overset{\text{def}}{=} 2\bar\rho$. We recall that

$$\rho < \frac{2-\beta}{2} - \frac{3}{q} \quad \text{and} \quad \bar\rho = \rho + \frac{3}{q}, \quad \text{so} \quad \alpha < 2 - \beta. \tag{174}$$

The *Riesz potential* of a function $\varphi : \mathbf{R}^d \to \mathbf{R}$ is defined by

$$(I_a\varphi)(x) = \frac{1}{\gamma(a)} \int_{\mathbf{R}^d} \frac{\varphi(y)}{|x-y|^{d-a}}\, dy, \qquad a \in\,]0, d[, \tag{175}$$

where $\gamma(a) = \pi^{d/2} 2^a \Gamma(a/2)/\Gamma(\frac{1}{2}(d-a))$. Riesz potentials have many interesting properties (see [17]), of which we mention the following:

(1) $I_{a+b}(\varphi) = I_a(I_b\,\varphi)$ if $a + b \in]0, d[$. Further, I_a can be seen as a "fractional integral of order a," in the sense that

$$\mathscr{F}(I_a\,\varphi)(\xi) = \mathscr{F}\varphi(\xi)\,\mathscr{F}\left(\frac{1}{|\cdot|^{d-a}}\right)(\xi) = \frac{\mathscr{F}\varphi(\xi)}{|\xi|^a}. \tag{176}$$

(2) Our covariance function $k_\beta(x) = |x|^{-\beta}$ is a *Riesz kernel*. These kernels have the following property:

$$|x|^{-d+a+b} = \int_{\mathbf{R}^d} dz\, k_{d-b}(x - z)\,|z|^{-d+a} \tag{177}$$

$$= I_b\left(|\cdot|^{-d+a}\right). \tag{178}$$

This equality can be viewed as saying that $|z|^{-d+a}$ is "bth derivative (or Laplacian)" of $|z|^{-d+a+b}$, in the sense that

$$(-\Delta)^{b/2}\left(|z|^{-d+a+b}\right) = |z|^{-d+a}. \tag{179}$$

Indeed, taking Fourier transforms, this equality becomes simply

$$|\xi|^b\,|\xi|^{-a-b} = |\xi|^{-a}. \tag{180}$$

Recall the notation

$$Df(u, y) = f(u + y) - f(u). \tag{181}$$

From (177), one can easily deduce (see [5, Lemma 2.6]) that

$$Dk_{d-a-b}(u, cx) = |c|^b \int_{\mathbf{R}^d} dw\, k_{d-a}(u - cw)\, Dk_{d-b}(w, x), \tag{182}$$

and

$$|D^2 k_{d-a-b}(u, x)| \le |x|^b \int_{\mathbf{R}^d} dw\, k_{d-a}(u - |x|w)\, D^2 k_{d-b}\left(w, \frac{x}{\|x\|}\right). \tag{183}$$

Set $b = \alpha = 2\bar\rho$ and $a = 3 - \alpha - \beta$, where $\alpha + \beta \in]0, 2[$. Looking back to (165), these two relations lead to the following estimate:

$$\mu_n(x, y)$$

$$\le \sup_{s \in [0,T]} \frac{1}{|x - y|^\alpha} \int_{\mathbf{R}^3} du \int_{\mathbf{R}^3} dv\, G_n(s, u)\, G_n(s, v)\, |x - y|^\alpha$$

$$\times \int_{\mathbf{R}^3} dw\, k_{\alpha+\beta}(v - u - |y - x|w) \times \left| D^2 k_{3-\alpha}\left(w, \frac{x}{|x|}\right)\right|$$

$$\le \sup_{s \in [0,T]} \left(\sup_{x,y,w} \int_{\mathbf{R}^3} du \int_{\mathbf{R}^3} dv\, G_n(s, u)\, G_n(s, v)\, k_{\alpha+\beta}(v - u - |y - x|w)\right)$$

$$\times \sup_x \int dw\, \left| D^2 k_{3-\alpha}\left(w, \frac{x}{\|x\|}\right)\right|. \tag{184}$$

The double integral above is finite since $\alpha + \beta < 2$. Indeed, taking Fourier transforms, the shift $-|y - x|w$ introduces a factor $e^{i\eta \cdot |y-x|w}$, which is of no consequence. The second integral is finite (and does not depend on x). For this calculation, see [5, Lemma 2.6]. This proves Theorem 7.3.   □

8 Hölder-Continuity in the 3-d Wave Equation

We consider the stochastic wave equation (113) for $d = 3$, driven by spatially homogeneous Gaussian noise with covariance $f(x) = |x|^{-\beta}$, where $0 < \beta < 2$.

The main idea for checking Hölder continuity of the solution is to go back to the Picard iteration scheme that was used to construct the solution, starting with a smooth function $u_0(t, x)$ (whose smoothness depends only on the regularity of the initial conditions), and then check that regularity is preserved at each iteration step and passes to the limit. The details are carried out in [5, Chapter 4]. The main result is the following.

Theorem 8.1. *Assume the following three properties:*

(a) the initial value v_0 is such that $v_0 \in C^2(\mathbf{R}^3)$ and Δv_0 is Hölder continuous with exponent γ_1;
(b) the initial velocity $\tilde{v}_0$ is Hölder continuous with exponent γ_2;
(c) the nonlinearities $\sigma, b : \mathbf{R} \to \mathbf{R}$ are Lipschitz continuous.

Then, for any $q \in [2, \infty[$ and

$$\alpha \in \left]0, \gamma_1 \wedge \gamma_2 \wedge \frac{2-\beta}{2}\right[, \tag{185}$$

there is $C > 0$ such that for all $(t, x), (s, y) \in [0, T] \times D$,

$$\mathrm{E}\left(|u(t, x) - u(s, y)|^q\right) \le C\left(|t - s| + |x - y|\right)^{\alpha q}. \tag{186}$$

In particular, $(t, x) \mapsto u(t, x)$ has a Hölder continuous version with exponent α.

We observe that the presence of $\gamma_1 \wedge \gamma_2$ in (185) can be interpreted by saying that the (ir)regularity of the initial conditions limits the possible regularity of the solution: there is no smoothing effect in the wave equation, contrary to the heat equation.

We note that this result is *sharp*. Indeed, if we consider the linear wave equation, in which we take $\sigma \equiv 1$ and $b \equiv 0$ in (113), with vanishing initial condition $v_0 \equiv \tilde{v}_0 \equiv 0$, then it is possible to show (see [5, Chapter 5] that

$$\mathrm{E}\left(|u(t, x) - u(t, y)|^2\right) \ge c_1 |x - y|^{2-\beta} \tag{187}$$

and

$$\mathrm{E}\left(|u(t, x) - u(s, x)|^2\right) \ge c_2 |t - s|^{2-\beta}. \tag{188}$$

This implies in particular that $t \mapsto u(t, x)$ and $x \mapsto u(t, x)$ are *not* γ-Hölder continuous, for $\gamma > \frac{2-\beta}{2}$ (see [5, Chapter 5]).

References

[1] R. C. Dalang: *Extending the martingale measure stochastic integral with applications to spatially homogeneous spde's.* Electronic J. of Probability, Vol 4, 1999.

[2] R. C. Dalang, N. E. Frangos: *The stochastic wave equation in two spatial dimensions.* Annals of Probab. 26, 1, 187–212, 1998.

[3] R. C. Dalang, O. Lévèque: *Second-order linear hyperbolic SPDEs driven by isotropic Gaussian noise on a sphere.* Annals of Probab. 32, 1068–1099, 2004.

[4] R. C. Dalang, C. Mueller: *Some non-linear SPDE's that are second order in time.* Electronic J. of Probability, Vol 8, 1, 1–21, 2003.

[5] R. C. Dalang, M. Sanz-Solé: *Hölder-Sobolev regularity of the solution to the stochastic wave equation in dimension 3.* Memoirs of the AMS (2009, to appear).

[6] J. L. Davis: *Wave Propagation in Solids and Fluids.* Springer Verlag, 1988.

[7] G. B. Folland: *Introduction to Partial Differential Equations.* Princeton Univ. Press, 1976.

[8] O. Gonzalez, J. H. Maddocks: *Extracting parameters for base-pair level models of DNA from molecular dynamics simulations.* Theoretical Chemistry Accounts, 106, 76–82, 2001.

[9] A. Hellemans: *SOHO Probes Sun's Interior by Tuning In to Its Vibrations.* Science May 31 1996: 1264–1265.

[10] D. Khoshnevisan: *A Primer on Stochastic Partial Differential Equations.* In: this volume, 2007.

[11] A. Karkzewska, J. Zabczyk: *Stochastic PDE's with function-valued solutions.* In: *Infinite-dimensional stochastic analysis* (Clément Ph., den Hollander F., van Neerven J. & de Pagter B., eds), pp. 197–216, Proceedings of the Colloquium of the Royal Netherlands Academy of Arts and Sciences, 1999, Amsterdam.

[12] M. Oberguggenberger & F. Russo: *Nonlinear stochastic wave equations.* In: Generalized functions—linear and nonlinear problems (Novi Sad, 1996). *Integral Transform. Spec. Funct.* 6, no. 1–4, 71–83, 1998.

[13] S. Peszat *The Cauchy problem for a non-linear stochastic wave equation in any dimension.* Journal of Evolution Equations, Vol. 2, no. 3, 383–394, 2002.

[14] M. Sanz-Solé, M. Sarrà: *Hölder continuity for the stochastic heat equation with spatially correlated noise.* In: *Stochastic analysis, random fields and applications* (R.C. Dalang, M. Dozzi & F. Russo, eds), pp. 259–268, Progress in Probability 52, Birkhäuser, Basel, 2002.

[15] L. Schwartz: *Théorie des distributions.* Hermann, Paris, 1966.

[16] N. Shimakura: *Partial differential operators of elliptic type.* Translations of Mathematical Monographs, 99. American Mathematical Society, 1992.

[17] E. M. Stein: *Singular Integrals and Differentiability Properties of Functions.* Princeton University Press, Princeton, 1970.

[18] F. Treves: *Basic Linear Partial Differential Equations.* Academic Press, 1975.

[19] J. B. Walsh: *An introduction to stochastic partial differential equations,* École d'été de Probabilités de Saint Flour XIV, Lecture Notes in Mathematics, Vol. 1180, Springer Verlag, 1986.

[20] J. Zabczyk: *A mini course on stochastic partial differential equations.* In: *Stochastic climate models* (P. Imkeller & J.-S. von Storch, eds), pp. 257–284, Progr. Probab., 49, Birkhäuser, 2001.

Application of Malliavin Calculus to Stochastic Partial Differential Equations

David Nualart

1 Introduction

The aim of these notes is to provide an introduction to the Malliavin calculus and its application to the regularity of the solutions of a class of stochastic partial differential equations. The Malliavin calculus is a differential calculus on a Gaussian space which has been developed from the probabilistic proof by Malliavin of Hörmander's hypoellipticity theorem (see [8]). In the next section we present an introduction to the Malliavin calculus, and we derive the main properties of the derivative and divergence operators. Section 3 is devoted to establish the main criteria for the existence and regularity of density for a random variable in a Gaussian space.

The last two sections are devoted to discuss the applications of the Malliavin calculus to stochastic partial differential equations. First we consider a one-dimensional heat equation driven by a space-time white noise on the time interval $[0, 1]$ with Dirichlet boundary conditions, and we show that for any $(t, x) \in (0, \infty) \times (0, 1)$, the solution $u(t, x)$ has an infinitely differentiable density if the coefficients are smooth and the diffusion coefficient is bounded away from zero. The last section deals with a class of stochastic partial differential equations perturbed by a Gaussian noise on $[0, \infty) \times \mathbf{R}^d$ with homogeneous spacial covariance, introduced by Dalang in [6]. We survey the results obtained in some recent papers [13; 16; 17] on the regularity of the density for the solution to this class of equations.

2 Malliavin Calculus

The Malliavin calculus is an infinite dimensional calculus on a Gaussian space, which is mainly applied to establish the regularity of the law of nonlinear functionals of the underlying Gaussian process.

D. Khoshnevisan and F. Rassoul-Agha (eds.) *A Minicourse on Stochastic Partial Differential Equations.*
Lecture Notes in Mathematics 1962.
© Springer-Verlag Berlin Heidelberg 2009

Suppose that H is a real separable Hilbert space with scalar product denoted by $\langle \cdot, \cdot \rangle_H$. Consider a Gaussian family of random variables $W = \{W(h), h \in H\}$ defined in a complete probability space $(\Omega, \mathscr{F}, \mathrm{P})$, with zero mean and covariance

$$\mathrm{E}(W(h)W(g)) = \langle h, g \rangle_H. \tag{1}$$

The mapping $h \to W(h)$ provides a linear isometry of H onto a closed subspace of H_1 of $L^2(\Omega)$.

Example 2.1. (Brownian motion) If $B = \{B_t, t \geq 0\}$ is a Brownian motion, we take $H = L^2([0, \infty))$ and

$$W(h) = \int_0^\infty h(t)dB_t. \tag{2}$$

Example 2.2. (White noise) Suppose that $H = L^2(T, \mathscr{B}, \mu)$, where μ is a σ-finite measure without atoms. In this case, for any set $A \in \mathscr{B}$ with $\mu(A) < \infty$ we make use of the notation $W(A) = W(\mathbf{1}_A)$. Then, $A \mapsto W(A)$ is a Gaussian measure with independent increments (Gaussian white noise). That is, if $A_1, \ldots, A_n$ are disjoint sets with finite measure, the random variables $W(A_1), \ldots, W(A_n)$ are independent, and for any $A \in \mathscr{B}$ with $\mu(A) < \infty$, $W(A)$ has the distribution $N(0, \mu(A))$. Then, any square integrable random variable $F \in L^2(\Omega, \mathscr{F}, \mathrm{P})$ (assuming that the σ-field $\mathscr{F}$ is generated by $\{W(h)\}$) admits the following Wiener chaos expansion

$$F = \mathrm{E}(F) + \sum_{n=1}^\infty I_n(f_n). \tag{3}$$

In this formula f_n is a symmetric function of $L^2(T^n)$ and I_n denotes the multiple stochastic integral introduced by Itô in [7]. In particular $I_1(f_1) = W(f_1)$. Furthermore (see Exercise 2.18),

$$\mathrm{E}(F^2) = \mathrm{E}(F)^2 + \sum_{n=1}^\infty n! \|f_n\|_{L^2(T^n)}^2. \tag{4}$$

2.1 Derivative Operator

Let $\mathscr{S}$ denote the class of smooth and cylindrical random variables of the form

$$F = f(W(h_1), \ldots, W(h_n)), \tag{5}$$

where f belongs to $C_p^\infty(\mathbf{R}^n)$ (f and all its partial derivatives have polynomial growth order), $h_1, \ldots, h_n$ are in H, and $n \geq 1$.

The *derivative* of F is the H-valued random variable given by

$$DF = \sum_{i=1}^{n} \frac{\partial f}{\partial x_i}(W(h_1), \ldots, W(h_n))h_i. \tag{6}$$

For example, $D(W(h)) = h$, and $D(W(h)^2) = 2W(h)h$.

The following result is an integration-by-parts formula.

Proposition 2.3. *Suppose that F is a smooth and cylindrical random variable and $h \in H$. Then*

$$\mathrm{E}(\langle DF, h \rangle_H) = \mathrm{E}(FW(h)). \tag{7}$$

Proof. We can restrict the proof to the case where there exist orthonormal elements of H, $e_1, \ldots, e_n$, such that $h = e_1$ and

$$F = f(W(e_1), \ldots, W(e_n)), \tag{8}$$

where $f \in C_p^{\infty}(\mathbf{R}^n)$. Let $\phi(x)$ denote the density of the standard normal distribution on $\mathbf{R}^n$, that is,

$$\phi(x) = (2\pi)^{-n/2} \exp\left(-\frac{1}{2}\sum_{i=1}^{n} x_i^2\right). \tag{9}$$

Then we have

$$\begin{aligned}
\mathrm{E}(\langle DF, h \rangle_H) &= \mathrm{E}\left(\frac{\partial f}{\partial x_1}(W(e_1), \ldots, W(e_n))\right) \\
&= \int_{\mathbf{R}^n} \frac{\partial f}{\partial x_1}(x)\phi(x)dx \\
&= \int_{\mathbf{R}^n} f(x)\phi(x)x_1 dx \\
&= \mathrm{E}(FW(e_1)),
\end{aligned} \tag{10}$$

which completes the proof. $\qquad\square$

Applying the previous result to a product FG, we obtain the following consequence.

Proposition 2.4. *Suppose that F and G are smooth and cylindrical random variables, and $h \in H$. Then we have*

$$\mathrm{E}(G\langle DF, h \rangle_H) = \mathrm{E}(-F\langle DG, h \rangle_H + FGW(h)). \tag{11}$$

Proof. Use the formula $D(FG) = FDG + GDF$. $\qquad\square$

The integration-by-parts formula (11) is a useful tool to show the closability of the derivative operator. In this sense, we have the following result.

Proposition 2.5. *The operator D is closable from $L^p(\Omega)$ to $L^p(\Omega; H)$ for any $p \geq 1$.*

Proof. Let $\{F_N, N \geq 1\}$ be a sequence of random variables in $\mathscr{S}$ such that F_N converges to 0 in $L^p(\Omega)$, and DF_N converges to η in $L^p(\Omega; H)$, as N tends to infinity. Then, we claim that $\eta = 0$. Indeed, for any $h \in H$ and for any random variable $F = f(W(h_1), , \ldots \ldots, W(h_n)) \in \mathscr{S}$ such that f and its partial derivatives are bounded, and $FW(h)$ is bounded, we have

$$
\begin{aligned}
\mathrm{E}(\langle \eta, h \rangle_H F) &= \lim_{N \to \infty} \mathrm{E}(\langle DF_N, h \rangle_H F) \\
&= \lim_{N \to \infty} \mathrm{E}(-F_N \langle DF, h \rangle_H + F_N F W(h)) \qquad (12) \\
&= 0.
\end{aligned}
$$

This implies that $\eta = 0$. ☐

For any $p \geq 1$ we denote by $\mathbf{D}^{1,p}$ the closure of $\mathscr{S}$ with respect to the seminorm

$$
\|F\|_{1,p} = [\mathrm{E}(|F|^p) + \mathrm{E}(\|DF\|_H^p)]^{1/p}. \qquad (13)
$$

For $p = 2$, the space $\mathbf{D}^{1,2}$ is a Hilbert space with the scalar product

$$
\langle F, G \rangle_{1,2} = \mathrm{E}(FG) + \mathrm{E}(\langle DF, DG \rangle_H). \qquad (14)
$$

We can define the iteration of the operator D in such a way that for a random variable $F \in \mathscr{S}$, the iterated derivative $D^k F$ is a random variable with values in $H^{\otimes k}$. For every $p \geq 1$ and any natural number $k \geq 1$ we introduce the seminorm on $\mathscr{S}$ defined by

$$
\|F\|_{k,p} = \left[\mathrm{E}(|F|^p) + \sum_{j=1}^{k} \mathrm{E}\left(\|D^j F\|_{H^{\otimes j}}^p\right) \right]^{1/p}. \qquad (15)
$$

We denote by $\mathbf{D}^{k,p}$ the closure of $\mathscr{S}$ with respect to the seminorm $\| \cdot \|_{k,p}$. For any $k \geq 1$ and $p > q$ we have $\mathbf{D}^{k,p} \subset \mathbf{D}^{k,q}$. We set $\mathbf{D}^\infty = \cap_{k,p} \mathbf{D}^{k,p}$.

Remark 2.6. The above definitions can be exended to Hilbert-space-valued random variables. That is, if V is a separable Hilbert space, then $\mathbf{D}^{k,p}(V)$ is the completion of the set $\mathscr{S}_V$ of V-valued smooth and cylindrical random variables of the form $F = \sum_{j=1}^{m} F_j v_j$, $F_j \in \mathscr{S}$, $v_j \in V$, with respect to the seminorm

$$
\|F\|_{k,p,V} = \left[\mathrm{E}\left(\|F\|_V^p\right) + \sum_{j=1}^{k} \mathrm{E}\left(\|D^j F\|_{H^{\otimes j} \otimes V}^p\right) \right]^{1/p}. \qquad (16)
$$

The following result is the chain rule, and its proof is contained in Exercise 2.20.

Proposition 2.7. *Let $\varphi : \mathbf{R}^m \to \mathbf{R}$ be a continuously differentiable function with bounded partial derivatives, and fix $p \geq 1$. Suppose that $F = (F^1, \ldots, F^m)$ is a random vector whose components belong to the space $\mathbf{D}^{1,p}$. Then $\varphi(F) \in \mathbf{D}^{1,p}$, and*

$$D(\varphi(F)) = \sum_{i=1}^{m} \frac{\partial \varphi}{\partial x_i}(F) DF^i. \tag{17}$$

The following Hölder inequality implies that $\mathbf{D}^{\infty}$ is closed under multiplication. We refer to Watanabe [19] and Exercise 2.21 for its proof.

Proposition 2.8. *Let $F \in \mathbf{D}^{k,p}$, $G \in \mathbf{D}^{k,q}$ for $k \geq 1$, $1 < p, q < \infty$ and let r be such that $p^{-1} + q^{-1} = r^{-1}$. Then, $FG \in \mathbf{D}^{k,r}$ and*

$$\|FG\|_{k,r} \leq c_{p,q,k} \|F\|_{k,p} \|G\|_{k,q}. \tag{18}$$

Consider now the white noise case, that is, $II = L^2(T, \mathscr{B}, \mu)$. Then, the derivative DF is a random element in $L^2(\Omega; H) \sim L^2(\Omega \times T, \mathscr{F} \otimes \mathscr{B}, \mathrm{P} \times \mu)$, that is, it is a stochastic process parameterized by T, that we denote by $\{D_t F, t \in T\}$. Suppose that F is a square integrable random variable having an orthogonal Wiener chaos expansion of the form

$$F = \mathrm{E}(F) + \sum_{n=1}^{\infty} I_n(f_n), \tag{19}$$

where the kernels f_n are symmetric functions from $L^2(T^n)$. The derivative $D_t F$ can be easily computed using this expression.

Proposition 2.9. *Let $F \in L^2(\Omega)$ be a square integrable random variable with a development of the form (19). Then F belongs to $\mathbf{D}^{1,2}$ if and only if*

$$\sum_{n=1}^{\infty} n n! \|f_n\|_{L^2(T^n)}^2 < \infty \tag{20}$$

and in this case we have

$$D_t F = \sum_{n=1}^{\infty} n I_{n-1}(f_n(\cdot, t)). \tag{21}$$

Proof. Suppose first that $F = I_n(f_n)$, where f_n is a symmetric and elementary function of the form

$$f(t_1, \ldots, t_n) = \sum_{i_1, \ldots, i_n = 1}^{m} a_{i_1 \cdots i_n} \mathbf{1}_{A_{i_1} \times \cdots \times A_{i_n}}(t_1, \ldots, t_n), \tag{22}$$

where $A_1, A_2, \ldots, A_m$ are pair-wise disjoint sets with finite measure, and the coefficients $a_{i_1 \cdots i_n}$ are zero if any two of the indices $i_1, \ldots, i_n$ are equal. Then

$$D_t F = \sum_{j=1}^{n} \sum_{i_1, \ldots, i_n = 1}^{m} a_{i_1 \cdots i_n} W(A_{i_1}) \cdots \mathbf{1}_{A_{i_j}}(t) \cdots W(A_{i_n}) \tag{23}$$
$$= n I_{n-1}(f_n(\cdot, t)).$$

Then the result follows easily. $\square$

The preceding proposition leads to an heuristic interpretation of the derivative operator on multiple stochastic integrals. Suppose that F is a multiple stochastic integral of the form $I_n(f_n)$, which can be denoted by

$$I_n(f_n) = \int_T \cdots \int_T f_n(t_1, \ldots, t_n) \, W(dt_1) \cdots W(dt_n). \tag{24}$$

Then, F belongs to the domain of the derivative operator and $D_t F$ is obtained simply by removing one of the stochastic integrals, letting the variable t be free, and multiplying by the factor n.

We will make use of the following result.

Lemma 2.10. *Let* $\{F_n, n \geq 1\}$ *be a sequence of random variables converging to* F *in* $L^p(\Omega)$ *for some* $p > 1$. *Suppose that* $\sup_n \|F_n\|_{k,p} < \infty$ *for some* $k \geq 1$. *Then,* F *belongs to* $\mathbf{D}^{k,p}$.

Proof. We do the proof only in the case $p = 2$, $k = 1$ and assuming that we are in the white noise context. There exists a subsequence $\{F_{n(k)}, k \geq 1\}$ such that the sequence of derivatives $DF_{n(k)}$ converges in the weak topology of $L^2(\Omega \times T)$ to some element $\alpha \in L^2(\Omega \times T)$. Then, for any $h \in H$ the projections of $\langle h, DF_{n(k)} \rangle_H$ on any Wiener chaos converge in the weak topology of $L^2(\Omega)$, as k tends to infinity, to those of $\langle h, \alpha \rangle_H$. Consequently, Proposition 2.9 implies $F \in \mathbf{D}^{1,2}$ and $\alpha = DF$. Moreover, for any weakly convergent subsequence the limit must be equal to α by the preceding argument, and this implies the weak convergence of the whole sequence. $\square$

Proposition 2.11. *Let* F *be a random variable of the space* $\mathbf{D}^{1,2}$ *such that* $DF = 0$. *Then* $F = \mathrm{E}(F)$.

Proof. In the white noise case, this proposition is obvious from the Wiener chaos expansion of the derivative provided in Proposition 2.9. In the general case the result is also true, even for random variables in $\mathbf{D}^{1,1}$ (see [12]). $\square$

Proposition 2.12. *Let* $A \in \mathscr{F}$. *Then the indicator function of* A *belongs to* $\mathbf{D}^{1,2}$ *if and only if* $\mathrm{P}(A)$ *is equal to zero or one.*

Proof. By the chain rule (Proposition 2.7) applied to a function $\varphi \in C_0^\infty(\mathbf{R})$, which is equal to x^2 on $[0, 1]$, we have

$$D1_A = D(1_A)^2 = 21_A D1_A \tag{25}$$

and, therefore, $D1_A = 0$ because from the above equality we get that this derivative is zero on A^c and equal to twice its value on A. So, by Proposition 2.11 we obtain $1_A = P(A)$. □

2.2 Divergence Operator

We denote by δ the adjoint of the operator D (*divergence operator*). That is, δ is an unbounded operator on $L^2(\Omega; H)$ with values in $L^2(\Omega)$ such that:

(i) The domain of δ, denoted by $Dom\,\delta$, is the set of H-valued square integrable random variables $u \in L^2(\Omega; H)$ such that

$$|\mathrm{E}(\langle DF, u \rangle_H)| \leq c\|F\|_2, \tag{26}$$

for all $F \in \mathbf{D}^{1,2}$, where c is some constant depending on u.

(ii) If u belongs to $Dom\,\delta$, then $\delta(u)$ is the element of $L^2(\Omega)$ characterized by the duality relationship

$$\mathrm{E}(F\delta(u)) = \mathrm{E}(\langle DF, u \rangle_H) \tag{27}$$

for any $F \in \mathbf{D}^{1,2}$.

Properties of the Divergence

1. $\mathrm{E}(\delta(u)) = 0$ (take $F = 1$ in the duality formula (27)).
2. Suppose that $u \in \mathscr{S}_H$ is an H-valued cylindrical and smooth random variable of the form $u = \sum_{j=1}^n F_j h_j$, where the $F_j \in \mathscr{S}$, and $h_j \in H$. Then, Proposition 2.4 implies that u belongs to the domain of δ and

$$\delta(u) = \sum_{j=1}^n F_j W(h_j) - \sum_{j=1}^n \langle DF_j, h_j \rangle_H. \tag{28}$$

We will make use of the notation $D_h F = \langle DF, h \rangle_H$, for any $h \in H$ and $F \in \mathbf{D}^{1,2}$.

Three Basic Formulas

Suppose that $u, v \in \mathscr{S}_H$, $F \in \mathscr{S}$ and $h \in H$. Then, if $\{e_i\}$ is a complete orthonormal system in H:

$$\mathrm{E}(\delta(u)\delta(v)) = \mathrm{E}(\langle u, v \rangle_H) + \mathrm{E}\left(\sum_{i,j=1}^\infty D_{e_i}\langle u, e_j \rangle_H D_{e_j}\langle v, e_i \rangle_H\right), \tag{29}$$

$$D_h(\delta(u)) = \delta(D_h u) + \langle h, u \rangle_H, \tag{30}$$

$$\delta(Fu) = F\delta(u) - \langle DF, u \rangle_H. \tag{31}$$

Proof (of (30)). Assume $u = \sum_{j=1}^{n} F_j h_j$. Then

$$
\begin{aligned}
D_h(\delta(u)) &= D_h \left(\sum_{j=1}^{n} F_j W(h_j) - \sum_{j=1}^{n} \langle DF_j, h_j \rangle_H \right) \\
&= \sum_{j=1}^{n} F_j \langle h, h_j \rangle_H + \sum_{j=1}^{n} \left(D_h F_j W(h_j) - \langle D_h (DF_j), h_j \rangle_H \right) \\
&= \langle u, h \rangle_H + \delta(D_h u).
\end{aligned}
\tag{32}
$$

This completes the proof. $\qquad\square$

Proof (of (29)). Using the duality formula (27) and (30) yields

$$
\begin{aligned}
\mathrm{E}\left(\delta(u)\delta(v)\right) &= \mathrm{E}\left(\langle v, D(\delta(u)) \rangle_H\right) \\
&= \mathrm{E}\left(\sum_{i=1}^{\infty} \langle v, e_i \rangle_H D_{e_i}(\delta(u)) \right) \\
&= \mathrm{E}\left(\sum_{i=1}^{\infty} \langle v, e_i \rangle_H \left(\langle u, e_i \rangle_H + \delta(D_{e_i}u) \right) \right) \\
&= \mathrm{E}\left(\langle u, v \rangle_H \right) + \mathrm{E}\left(\sum_{i,j=1}^{\infty} D_{e_i} \langle u, e_j \rangle_H D_{e_j} \langle v, e_i \rangle_H \right).
\end{aligned}
\tag{33}
$$

Hence follows (29). $\qquad\square$

Proof (of (31)). For any smooth random variable $G \in \mathscr{S}$ we have

$$
\begin{aligned}
\mathrm{E}\left(\langle DG, Fu \rangle_H\right) &= \mathrm{E}\left(\langle u, D(FG) - GDF \rangle_H\right) \\
&= \mathrm{E}\left((\delta(u)F - \langle u, DF \rangle_H)G\right).
\end{aligned}
\tag{34}
$$

This verifies (31). $\qquad\square$

Remark 2.13. Property (29) implies the estimate

$$
\mathrm{E}\left(\delta(u)^2\right) \leq \mathrm{E}\left(\|u\|_H^2\right) + \mathrm{E}\left(\|Du\|_{H \otimes H}^2\right) = \|u\|_{1,2,H}^2.
\tag{35}
$$

As a consequence, $\mathbf{D}^{1,2}(H) \subset \mathrm{Dom}\,\delta$.

Remark 2.14. Properties (29), (30) and (31) hold under more general conditions:

1. $u \in \mathbf{D}^{1,2}(H)$ for (29).
2. $u \in \mathbf{D}^{1,2}(H)$ and $D_h u$ belongs to $\mathrm{Dom}\,\delta$ for (30).
3. $F \in \mathbf{D}^{1,2}$, $u \in \mathrm{Dom}\,\delta$, $Fu \in L^2(\Omega; H)$, and $F\delta(u) - \langle DF, u \rangle_H \in L^2(\Omega)$ for (31).

Consider the case of a Gaussian white noise $H = L^2(T, \mathcal{B}, \mu)$. The next proposition tells us that the second summand in the right-hand side of formula (31) cancels when F and u are independent.

Proposition 2.15. *Fix a set $A \in \mathcal{B}$ with finite measure. Let $\mathscr{F}_{A^c}$ be the σ-field generated by the random variables $\{W(B), B \subset A^c\}$. Suppose that $F \in L^2(\Omega, \mathscr{F}_{A^c}, P)$. Then $F\mathbf{1}_A$ belongs to the domain of the divergence and*

$$\delta(F\mathbf{1}_A) = FW(A). \tag{36}$$

Proof. If F is a cylindrical and smooth random variable, then

$$\delta(F\mathbf{1}_A) = FW(A) - \langle DF, \mathbf{1}_A \rangle_H$$

$$= FW(A) - \int_A D_t F \mu(dt) \tag{37}$$

$$= FW(A),$$

because $D_t F = 0$ if $t \in A$. The general case follows easily. $\qquad\square$

Consider the particular case $T = [0, \infty)$. Then $B_t = W(\mathbf{1}_{[0,t]})$ is a Brownian motion. Let $\mathscr{F}_t$ be the σ-field generated by the random variables $\{B_s, 0 \le s \le t\}$. We say that a stochastic process $\{u_t, t \ge 0\}$ is adapted if for all $t \ge 0$ the random variable u_t is $\mathscr{F}_t$ measurable. Then, the class L_a^2 of adapted stochastic processes such that $E\left(\int_0^\infty u_t^2 \, dt\right) < \infty$ is included in the domain of the divergence and $\delta(u)$ coincides with the Itô stochastic integral:

$$\delta(u) = \int_0^\infty u_t \, dB_t. \tag{38}$$

This is a consequence of Proposition 2.15 and the fact that the operator δ is closed.

The following theorem is based on Meyer inequalities and it is a central result in Malliavin Calculus. It tells us that the operator δ can be extended continuously from the space $\mathbf{D}^{k,p}(H)$ to the space $\mathbf{D}^{k-1,p}$ for all $p > 1$ and $k \ge 1$. We refer to Watanabe [20] and Nualart [12, Proposition 1.5.7] for its proof.

Theorem 2.16. *The operator δ is continuous from $\mathbf{D}^{k,p}(H)$ into $\mathbf{D}^{k-1,p}$ for all $p > 1$ and $k \ge 1$. That is,*

$$\|\delta(u)\|_{k-1,p} \le C_{k,p}\|u\|_{k,p}. \tag{39}$$

In the particular case $k = 1$, the p norm of the divergence $\delta(u)$ can be dominated by $\|u\|_{1,p}$. Actually, this domination can be made more precise as it is shown in the next proposition (see [12] for the proof).

Proposition 2.17. *Let u be an element of $\mathbf{D}^{1,p}(H)$, $p > 1$. Then we have*

$$\|\delta(u)\|_p \le c_p \left(\|E(u)\|_H + \|Du\|_{L^p(\Omega; H \otimes H)} \right). \tag{40}$$

Exercise 2.18. Show Equation (4). *Hint*: Use the orthogonality of multiple stochastic integrals of different order and the variance formula (see Itô [7]): $E(I_n(f_n)I_n(g_n)) = n!\langle f_n, g_n \rangle_{L^2(T^n)}$, for all $n \geq 0$.

Exercise 2.19. Show that if $H = \mathbf{R}^n$, then the spaces $\mathbf{D}^{k,p}$ can be identified as weighted Sobolev spaces of functions on $\mathbf{R}^n$ such that that together with their k first partial derivatives have moments of order p with respect to the standard normal law.

Exercise 2.20. Prove Proposition 2.7. *Hint*: Approximate the components of the random vector F by smooth and cylindrical random variables and the function φ by $\varphi * \psi_n$, where ψ_n is an approximation of the identity.

Exercise 2.21. Prove Proposition 2.8.

Exercise 2.22. Let $F \in \mathbf{D}^{k,2}$ be given by the expansion $F = E(F) + \sum_{n=1}^{\infty} I_n(f_n)$. Show that for all $k \geq 1$,

$$D_{t_1,\ldots,t_k}^k F = \sum_{n=k}^{\infty} n(n-1)\cdots(n-k+1)I_{n-k}(f_n(\cdot, t_1, \ldots, t_k)), \qquad (41)$$

and

$$E\left(\|D^k F\|_{L^2(T^k)}^2\right) = \sum_{n=k}^{\infty} \frac{(n!)^2}{(n-k)!}\|f_n\|_{L^2(T^n)}^2. \qquad (42)$$

Exercise 2.23 (Stroock's formula, see Stroock [18]). Suppose that $F = E(F) + \sum_{n=1}^{\infty} I_n(f_n)$ is a random variable belonging to the space $\mathbf{D}^{\infty,2} = \cap_k \mathbf{D}^{k,2}$. Show that $f_n = E(D^n F)/n!$ for every $n \geq 1$.

Exercise 2.24. Let $F = \exp(W(h) - \frac{1}{2}\int_T h_s^2 \mu(ds))$, $h \in L^2(T)$. Show that the iterated derivatives of F are given by

$$D_{t_1,\ldots,t_n}^n F = F h(t_1) \cdots h(t_n). \qquad (43)$$

Using Exercise 2.23, show that the kernels of the expansion of F into the Wiener chaos are $f_n(t_1, \ldots, t_n) = \frac{1}{n!}h(t_1)\cdots h(t_n)$.

Exercise 2.25. Let $F \in \mathbf{D}^{1,2}$ be a random variable such that $E(|F|^{-2}) < \infty$. Then $P\{F > 0\}$ is zero or one.

Exercise 2.26. Suppose that $H = L^2(T)$. Let δ^k be the adjoint of the operator D^k. That is, a multiparameter process $u \in L^2(\Omega \times T^k)$ belongs to the domain of δ^k if and only if there exists a random variable $\delta^k(u)$ such that

$$E\left(F\delta^k(u)\right) = E\left(\langle u, D^k F \rangle_{L^2(T^k)}\right) \qquad (44)$$

for all $F \in \mathbf{D}^{k,2}$. Show that a process $u \in L^2(\Omega \times T^k)$ with an expansion

$$u_t = \mathrm{E}(u_t) + \sum_{n=1}^{\infty} I_n(f_n(\cdot,t)), \quad t \in T^k, \tag{45}$$

belongs to the domain of δ^k if and only if the series

$$\delta^k(u) = \int_T \mathrm{E}(u_t)\, dW_t + \sum_{n=1}^{\infty} I_{n+k}(f_n) \tag{46}$$

converges in $L^2(\Omega)$. For more details, see Nualart and Zakai [14].

Exercise 2.27. Let $\{W_t, t \in [0,1]\}$ be a one-dimensional Brownian motion. Using Exercise 2.23 find the Wiener chaos expansion of the random variables

$$F_1 = \int_0^1 \left(t^3 W_t^3 + 2t W_t^2\right) dW_t, \quad F_2 = \int_0^1 t e^{W_t}\, dW_t. \tag{47}$$

Answer: The kernels of the Wiener chaos expansion of F_1 are

$$f_1(t_1) = 2t_1$$
$$f_2(t_1,t_2) = \frac{3}{2}\max(t_1,t_2)^3$$
$$f_3(t_1,t_2,t_3) = \max(t_1,t_2,t_3)^3 + \frac{2}{3}\max(t_1,t_2,t_3)$$
$$f_4(t_1,t_2,t_3,t_4) = \frac{1}{4}\max(t_1,t_2,t_3,t_4)^3,$$

and those of F_2 are

$$f_n(t_1,\ldots,t_n) = \frac{1}{n!}\max(t_1,\ldots,t_n)e^{\frac{1}{2}\max(t_1,\ldots,t_n)}.$$

3 Application of Malliavin Calculus to Regularity of Probability Laws

The integration-by-parts formula leads to the following explicit expression for the density of a one-dimensional random variable.

Proposition 3.1. *Let F be a random variable in the space $\mathbf{D}^{1,2}$. Suppose that $DF/\|DF\|_H^2$ belongs to the domain of the operator δ in $L^2(\Omega)$. Then the law of F has a continuous and bounded density given by*

$$p(x) = \mathrm{E}\left[\mathbf{1}_{\{F>x\}}\delta\left(\frac{DF}{\|DF\|_H^2}\right)\right]. \tag{48}$$

Proof. Let ψ be a nonnegative smooth function with compact support, and set $\varphi(y) = \int_{-\infty}^{y} \psi(z)dz$. We know that $\varphi(F)$ belongs to $\mathbf{D}^{1,2}$, and making the scalar product of its derivative with DF yields

$$\langle D(\varphi(F)), DF \rangle_H = \psi(F)\|DF\|_H^2. \tag{49}$$

Using the duality formula (27) we obtain

$$
\begin{aligned}
\mathrm{E}[\psi(F)] &= \mathrm{E}\left[\left\langle D(\varphi(F)), \frac{DF}{\|DF\|_H^2} \right\rangle_H \right] \\
&= \mathrm{E}\left[\varphi(F)\delta\left(\frac{DF}{\|DF\|_H^2} \right) \right].
\end{aligned}
\tag{50}
$$

By an approximation argument, Equation (50) holds for $\psi(y) = \mathbf{1}_{[a,b]}(y)$, where $a < b$. As a consequence, we apply Fubini's theorem to get

$$
\begin{aligned}
\mathrm{P}(a \le F \le b) &= \mathrm{E}\left[\left(\int_{-\infty}^{F} \psi(x)dx \right) \delta\left(\frac{DF}{\|DF\|_H^2} \right) \right] \\
&= \int_a^b \mathrm{E}\left[\mathbf{1}_{\{F>x\}}\delta\left(\frac{DF}{\|DF\|_H^2} \right) \right] dx,
\end{aligned}
\tag{51}
$$

which implies the desired result. $\qquad\square$

Notice that Equation (48) still holds under the hypotheses $F \in \mathbf{D}^{1,p}$ and $DF/\|DF\|_H^2 \in \mathbf{D}^{1,p'}(H)$ for some $p, p' > 1$.

From expression (48) we can deduce estimates for the density. Fix p and q such that $p^{-1} + q^{-1} = 1$. By Hölder's inequality we obtain

$$p(x) \le (\mathrm{P}(F > x))^{1/q} \left\| \delta\left(\frac{DF}{\|DF\|_H^2} \right) \right\|_p. \tag{52}$$

In the same way, taking into account the relation $\mathrm{E}[\delta(DF/\|DF\|_H^2)] = 0$ we can deduce the inequality

$$p(x) \le (\mathrm{P}(F < x))^{1/q} \left\| \delta\left(\frac{DF}{\|DF\|_H^2} \right) \right\|_p. \tag{53}$$

As a consequence, we obtain

$$p(x) \le (\mathrm{P}(|F| > |x|))^{1/q} \left\| \delta\left(\frac{DF}{\|DF\|_H^2} \right) \right\|_p, \tag{54}$$

for all $x \in \mathbf{R}$. Now using the $L^p(\Omega)$ estimate of the operator δ established in Proposition 2.17 we obtain

$$
\begin{aligned}
&\left\| \delta\left(\frac{DF}{\|DF\|_H^2} \right) \right\|_p \\
&\le c_p \left(\left\| \mathrm{E}\left(\frac{DF}{\|DF\|_H^2} \right) \right\|_H + \left\| D\left(\frac{DF}{\|DF\|_H^2} \right) \right\|_{L^p(\Omega;H\otimes H)} \right).
\end{aligned}
\tag{55}
$$

We have

$$D\left(\frac{DF}{\|DF\|_H^2}\right) = \frac{D^2F}{\|DF\|_H^2} - 2\frac{\langle D^2F, DF \otimes DF\rangle_{H\otimes H}}{\|DF\|_H^4}, \tag{56}$$

and, hence,

$$\left\|D\left(\frac{DF}{\|DF\|_H^2}\right)\right\|_{H\otimes H} \le \frac{3\,\|D^2F\|_{H\otimes H}}{\|DF\|_H^2}. \tag{57}$$

Finally, from the inequalities (54), (55) and (57) we deduce the following estimate.

Proposition 3.2. *Let q, α, β be three positive real numbers such that $q^{-1} + \alpha^{-1} + \beta^{-1} = 1$. Let F be a random variable in the space $\mathbf{D}^{2,\alpha}$, such that $\mathrm{E}(\|DF\|_H^{-2\beta}) < \infty$. Then the density $p(x)$ of F can be estimated as follows*

$$p(x) \le c_{q,\alpha,\beta}\,(\mathrm{P}(|F| > |x|))^{1/q}$$

$$\times \left(\mathrm{E}(\|DF\|_H^{-1}) + \|D^2F\|_{L^\alpha(\Omega;H\otimes H)}\left\|\|DF\|_H^{-2}\right\|_\beta\right). \tag{58}$$

Suppose now that $F = (F^1, \ldots, F^m)$ is a random vector whose components belong to the space $\mathbf{D}^{1,1}$. We associate to F the following random symmetric nonnegative definite matrix:

$$\gamma_F = (\langle DF^i, DF^j\rangle_H)_{1\le i,j\le m}. \tag{59}$$

This matrix will be called the Malliavin matrix of the random vector F. The basic condition for the absolute continuity of the law of F will be that the matrix γ_F is invertible a.s. In this sense we have the following criterion which was proved by Bouleau and Hirsch (see [2]) using techniques of geometric measure theory.

Theorem 3.3. *Let $F = (F^1, \ldots, F^m)$ be a random vector verifying the following conditions:*

(i) $F^i \in \mathbf{D}^{1,2}$ for all $i = 1, \ldots, m$.
(ii) The matrix γ_F safistifes $\det \gamma_F > 0$ almost surely.

Then the law of F is absolutely continuous with respect to the Lebesgue measure on $\mathbf{R}^m$.

Condition (i) in Theorem 3.3 implies that the measure $(\det(\gamma_F)\cdot\mathrm{P})\circ F^{-1}$ is absolutely continuous with respect to the Lebesgue measure on $\mathbf{R}^m$. In other words, the random vector F has an absolutely continuous law conditioned by the set $\{\det(\gamma_F) > 0\}$; that is,

$$\mathrm{P}\{F \in B\,, \det(\gamma_F) > 0\} = 0 \tag{60}$$

for any Borel subset B of $\mathbf{R}^m$ of zero Lebesgue measure.

The regularity of the density requires under stronger conditions, and for this we introduce the following definition.

Definition 3.4. *We say that a random vector* $F = (F^1, \ldots, F^m)$ *is nonde-generate if it satisfies the following conditions:*

(i) $F^i \in \mathbf{D}^\infty$ *for all* $i = 1, \ldots, m$.
(ii) The matrix γ_F *satisfies* $\mathrm{E}[(\det \gamma_F)^{-p}] < \infty$ *for all* $p \geq 2$.

Using the techniques of Malliavin calculus we will establish the following general criterion for the smoothness of densities.

Theorem 3.5. *Let* $F = (F^1, \ldots, F^m)$ *be a nondegenerate random vector in the sense of Definition 3.4. Then the law of* F *possesses an infinitely differentiable density.*

In order to prove Theorem 3.5 we need some technical results. Set $\partial_i = \partial/\partial x_i$, and for any multiindex $\alpha \in \{1, \ldots, m\}^k$, $k \geq 1$, we denote by ∂_α the partial derivative $\partial^k/(\partial x_{\alpha_1} \cdots \partial x_{\alpha_k})$.

Let γ be an $m \times m$ random matrix that is invertible a.s., and whose components belong to $\mathbf{D}^{1,p}$ for all $p \geq 2$. Denote by $A(\gamma)$ the adjoint matrix of γ. Applying the derivative operator to the expression $A(\gamma)\gamma = \det \gamma I$, where I is the identity matrix, we obtain

$$D(\det \gamma)I = D(A(\gamma))\gamma + A(\gamma)D\gamma, \tag{61}$$

which implies

$$D(A(\gamma)) = D(\det \gamma)\gamma^{-1} - A(\gamma)D\gamma\gamma^{-1}. \tag{62}$$

Lemma 3.6. *Suppose that* γ *is an* $m \times m$ *random matrix that is invertible a.s. and such that* $|\det \gamma|^{-1} \in L^p(\Omega)$ *for all* $p \geq 1$. *Suppose that the entries* γ^{ij} *of* γ *are in* $\mathbf{D}^\infty$. *Then* $(\gamma^{-1})^{ij}$ *belongs to* $\mathbf{D}^\infty$ *for all* i, j, *and*

$$D \left(\gamma^{-1}\right)^{ij} = -\sum_{k,l=1}^{m} \left(\gamma^{-1}\right)^{ik} \left(\gamma^{-1}\right)^{lj} D\gamma^{kl}. \tag{63}$$

Proof. First notice that the event $\{\det \gamma > 0\}$ has probability zero or one (see Exercise 2.25). We will assume that $\det \gamma > 0$ a.s. For any $\epsilon > 0$ define

$$\gamma_\epsilon^{-1} = (\det \gamma + \epsilon)^{-1} A(\gamma). \tag{64}$$

Note that $(\det \gamma + \epsilon)^{-1}$ belongs to $\mathbf{D}^\infty$ because it can be expressed as the composition of $\det \gamma$ with a function in $C_p^\infty(\mathbf{R})$. Therefore, the entries of γ_ϵ^{-1} belong to $\mathbf{D}^\infty$. Furthermore, for any i, j, $(\gamma_\epsilon^{-1})^{ij}$ converges in $L^p(\Omega)$ to $(\gamma^{-1})^{ij}$ as ϵ tends to zero. Then, in order to check that the entries of γ^{-1} belong to $\mathbf{D}^\infty$, it suffices to show (taking into account Lemma 2.10) that the iterated derivatives of $(\gamma_\epsilon^{-1})^{ij}$ are bounded in $L^p(\Omega)$, uniformly with respect to ϵ, for any $p \geq 1$. This boundedness in $L^p(\Omega)$ holds from Leibnitz rule for the operator D^k, because $(\det \gamma)\gamma^{-1}$ belongs to $\mathbf{D}^\infty$, and the fact that $(\det \gamma + \epsilon)^{-1}$ has bounded $\| \cdot \|_{k,p}$ norms for all k, p, due to our hypotheses.

Finally, from (64) we deduce (63) by first applying the derivative operator D using (62), and then letting ϵ tend to zero. □

Proposition 3.7. *Let $F = (F^1, \ldots, F^m)$ be a nondegenerate random vector. Let $G \in \mathbf{D}^\infty$ and let φ be a function in the space $C_p^\infty(\mathbf{R}^m)$. Then for any multiindex $\alpha \in \{1, \ldots, m\}^k$, $k \geq 1$, there exists an element $H_\alpha(F, G) \in \mathbf{D}^\infty$ such that*

$$\mathrm{E}\left[\partial_\alpha \varphi(F) G\right] = \mathrm{E}\left[\varphi(F) H_\alpha(F, G)\right], \tag{65}$$

where the elements $H_\alpha(F, G)$ are recursively given by

$$H_{(i)}(F, G) = \sum_{j=1}^m \delta\left(G\left(\gamma_F^{-1}\right)^{ij} DF^j\right), \tag{66}$$

$$H_\alpha(F, G) = H_{\alpha_k}(F, H_{(\alpha_1, \ldots, \alpha_{k-1})}(F, G)). \tag{67}$$

Proof. By the chain rule (Proposition 2.7) we have

$$
\begin{aligned}
\langle D(\varphi(F)), DF^j \rangle_H &= \sum_{i=1}^m \partial_i \varphi(F) \langle DF^i, DF^j \rangle_H \\
&= \sum_{i=1}^m \partial_i \varphi(F) \gamma_F^{ij},
\end{aligned}
\tag{68}
$$

and, consequently,

$$\partial_i \varphi(F) = \sum_{j=1}^m \langle D(\varphi(F)), DF^j \rangle_H (\gamma_F^{-1})^{ji}. \tag{69}$$

Taking expectations and using the duality relationship (27) between the derivative and the divergence operators we get

$$\mathrm{E}\left[\partial_i \varphi(F) G\right] = \mathrm{E}\left[\varphi(F) H_{(i)}(F, G)\right], \tag{70}$$

where $H_{(i)}$ equals to the right-hand side of Equation (66). Notice that the continuity of the operator δ (Theorem 2.16), and Lemma 3.6 imply that $H_{(i)}$ belongs to $\mathbf{D}^\infty$. Equation (67) follows by recurrence. □

As a consequence, there exist constants $\beta, \gamma > 1$ and integers n, m such that

$$\|H_\alpha(F, G)\|_p \leq c_{p,q} \left\|\det \gamma_F^{-1}\right\|_\beta^m \|DF\|_{k,\gamma}^n \|G\|_{k,q}. \tag{71}$$

Proof (Theorem 3.5). Equality (65) applied to the multiindex $\alpha = (1, 2, \ldots, m)$ yields

$$\mathrm{E}\left[G \partial_\alpha \varphi(F)\right] = \mathrm{E}[\varphi(F) H_\alpha(F, G)]. \tag{72}$$

Notice that

$$\varphi(F) = \int_{-\infty}^{F^1} \cdots \int_{-\infty}^{F^m} \partial_\alpha \varphi(x)\, dx. \tag{73}$$

Hence, by Fubini's theorem we can write

$$\mathrm{E}\left[G\partial_\alpha\varphi(F)\right] = \int_{\mathbf{R}^m} \partial_\alpha\varphi(x)\mathrm{E}\left[\mathbf{1}_{\{F>x\}}H_\alpha(F,G)\right]\, dx. \tag{74}$$

We can take as $\partial_\alpha\varphi$ any function in $C_0^\infty(\mathbf{R}^m)$. Then Equation (74) implies that the random vector F has a density given by

$$p(x) = \mathrm{E}\left[\mathbf{1}_{\{F>x\}}H_\alpha(F,1)\right]. \tag{75}$$

Moreover, for any multiindex β we have

$$\mathrm{E}\left[\partial_\beta\partial_\alpha\varphi(F)\right] = \mathrm{E}[\varphi(F)H_\beta(F,H_\alpha(F,1)))]$$
$$= \int_{\mathbf{R}^m} \partial_\alpha\varphi(x)\mathrm{E}\left[\mathbf{1}_{\{F>x\}}H_\beta(H_\alpha)\right]\, dx. \tag{76}$$

Hence, for any $\xi \in C_0^\infty(\mathbf{R}^m)$

$$\int_{\mathbf{R}^m} \partial_\beta\xi(x)p(x)dx = \int_{\mathbf{R}^m} \xi(x)\mathrm{E}\left[\mathbf{1}_{\{F>x\}}H_\beta(F,H_\alpha(F,1))\right]\, dx. \tag{77}$$

Therefore $p(x)$ is infinitely differentiable, and for any multiindex β we have

$$\partial_\beta p(x) = (-1)^{|\beta|}\mathrm{E}\left[\mathbf{1}_{\{F>x\}}H_\beta(F,(H_\alpha(F,1)))\right]. \tag{78}$$

This completes the proof.   □

3.1 Properties of the Support of the Law

The Malliavin calculus is also useful to derive properties of the topological support of the law of a random variable. We will illustrate this application in this section. Given a random vector $F : \Omega \to \mathbf{R}^m$, the topological support of the law of F is defined as the set of points $x \in \mathbf{R}^m$ such that $\mathrm{P}(|x-F| < \varepsilon) > 0$ for all $\varepsilon > 0$. The support is a closed set because clearly its complement is open. On the other hand, the support of the law of F coincides with the intersection of all closed sets C in $\mathbf{R}^m$ such that $P(F \in C) = 1$.

   The following result asserts the connectivity property of the support of a smooth random vector.

Proposition 3.8. *Let $F = (F^1, \ldots, F^m)$ be a random vector whose components belong to $\mathbf{D}^{1,2}$. Then, the topological support of the law of F is a closed connected subset of $\mathbf{R}^m$.*

Proof. If the support of F is not connected, it can be decomposed as the union of two nonempty disjoint closed sets A and B. For each integer $M \geq 2$ let $\psi_M : \mathbf{R}^m \to \mathbf{R}$ be an infinitely differentiable function such that $0 \leq \psi_M \leq 1$, $\psi_M(x) = 0$ if $|x| \geq M$, $\psi_M(x) = 1$ if $|x| \leq M - 1$, and $\sup_{x,M} |\nabla \psi_M(x)| < \infty$. Set $A_M = A \cap \{|x| \leq M\}$ and $B_M = B \cap \{|x| \leq M\}$. For M large enough we have $A_M \neq \emptyset$ and $B_M \neq \emptyset$, and there exists an infinitely differentiable function f_M such that $0 \leq f_M \leq 1$, $f_M = 1$ in a neighborhood of A_M, and $f_M = 0$ in a neighborhood of B_M. The sequence $(f_M \psi_M)(F)$ converges a.s. and in $L^2(\Omega)$ to $\mathbf{1}_{\{F \in A\}}$ as M tends to infinity. On the other hand, we have

$$
D\left[(f_M \psi_M)(F)\right] = \sum_{i=1}^{m} \left[(\psi_M \partial_i f_M)(F) DF^i + (f_M \partial_i \psi_M)(F) DF^i\right]
$$
$$
= \sum_{i=1}^{m} (f_M \partial_i \psi_M)(F) DF^i.
$$

Hence,

$$
\sup_M \left\| D\left[(f_M \psi_M)(F)\right] \right\|_H \leq \sum_{i=1}^{m} \sup_M \left\| \partial_i \psi_M \right\|_\infty \left\| DF^i \right\|_H \in L^2(\Omega). \tag{79}
$$

By Lemma 2.10 we get that $\mathbf{1}_{\{F \in A\}}$ belongs to $\mathbf{D}^{1,2}$, and by Proposition 2.12 this is contradictory because $0 < P(F \in A) < 1$. □

As a consequence, the support of the law of a random variable $F \in \mathbf{D}^{1,2}$ is a closed interval. On the other hand, a random variable F which takes only finitely many different values has a disconnected support, and, therefore, it cannot belong to $\mathbf{D}^{1,2}$.

The next result provides sufficient conditions for the density of F to be nonzero in the interior of the support.

Proposition 3.9. *Let $F \in \mathbf{D}^{1,p}$, $p > 2$, and suppose that F possesses a density $p(x)$ which is locally Lipschitz in the interior of the support of the law of F. Let a be a point in the interior of the support of the law of F. Then $p(a) > 0$.*

Proof. Suppose $p(a) = 0$. Set $r = 2p/(p+2) > 1$. From Proposition 2.12 we know that $1_{\{F > a\}} \notin \mathbf{D}^{1,r}$ because $0 < P(F > a) < 1$. Fix $\epsilon > 0$ and set

$$
\varphi_\epsilon(x) = \int_{-\infty}^{x} \frac{1}{2\epsilon} \mathbf{1}_{[a-\epsilon, a+\epsilon]}(y) \, dy. \tag{80}
$$

Then $\varphi_\epsilon(F)$ converges to $\mathbf{1}_{\{F > a\}}$ in $L^r(\Omega)$ as $\epsilon \downarrow 0$. Moreover, $\varphi_\epsilon(F) \in \mathbf{D}^{1,r}$ and

$$
D(\varphi_\epsilon(F)) = \frac{1}{2\epsilon} \mathbf{1}_{[a-\epsilon, a+\epsilon]}(F) DF. \tag{81}
$$

We have

$$\mathrm{E}\left(\|D(\varphi_\epsilon(F))\|_H^r\right)$$

$$\leq (\mathrm{E}(\|DF\|_H^p)^{2/(p+2)} \left(\frac{1}{(2\epsilon)^2} \int_{a-\epsilon}^{a+\epsilon} p(x)dx\right)^{p/(p+2)}. \qquad (82)$$

The local Lipschitz property of p implies that $p(x) \leq K|x-a|$, and we obtain

$$\mathrm{E}\left(\|D(\varphi_\epsilon(F))\|_H^r\right) \leq (\mathrm{E}(\|DF\|_H^p)^{2/(p+2)} 2^{-r} K^{p/(p+2)}. \qquad (83)$$

By Lemma 2.10 this implies $1_{\{F>a\}} \in \mathbf{D}^{1,r}$, resulting in a contradiction. $\square$

The following example shows that, unlike the one-dimensional case, in dimension $m > 1$ the density of a nondegenerate random vector may vanish in the interior of the support.

Example 3.10. Let h_1 and h_2 be two orthonormal elements of H. Define $X = (X_1, X_2)$, $X_1 = \arctan W(h_1)$, and $X_2 = \arctan W(h_2)$. Then, $X_i \in \mathbf{D}^\infty$ and

$$DX_i = (1 + W(h_i)^2)^{-1} h_i, \qquad (84)$$

for $i = 1, 2$, and

$$\det \gamma_X = \left[(1 + W(h_1)^2)(1 + W(h_2)^2)\right]^{-2}. \qquad (85)$$

The support of the law of the random vector X is the rectangle $\left[-\frac{\pi}{2}, \frac{\pi}{2}\right]^2$, and the density of X is strictly positive in the interior of the support. Now consider the vector $Y = (Y_1, Y_2)$ given by

$$Y_1 = \left(X_1 + \frac{3\pi}{2}\right)\cos(2X_2 + \pi),$$

$$Y_2 = \left(X_1 + \frac{3\pi}{2}\right)\sin(2X_2 + \pi).$$

We have that $Y_i \in \mathbf{D}^\infty$ for $i = 1, 2$, and

$$\det \gamma_Y = 4\left(X_1 + \frac{3\pi}{2}\right)^2 \left[(1 + W(h_1)^2)(1 + W(h_2)^2)\right]^{-2}. \qquad (86)$$

This implies that Y is a nondegenerate random vector. Its support is the set $\{(x, y) : \pi^2 \leq x^2 + y^2 \leq 4\pi^2\}$, and the density of Y vanishes on the points (x, y) in the support such that $-2\pi < y < -\pi$ and $y = 0$.

Exercise 3.11. Prove Proposition 3.2.

Exercise 3.12. Show that if $F \in \mathbf{D}^{2,4}$ satisfies $\mathrm{E}(\|DF\|^{-8}) < \infty$, then $DF/\|DF\|^2 \in \mathrm{Dom}\,\delta$ and

$$\delta\left(\frac{DF}{\|DF\|_H^2}\right) = \frac{\delta(DF)}{\|DF\|_H^2} + 2\frac{\langle DF \otimes DF, D^2F\rangle_{H \otimes H}}{\|DF\|_H^4}. \tag{87}$$

Exercise 3.13. Let F be a random variable in $\mathbf{D}^{1,2}$ such that $G_k DF/\|DF\|_H^2$ belongs to $\mathrm{Dom}\,\delta$ for any $k = 0, \dots, n$, where $G_0 = 1$ and

$$G_k = \delta\left(G_{k-1}\frac{DF}{\|DF\|_H^2}\right) \tag{88}$$

if $1 \le k \le n+1$. Show that F has a density of class C^n and

$$f^{(k)}(x) = (-1)^k \mathrm{E}\left[\mathbf{1}_{\{F>x\}} G_{k+1}\right], \tag{89}$$

$0 \le k \le n$.

Exercise 3.14. Set $M_t = \int_0^t u(s)\,dW_s$, where $W = \{W(t), t \in [0,T]\}$ is a Brownian motion and $u = \{u(t), t \in [0,T]\}$ is an adapted process such that $|u(t)| \ge \rho > 0$ for some constant ρ, $E\left(\int_0^T u(t)^2\,dt\right) < \infty$, $u(t) \in \mathbf{D}^{2,2}$ for each $t \in [0,T]$, and

$$\lambda := \sup_{s,t\in[0,T]} \mathrm{E}(|D_s u_t|^p) + \sup_{r,s\in[0,T]} \mathrm{E}\left[\left(\int_0^T |D_{r,s}^2 u_t|^p\,dt\right)^{p/2}\right] < \infty, \tag{90}$$

for some $p > 3$. Show that the density of M_t, denoted by $p_t(x)$ satisfies

$$p_t(x) \le \frac{c}{\sqrt{t}} \mathrm{P}(|M_t| > |x|)^{\frac{1}{q}}, \tag{91}$$

for all $t > 0$, where $q > p/(p-3)$ and the constant c depends on λ, ρ and p.

Exercise 3.15. Let $F \in \mathbf{D}^{3,\alpha}$, $\alpha > 4$, be a random variable such that $\mathrm{E}(\|DF\|_H^{-p}) < \infty$ for all $p \ge 2$. Show that the density $p(x)$ of F is continuously differentiable, and compute $p'(x)$.

Exercise 3.16. Show that the random variable $F = \int_0^1 t^2 \arctan(W_t)\,dt$, where W is a Brownian motion, has a C^∞ density.

Exercise 3.17. Let $W = \{W(s,t), (s,t) \in [0,\infty)^2\}$ be a two-parameter Wiener process. That is, W is a zero mean Gaussian field with covariance

$$\mathrm{E}\left(W(s,t)W(s',t')\right) = (s \wedge s')(t \wedge t'). \tag{92}$$

Show that $F = \sup_{(s,t)\in[0,1]^2} W(s,t)$ has an absolutely continuous distribution. Show also that the density of F is strictly positive in $(0, +\infty)$.

4 Stochastic Heat Equation

Suppose that $W = \{W(A), A \in \mathcal{B}(\mathbf{R}^2), |A| < \infty\}$ is a Gaussian family of random variables with zero mean and covariance

$$E(W(A)W(B)) = |A \cap B|. \tag{93}$$

That is, W is a Gaussian white noise on the plane. Then, if we set $W(t, x) = W([0, t] \times [0, x])$, for $t, x \geq 0$, $W = \{W(t, x), (t, x) \in [0, \infty)^2\}$ is a two-parameter Wiener process (see Exercise 3.17).

For each $t \geq 0$ we will denote by $\mathcal{F}_t$ the σ-field generated by the random variables $\{W(s, x), s \in [0, t], x \geq 0\}$ and the P-null sets. We say that a random field $\{u(t, x), t \geq 0, x \geq 0)\}$ is adapted if for all (t, x) the random variable $u(t, x)$ is $\mathcal{F}_t$-measurable.

Consider the following stochastic partial differential equation on $[0, \infty) \times [0, 1]$:

$$\frac{\partial u}{\partial t} = \frac{\partial^2 u}{\partial x^2} + b(u(t, x)) + \sigma(u(t, x))\frac{\partial^2 W}{\partial t \partial x} \tag{94}$$

with initial condition $u(0, x) = u_0(x)$, and Dirichlet boundary conditions $u(t, 0) = u(t, 1) = 0$. We will assume that $u_0 \in C([0, 1])$ satisfies $u_0(0) = u_0(1) = 0$.

Equation (94) is formal because the derivative $\partial^2 W/(\partial t \partial x)$ does not exist, and we will replace it by the following integral equation:

$$u(t, x) = \int_0^1 G_t(x, y)u_0(y)\, dy + \int_0^t \int_0^1 G_{t-s}(x, y)b(u(s, y))\, dy\, ds$$
$$+ \int_0^t \int_0^1 G_{t-s}(x, y)\sigma(u(s, y))\, W(dy, ds), \tag{95}$$

where $G_t(x, y)$ is the fundamental solution of the heat equation on $[0, 1]$ with Dirichlet boundary conditions:

$$\frac{\partial G}{\partial t} = \frac{\partial^2 G}{\partial y^2}, \quad G_0(x, y) = \delta_x(y). \tag{96}$$

The kernel $G_t(x, y)$ has the following explicit formula:

$$G_t(x, y) = \frac{1}{\sqrt{4\pi t}} \sum_{n=-\infty}^{\infty} \left\{ \exp\left(-\frac{(y - x - 2n)^2}{4t} \right) \right.$$
$$\left. - \exp\left(-\frac{(y + x - 2n)^2}{4t} \right) \right\}. \tag{97}$$

On the other hand, $G_t(x, y)$ coincides with the probability density at point y of a Brownian motion with variance $2t$ starting at x and killed if it leaves the interval $[0, 1]$:

$$G_t(x\,,y) = \frac{d}{dy}\mathrm{E}^x\{B_t \in dy, B_s \in (0\,,1) \quad \forall s \in [0\,,t]\}. \tag{98}$$

This implies that

$$G_t(x\,,y) \le \frac{1}{\sqrt{4\pi t}}\exp\left(-\frac{|x-y|^2}{4t}\right). \tag{99}$$

Therefore, for any $\beta > 0$ we have

$$\int_0^1 G_t(x\,,y)^\beta\,dy \le (4\pi t)^{-\beta/2}\int_{\mathbf{R}} e^{-\beta|x|^2/(4t)}\,dx = C_\beta t^{(1-\beta)/2}. \tag{100}$$

The solution in the the particular case $u_0 = 0$, $b = 0$, $\sigma = 1$ is

$$u(t\,,x) = \int_0^t \int_0^1 G_{t-s}(x\,,y)\,W(ds\,,dy). \tag{101}$$

This stochastic integral exists and it is a Gaussian centered random variable with variance

$$\int_0^t \int_0^1 G_{t-s}(x\,,y)^2\,dy\,ds = \int_0^t G_{2s}(x\,,x)\,ds < \infty, \tag{102}$$

because (99) implies that $G_{2s}(x\,,x) \le Cs^{-1/2}$. Notice that in dimension $d \ge 2$, $G_{2s}(x\,,x) \sim Cs^{-1}$ and the variance is infinite. For this reason, the study of space-time white noise driven parabolic equations is restricted to the one-dimensional case.

The following well-known result asserts that Equation (95) has a unique solution if the coefficients are Lipschitz continuous (see Walsh [19, Theorem 3.2], and Theorem 6.4 in the first chapter).

Theorem 4.1. *Suppose that the coefficients b and σ are globally Lipschitz functions. Let $u_0 \in C([0\,,1])$ be such that $u_0(0) = u_0(1) = 0$. Then there is a unique adapted process $u = \{u(t\,,x), t \ge 0, x \in [0\,,1]\}$ such that for all $T > 0$*

$$\mathrm{E}\left(\int_0^T \int_0^1 u(t\,,x)^2\,dx\,dt\right) < \infty, \tag{103}$$

and satisfies (95). Moreover, the solution u satisfies

$$\sup_{(t,x)\in[0,T]\times[0,1]} \mathrm{E}\left(|u(t\,,x)|^p\right) < \infty. \tag{104}$$

Furthermore, one can obtain the following moment estimates for the increments of the solution (see Walsh [19, Corollary 3.4], Nualart [12, Proposition 2.4.3], and Theorem 6.7 in the first chapter):

$$\mathrm{E}\left(|u(t\,,x) - u(s\,,y)|^p\right) \le C_{T,p}\left(|t-s|^{(p-6)/4} + |x-y|^{(p-6)/2}\right), \tag{105}$$

for all $s,t \in [0\,,T]$, $x,y \in [0\,,1]$, $p \ge 2$, assuming the that initial condition is Hölder continuous of order $\frac{1}{2}$. As a consequence, for any $\epsilon > 0$ the trajectories of the process $u(t\,,x)$ are Hölder continuous of order $\frac{1}{4} - \epsilon$ in the variable t and Hölder continuous of order $\frac{1}{2} - \epsilon$ in the variable x.

4.1 Regularity of the Probability Law of the Solution

The aim of this section is to show that under a suitable nondegeneracy condition on the coefficient σ, the solution $u(t,x)$ of Equation (95) has a regular density. First we will discuss the differentiability in the sense of Malliavin calculus of the random variable $u(t,x)$, and we will show that it belongs to the space $\mathbf{D}^{1,p}$ for all p. Here the underlying Hilbert space is $H = L^2([0,\infty) \times [0,1])$, and the derivative of a random variable $F \in \mathbf{D}^{1,p}$ is a two-parameter stochastic process $\{D_{s,y}F, (s,y) \in [0,\infty) \times [0,1]\}$.

Remark 4.2. Consider an adapted stochastic process $v = \{v(t,x), (t,x) \in [0,\infty) \times [0,1]\}$ satisfying $E(\int_0^\infty \int_0^1 v(t,x)^2 \, dx \, dt) < \infty$. Then, v belongs to the domain of the divergence operator and $\delta(v)$ equals to the Itô stochastic integral $\int_0^\infty \int_0^1 v(t,x) \, W(dt,dx)$.

Proposition 4.3. *Let b and σ be C^1 functions with bounded derivatives. Then, for all $p \geq 2$, $u(t,x) \in \mathbf{D}^{1,p}$, and the derivative $D_{s,y}u(t,x)$ satisfies*

$$D_{s,y}u(t,x) = G_{t-s}(x,y)\sigma(u(s,y))$$
$$+ \int_s^t \int_0^1 G_{t-r}(x,z)b'(u(r,z))D_{s,y}u(r,z) \, dz \, dr \quad (106)$$
$$+ \int_s^t \int_0^1 G_{t-r}(x,z)\sigma'(u(r,z))D_{s,y}u(r,z) \, W(dr,dz)$$

if $s < t$, and $D_{s,y}u(t,x) = 0$ if $s > t$. Moreover, for all $t \in [0,T]$, we have

$$\sup_{(r,z)\in[0,t]\times[0,1]} E\left(\|D(r,z)\|_H^p\right) < C_{T,p}t^{p/4}. \quad (107)$$

Furthermore, if the coefficients b and σ are infinitely differentiable with bounded derivatives of all orders, then $u(t,x)$ belongs to the space $\mathbf{D}^\infty$.

Notice that for each fixed (s,y), $\{D_{s,y}u(t,x), t \geq s\}$ is the solution of the stochastic heat equation

$$\frac{\partial D_{s,y}u}{\partial t} = \frac{\partial^2 D_{s,y}u}{\partial x^2} + b'(u)D_{s,y}u + \sigma'(u)D_{s,y}u\frac{\partial^2 W}{\partial t \partial x} \quad (108)$$

on $[s,\infty) \times [0,1]$, with Dirichlet boundary conditions and initial condition $\sigma(u(s,y))\delta_0(x-y)$.

On the other hand, if the coefficients are Lipschitz continuous it is also true that $u(t,x)$ belongs to the space $\mathbf{D}^{1,p}$ for all $p \geq 2$, and Equation (106) holds but replacing $b'(u(t,x))$ and $\sigma'(u(t,x))$ by some bounded and adapted processes.

Proof. Consider the Picard iteration scheme defined by

$$u_0(t\,,x) = \int_0^1 G_t(x\,,y)u_0(y)\,dy \qquad (109)$$

and

$$u_{n+1}(t\,,x) = u_0(t\,,x) + \int_0^t \int_0^1 G_{t-r}(x\,,z)b(u_n(r\,,z))\,dz\,dr$$
$$+ \int_0^t \int_0^1 G_{t-r}(x\,,z)\sigma(u_n(r\,,z))\,W(dz\,,dr), \qquad (110)$$

$n \geq 0$. It holds that $u_n(t\,,x)$ converges in L^p to $u(t\,,x)$ for all $p \geq 2$, and

$$\sup_n \sup_{(t,x)\in[0,T]\times[0,1]} \mathrm{E}\left(|u(t\,,x)|^p\right) < \infty. \qquad (111)$$

Fix a time interval $[0\,,T]$. Suppose that for all $(t\,,x) \in [0\,,T] \times [0\,,1]$ and $p \geq 2$ we have $u_n(t\,,x) \in \mathbf{D}^{1,p}$ and

$$V_n(t) := \sup_{x\in[0,1]} \mathrm{E}\left[\left(\int_0^t \int_0^1 |D_{s,y}u_n(t\,,x)|^2\,dy\,ds\right)^{p/2}\right] < \infty. \qquad (112)$$

By Remark 4.2, the Itô stochastic integral

$$\int_0^t \int_0^1 G_{t-r}(x\,,z)\sigma(u_n(r\,,z))\,W(dz\,,dr) \qquad (113)$$

coincides with the divergence of the adapted stochastic process

$$\left\{G_{t-r}(x\,,z)\sigma(u_n(r\,,z))\mathbf{1}_{[0,t]}(r), (r\,,z) \in [0,\infty) \times [0\,,1]\right\}. \qquad (114)$$

As a consequence, we can compute the derivative of this stochastic integral using property (30) and we obtain for $s < t$

$$D_{s,y}\left(\int_0^t \int_0^1 G_{t-r}(x\,,z)\sigma(u_n(r\,,z))\,W(dz\,,dr)\right)$$
$$= G_{t-s}(x\,,y)\sigma(u_n(s\,,y)) \qquad (115)$$
$$+ \int_0^t \int_0^1 G_{t-r}(x\,,z)\sigma'(u_n(r\,,z))D_{s,y}u_n(r\,,z)\,W(dr\,,dz).$$

Notice that the stochastic integral in the right-hand side of Equation (115) vanishes if $r \leq s$, because $u(r\,,z)$ is $\mathscr{F}_r$-measurable.

Hence, applying the operator D to Eq. (110), and using that $u_n(t\,,x) \in \mathbf{D}^{1,p}$ and (112), we obtain that $u_{n+1}(t\,,x) \in \mathbf{D}^{1,p}$ and that

$$D_{s,y}u_{n+1}(t\,,x)$$
$$= G_{t-s}(x\,,y)\sigma(u_n(s\,,y))$$
$$+ \int_s^t \int_0^1 G_{t-r}(x\,,z)b'(u_n(r\,,z))D_{s,y}u_n(r\,,z)\,dz\,dr \qquad (116)$$
$$+ \int_s^t \int_0^1 G_{t-r}(x\,,z)\sigma'(u_n(r\,,z))D_{s,y}u_n(r\,,z)\,W(dr\,,dz).$$

By Hölder's inequality, (100) and (111) we obtain

$$
\begin{aligned}
&\mathrm{E}\left[\left(\int_0^t \int_0^1 G_{t-s}(x,y)^2 \sigma(u_n(s,y))^2\, dy\, ds\right)^{p/2}\right] \\
&\leq \left(\int_0^t \int_0^1 G_{t-s}(x,y)^2\, dy\, ds\right)^{p/2} \sup_{(t,x)\in[0,T]\times[0,1]} \mathrm{E}\left(|\sigma(u_n(t,x))|^p\right) \\
&\leq C_1 t^{p/4},
\end{aligned}
\tag{117}
$$

for some constant $C_1 > 0$. On the other hand, Burkholder's inequality for Hilbert-space-valued stochastic integrals, and Hölder inequality yield

$$
\begin{aligned}
&\mathrm{E}\left[\left(\int_0^t \int_0^1 \left(\int_s^t \int_0^1 G_{t-r}(x,z)\sigma'(u_n(r,z))D_{s,y}u_n(r,z)\,W(dr,dz)\right)^2 dy\, ds\right)^{p/2}\right] \\
&\leq c_p \mathrm{E}\left[\left(\int_0^t \int_0^1 \int_0^r \int_0^1 G_{t-r}(x,z)^2 \sigma'(u_n(r,z))^2\, |D_{s,y}u_n(r,z)|^2\, dy\, ds\, dz\, dr\right)^{p/2}\right] \\
&\leq c_p \,\|\sigma'\|_\infty^p\, \mathrm{E}\left[\left(\int_0^t \int_0^1 G_{t-r}(x,z)^2 \left(\int_0^r \int_0^1 |D_{s,y}u_n(r,z)|^2\, dy\, ds\right) dz\, dr\right)^{p/2}\right] \\
&\leq c_p \,\|\sigma'\|_\infty^p \left(\int_0^t \int_0^1 G_{t-r}(x,z)^2\, dz\, dr\right)^{(p/2)-1} \\
&\quad \times \int_0^t \int_0^1 G_{t-r}(x,z)^2 \mathrm{E}\left[\left(\int_0^r \int_0^1 |D_{s,y}u_n(r,z)|^2\, dy\, ds\right)^{p/2}\right] dr\, dz.
\end{aligned}
\tag{118}
$$

A similar estimate can be obtained for the second summand in the right-hand side of Equation (116). Then, from (117) and (118) we obtain

$$
\begin{aligned}
&\mathrm{E}\left[\left(\int_0^t \int_0^1 |D_{s,y}u_{n+1}(t,x)|^2\, dy\, ds\right)^{p/2}\right] \\
&\leq C_2 \left(t^{p/4} + t^{\frac{p}{4}-\frac{1}{2}} \int_0^t \int_0^1 G_{t-r}(x,z)^2 \mathrm{E}\left[\left(\int_0^r \int_0^1 |D_{s,y}u_n(r,z)|^2\, dy\, ds\right)^{p/2}\right] dr\, dz\right). \\
&\leq C_3 \left(t^{p/4} + t^{\frac{p}{4}-\frac{1}{2}} \int_0^t (t-r)^{-\frac{1}{2}} \sup_{z\in[0,1]} \mathrm{E}\left[\left(\int_0^r \int_0^1 |D_{s,y}u_n(r,z)|^2\, dy\, ds\right)^{p/2}\right] dr\right).
\end{aligned}
$$

Then, using (112) we get

$$
\begin{aligned}
V_{n+1}(t) &\leq C_3 \left(t^{p/4} + t^{\frac{p}{4}-\frac{1}{2}} \int_0^t V_n(\theta)(t-\theta)^{-\frac{1}{2}}\, d\theta\right) \\
&\leq C_4 \left(t^{p/4} + t^{\frac{p}{4}-\frac{1}{2}} \int_0^t \int_0^\theta V_{n-1}(u)(t-\theta)^{-\frac{1}{2}}(\theta-u)^{-\frac{1}{2}}\, du\, d\theta\right) \\
&\leq C_5 \left(t^{p/4} + t^{\frac{p}{4}-\frac{1}{2}} \int_0^t V_{n-1}(u)\, du\right) < \infty.
\end{aligned}
$$

By iteration this implies that

$$V_n(t) < Ct^{p/4}, \tag{119}$$

where the constant C does not depend on n. Taking into account that $u_n(t,x)$ converges to $u(t,x)$ in $L^p(\Omega)$ for all $p \geq 2$, we deduce that $u(t,x) \in \mathbf{D}^{1,p}$, and $Du_n(t,x)$ converges to $Du(t,x)$ in the weak topology of $L^p(\Omega; H)$ (see Lemma 2.10). Finally, applying the operator D to both members of Eq. (95), we deduce the desired result. $\qquad\square$

In order to show the existence of negative moments, we will make use of the following lemma.

Lemma 4.4. *Let F be a nonnegative random variable. Then, $\mathrm{E}(F^{-p}) < \infty$ for all $p \geq 2$ if an only if for all $q \geq 2$ there exists an $\varepsilon_0(q) > 0$ such that for $\mathrm{P}(F < \varepsilon) \leq C\varepsilon^q$, for all $\varepsilon \leq \varepsilon_0$.*

We can now state and prove the following result on the regularity of the density of the random variable $u(t,x)$.

Theorem 4.5. *Assume that the coefficients b and σ are infinitely differentiable functions with bounded derivatives. Suppose that u_0 is a Hölder continuous function of order $\frac{1}{2}$ which satisfies the Dirichlet boundary conditions. Then, if $|\sigma(r)| > c > 0$ for all $r \in \mathbf{R}$, $u(t,x)$ has a C^∞ density for all (t,x) such that $t > 0$ and $x \in (0,1)$.*

Proof. From the general criterion for smoothness of densities (see Theorem 3.5), it suffices to show that $\mathrm{E}(\gamma_{u(t,x)}^{-p}) < \infty$ for all $p \geq 2$, where $\gamma_{u(t,x)}$ denote the Malliavin matrix of the random variable $u(t,x)$. By Lemma 4.4 it is enough to show that for any $q \geq 2$ there exists an $\varepsilon_0(q) > 0$ such that for all $\varepsilon \leq \varepsilon_0$

$$\mathrm{P}\left(\int_0^t \int_0^1 (D_{s,y}u(t,x))^2 \, dy \, ds < \varepsilon \right) \leq C\varepsilon^q. \tag{120}$$

We fix $\delta > 0$ sufficiently small and write

$$\int_0^t \int_0^1 (D_{s,y}u(t,x))^2 \, dy \, ds$$
$$\geq \frac{1}{2} \int_{t-\delta}^t \int_0^1 |G_{t-s}(x,y)\sigma(u(s,y))|^2 \, dy \, ds - I_\delta, \tag{121}$$

where

$$I_\delta = \int_{t-\delta}^t \int_0^1 \left| \int_s^t \int_0^1 G_{t-r}(x,z)\sigma'(u(r,z))D_{s,y}u(r,z) \, W(dr,dz) \right.$$
$$\left. + \int_s^t \int_0^1 G_{t-r}(x,z)b'(u(r,z))D_{s,y}u(r,z) \, dz \, dr \right|^2 \, dy \, ds. \tag{122}$$

Using (97) yields

$$\int_{t-\delta}^{t} \int_0^1 |G_{t-s}(x,y)\sigma(u(s,y))|^2 \, dy \, ds \geq c^2 \int_{t-\delta}^{t} \int_0^1 G_{t-s}^2(x,y) \, dy \, ds$$

$$= c^2 \int_0^\delta G_{2s}(x,x) \, ds \qquad (123)$$

$$\geq \frac{c^2}{\sqrt{2\pi}} \sqrt{\delta}.$$

Hence,

$$\mathrm{P}\left(\int_0^t \int_0^1 (D_{s,y}u(t,x))^2 \, dy \, ds < \varepsilon \right)$$

$$\leq \mathrm{P}\left(I_\delta \geq \frac{c^2}{\sqrt{2\pi}} \sqrt{\delta} - \varepsilon \right) \qquad (124)$$

$$\leq \left(\frac{c^2}{\sqrt{2\pi}} \sqrt{\delta} - \varepsilon \right)^{-p} \mathrm{E}\left(|I_\delta|^p \right),$$

for any $p > 0$. The term I_δ can be bounded by $2(I_{\delta,1} + I_{\delta,2})$, where

$$I_{\delta,1} \qquad (125)$$

$$= \int_{t-\delta}^t \int_0^1 \left| \int_s^t \int_0^1 G_{t-r}(x,z)\sigma'(u(r,z)) D_{s,y}u(r,z) \, W(dr,dz) \right|^2 \, dy \, ds,$$

and

$$I_{\delta,2}$$

$$= \int_{t-\delta}^t \int_0^1 \left| \int_s^t \int_0^1 G_{t-r}(x,z)b'(u(r,z)) D_{s,y}u(r,z) \, dz \, dr \right|^2 \, dy \, ds. \qquad (126)$$

Therefore, by Burkholder's inequality for Hilbert-space-valued stochastic integrals with respect to two-parameter processes and Hölder's inequality we obtain

$$\mathrm{E}\left(|I_{\delta,1}|^p \right)$$

$$\leq c_p \mathrm{E}\left[\left(\int_{t-\delta}^t \int_0^1 \int_s^t \int_0^1 [G_{t-r}(x,z)\sigma'(u(r,z)) D_{s,y}u(r,z)]^2 \, dz \, dr \, dy \, ds \right)^p \right]$$

$$\leq c_p \|\sigma'\|_\infty^p \left(\int_0^\delta \int_0^1 G_r(x,z)^2 \, dz \, dr \right)^p \sup_{(r,z) \in [0,\delta] \times [0,1]} \mathrm{E}\left(\|D(r,z)\|_H^p \right)$$

$$\leq C c_p \|\sigma'\|_\infty^p \left(\int_0^\delta G_{2r}(x,x) \, dr \right)^p \delta^{p/4}$$

$$\leq C \delta^{(p/2)+(p/4)}. \qquad (127)$$

As a consequence, choosing $\delta = 4\pi\varepsilon^2/c^4$ and using (124) and (127) yields

$$P\left(\int_0^t \int_0^1 (D_{s,y}u(t,x))^2 \, dy \, ds < \varepsilon\right) \le C\varepsilon^{p/2}, \tag{128}$$

which allows us to conclude the proof. $\square$

In [15] Pardoux and Zhang proved that $u(t,x)$ has an absolutely continuous distribution for all (t,x) such that $t > 0$ and $x \in (0,1)$, if the coefficients b and σ are Lipschitz continuous and $\sigma(u_0(y)) \ne 0$ for some $y \in (0,1)$. The regularity of the density under this more general nondegeneracy assumption is an open problem.

Bally and Pardoux considered in [3] the Equation (95) with Neumann boundary conditions on $[0,1]$, assuming that the coefficients b and σ are infinitely differentiable functions, which are bounded together with their derivatives. The main result of this paper says that the law of any vector of the form $(u(t,x_1),\ldots,u(t,x_d))$, $0 \le x_1 \le \cdots \le x_d \le 1$, $t > 0$, has a smooth and strictly positive density with respect to the Lebesgue measure on the set $\{\sigma > 0\}^d$.

Exercise 4.6. Prove Lemma 4.4.

Exercise 4.7. Let u be the solution to the linear stochastic differential equation

$$\frac{\partial u}{\partial t} = \frac{\partial^2 u}{\partial x^2} + u\frac{\partial^2 W}{\partial t \partial x}, \tag{129}$$

with initial condition $u_0(x)$ and Dirichler boundary conditions on $[0,1]$. Find the Wiener chaos expansion of $u(t,x)$.

5 Spatially Homogeneous SPDEs

We are interested in the following general class of stochastic partial differential equations

$$Lu(t,x) = \sigma(u(t,x))\dot{W}(t,x) + b(u(t,x)), \tag{130}$$

$t \ge 0$, $x \in \mathbf{R}^d$, where L denotes a second order differential operator, and we impose zero initial conditions.

We assume that the noise W is a zero mean Gaussian family $W = \{W(\varphi), \varphi \in C_0^\infty(\mathbf{R}^{d+1})\}$ with covariance

$$E(W(\varphi)W(\psi)) = \int_0^\infty \int_{\mathbf{R}^d} \int_{\mathbf{R}^d} \varphi(t,x)f(x-y)\psi(t,y) \, dx \, dy \, dt, \tag{131}$$

where f is a non-negative continuous function of $\mathbf{R}^d\backslash\{0\}$ such that it is the Fourier transform of a non-negative definite tempered measure μ on $\mathbf{R}^d$. That is,

$$f(x) = \int_{\mathbf{R}^d} \exp(-x \cdot \xi)\, \mu(d\xi), \tag{132}$$

and there is an integer $m \geq 1$ such that

$$\int_{\mathbf{R}^d} (1 + |\xi|^2)^{-m} \mu(d\xi) < \infty. \tag{133}$$

Then, the covariance (131) can also be written, using Fourier transform, as

$$E(W(\varphi)W(\psi)) = \int_0^\infty \int_{\mathbf{R}^d} \mathscr{F}\varphi(s)(\xi)\, \overline{\mathscr{F}\psi(s)(\xi)}\, \mu(d\xi)\, ds. \tag{134}$$

This class of equations has been studied, among others, by Dalang (see ([6]) and the references therein). They include the stochastic heat equation on $\mathbf{R}^d$, and the wave equation in dimension less or equal than 3. We are going to make the following assumption on the differential operator L:

Hypothesis H. *The fundamental solution of the operator L, denoted by G, is a non-negative distribution with rapid decrease such that such that for all $T > 0$*

$$\sup_{0 \leq t \leq T} G(t, \mathbf{R}^d) \leq C_T < \infty \tag{135}$$

and

$$\int_0^T \int_{\mathbf{R}^d} |\mathscr{F}G(t)(\xi)|^2\, \mu(d\xi)\, dt < \infty. \tag{136}$$

Here are two basic examples where condition (136) holds:

Example 5.1. (*The wave equation*) Let G_d be the fundamental solution of the wave equation

$$\frac{\partial^2 u}{\partial t^2} - \Delta u = 0. \tag{137}$$

We know that

$$G_1(t) = \frac{1}{2}\mathbf{1}_{\{|x|<1\}},$$
$$G_2(t) = c_2(t^2 - |x|^2)_+^{-1/2}, \tag{138}$$
$$G_3(t) = \frac{1}{4\pi}\sigma_t,$$

where σ_t denotes the surface measure on the 3-dimensional sphere of radius t. Furthermore, for all dimensions d

$$\mathscr{F}G_d(t)(\xi) = \frac{\sin(t|\xi|)}{|\xi|}. \tag{139}$$

Notice that only in dimensions $d = 1, 2, 3$, G_d is a measure. We can show that there are positive constants c_1 and c_2 depending on T such that

$$\frac{c_1}{1 + |\xi|^2} \leq \int_0^T \frac{\sin^2(t|\xi|)}{|\xi|^2} dt \leq \frac{c_2}{1 + |\xi|^2}. \tag{140}$$

Therefore, G_d satisfies Hypothesis **H** if and only if

$$\int_{\mathbf{R}^d} \frac{\mu(d\xi)}{1 + |\xi|^2} < \infty. \tag{141}$$

Example 5.2. (*The heat equation*) Let G be the fundamental solution of the heat equation

$$\frac{\partial u}{\partial t} - \frac{1}{2}\Delta u = 0. \tag{142}$$

Then,

$$G(t, x) = (2\pi t)^{-d/2} \exp\left(-\frac{|x|^2}{2t}\right) \tag{143}$$

and

$$\mathscr{F}G(t)(\xi) = \exp(-t|\xi|^2). \tag{144}$$

Because

$$\int_0^T \exp(-t|\xi|^2)\, dt = \frac{1}{|\xi|^2}(1 - \exp(-T|\xi|^2)), \tag{145}$$

we conclude that Hypothesis **H** holds if an only if (136) holds. We can also express condition (136) in terms of the covariance function f (see Exercise 5.9).

By definition, the solution to (130) on $[0, T]$ is an adapted stochastic process $u = \{u(t, x), (t, x) \in [0, T] \times \mathbf{R}^d\}$ satisfying

$$u(t, x) = \int_0^t \int_{\mathbf{R}^d} G(t - s, x - y)\sigma(u(s, y))\, W(ds, dy)$$
$$+ \int_0^t \int_{\mathbf{R}^d} b(u(t - s, x - y))\, G(s, dy). \tag{146}$$

The stochastic integral appearing in formula (146) requires some care because the integrand is a measure. In [6] Dalang constructed stochastic integrals of measure-valued processes on this type using the techniques of martingale measures. Actually, this stochastic integral is also a particular case of the integral with respect to a cylindrical Wiener proces. We will describe the construction of these integrals in the next section.

5.1 Existence and Uniqueness of Solutions

Fix a time interval $[0, T]$. The completion of the Schwartz space $\mathscr{S}(\mathbf{R}^d)$ of rapidly decreasing C^∞ functions, endowed with the inner product

$$\langle \varphi, \psi \rangle_{\mathscr{H}} = \int_{\mathbf{R}^d} \mathscr{F}\varphi(\xi) \, \overline{\mathscr{F}\psi(\xi)} \, \mu(d\xi)$$

$$= \int_{\mathbf{R}^d} \int_{\mathbf{R}^d} \varphi(x) f(x - y) \psi(y) \, dx \, dy \tag{147}$$

is denoted by $\mathscr{H}$. Notice that $\mathscr{H}$ may contain distributions. Set $\mathscr{H}_T = L^2([0,T]; \mathscr{H})$.

The Gaussian family $\{W(\varphi), \varphi \in C_0^\infty([0,T] \times \mathbf{R}^d)\}$ can be extended to the completion $\mathscr{H}_T = L^2([0,T]; \mathscr{H})$ of the space $C_0^\infty([0,T] \times \mathbf{R}^d)$ under the scalar product

$$\langle \varphi, \psi \rangle_{\mathscr{H}_T} = \int_0^T \int_{\mathbf{R}^d} \mathscr{F}\varphi(t)(\xi) \, \overline{\mathscr{F}\psi(t)(\xi)} \, \mu(d\xi) \, dt. \tag{148}$$

We will also denote by $W(g)$ the Gaussian random variable associated with an element $g \in L^2([0,T]; \mathscr{H})$.

Set $W_t(h) = W(1_{[0,t]}h)$ for any $t \geq 0$ and $h \in \mathscr{H}$. Then, $\{W_t, t \in [0,T]\}$ is a cylindrical Wiener process in the Hilbert space $\mathscr{H}$. That is, for any $h \in \mathscr{H}$, $\{W_t(h), t \in [0,T]\}$ is a Brownian motion with variance $\|h\|_{\mathscr{H}}^2$, and

$$\mathrm{E}(W_t(h) W_s(g)) = (s \wedge t) \, \langle h, g \rangle_{\mathscr{H}}. \tag{149}$$

Then (see, for instance, [5]), we can define the stochastic integral of $\mathscr{H}$-valued square integrable predictable processes. We denote by $\mathscr{F}_t$ the σ-field generated by the random variables $\{W_s(h), 0 \leq s \leq t, h \in \mathscr{H}\}$. The σ-field on $\Omega \times [0,T]$ generated by elements of the form $1_{(s,t]}X$, where X is a bounded $\mathscr{F}_s$-measurable random variable, is called the predictable σ-field and denoted by $\mathscr{P}$.

For any predictable process $g \in L^2(\Omega \times [0,T]; \mathscr{H})$ we denote its integral with respect to the cylindrical Wiener process W by

$$\int_0^T \int_{\mathbf{R}^d} g \, dW = g \cdot W, \tag{150}$$

and we have the isometry property

$$\mathrm{E}\left(|g \cdot W|^2\right) = \mathrm{E}\left(\int_0^T \|g_t\|_{\mathscr{H}}^2 \, dt\right). \tag{151}$$

Then, $M_t(A) = W_t(1_A)$ defines a martingale measure associated to the noise W in the sense of Walsh (see [19] and [6]) and the stochastic integral (150) coincides with the integral defined in the work of Dalang [6].

The following result provides examples of random distributions that can be integrated with respect to W.

Proposition 5.3. *Let* $Z = \{Z(t,x), (t,x) \in [0,T] \times \mathbf{R}^d\}$ *be a predictable process such that*

$$C_Z := \sup_{(t,x)\in[0,T]\times\mathbf{R}^d} \mathrm{E}\left(|Z(t,x)|^2\right) < \infty. \tag{152}$$

Then, the random element $G = G(t,dx) = Z(t,x)\,G(t,dx)$ is a predictable process in the space $L^2(\Omega \times [0,T];\mathscr{H})$, and

$$\mathrm{E}(\|G\|_{\mathscr{H}_T}^2) \leq C_Z \int_0^T \int_{\mathbf{R}^d} |\mathscr{F}G(t)(\xi)|^2\,\mu(d\xi)\,dt. \tag{153}$$

Proof. First notice that G belongs to $\mathscr{H}_T$, and

$$\|G\|_{\mathscr{H}_T}^2 = \int_0^T \int_{\mathbf{R}^d} |\mathscr{F}G(t)(\xi)|^2\,\mu(d\xi)\,dt. \tag{154}$$

We can assume that Z is nonnegative. For any $k > 0$ and any $t \in [0,T]$ we have

$$\int_{\mathbf{R}^d} (1 + |x|^2)^{k/2} Z(t,y)\,G(t,dy) < \infty. \tag{155}$$

This implies that G belongs to the space of distributions with rapid decrease, almost surely. Finally, the result follows by a regularization argument (see Proposition 3.3 in [13]).   □

Under the assumptions of Proposition 5.3, suppose in addition that

$$\sup_{(t,x)\in[0,T]\times\mathbf{R}^d} \mathrm{E}(|Z(t,x)|^p) < \infty, \tag{156}$$

for some $p \geq 2$. Then one can show the following estimate for the p moment of the stochastic integral, using the Burkholder and Hölder inequalities:

$$\begin{aligned}
&\mathrm{E}\left(\left|\int_0^T \int_{\mathbf{R}^d} G(t,dx)Z(t,x)\,W(dt,dx)\right|^p\right) \\
&\leq c_p(\nu_t)^{(p/2)-1} \int_0^T \left(\sup_{x\in\mathbf{R}^d} \mathrm{E}\left(|Z(t,x)|^p\right)\right) \int_{\mathbf{R}^d} |\mathscr{F}G(s)(\xi)|^2\,\mu(d\xi)\,ds,
\end{aligned} \tag{157}$$

where

$$\nu_t = \int_0^t \int_{\mathbf{R}^d} |\mathscr{F}G(s)(\xi)|^2\,\mu(d\xi)\,ds. \tag{158}$$

The construction of the stochastic integral can be extended to processes taking values in a Hilbert space. Let $\mathscr{A}$ be a separable real Hilbert space with inner-product and norm denoted by $\langle \cdot,\cdot \rangle_{\mathscr{A}}$ and $\|\cdot\|_{\mathscr{A}}$, respectively. Let $K = \{K(t,x),(t,x) \in [0,T] \times \mathbf{R}^d\}$ be an $\mathscr{A}$-valued predictable process satisfying the following condition:

$$\sup_{(t,x)\in[0,T]\times\mathbf{R}^d} \mathrm{E}\left(\|K(t,x)\|_{\mathscr{A}}^2\right) < \infty. \tag{159}$$

Set $\Gamma(t,dx) = G(t,dx)K(t,x) \in L^2(\Omega \times [0,T]; \mathscr{H} \otimes \mathscr{A})$, where G satisfies the hypothesis of Proposition 5.3. Then, if $\{e_j, j \geq 0\}$ is a complete orthonormal system of $\mathscr{A}$ we define

$$\Gamma \cdot W = \sum_{j \geq 0} \left(\int_0^T \int_{\mathbf{R}^d} G(s,y)\langle K(s,x), e_j \rangle_{\mathscr{A}} W(ds,dy) \right) e_j. \tag{160}$$

Condition (159) and Proposition 5.3 imply that the above series is convergent and $G \cdot W$ defines an element of $L^2(\Omega; \mathscr{A})$. Moreover, the following estimate for the moments of $G \cdot W$ is analogous to (157):

$$
\begin{aligned}
&\mathrm{E}\left(\|\Gamma \cdot W\|_{\mathscr{A}}^p \right) \\
&\qquad \leq C_p(\nu_T)^{\frac{p}{2}-1} \int_0^T \sup_{x \in \mathbf{R}^d} \mathrm{E}(\|K(s,x)\|_{\mathscr{A}}^p) \int_{\mathbf{R}^d} |\mathscr{F}G(s)(\xi)|^2 \, \mu(d\xi) \, ds,
\end{aligned} \tag{161}
$$

for all $p \geq 2$.

The following theorem gives the existence and uniqueness of a solution for Equation (130) (see Dalang [6]).

Theorem 5.4. *Suppose that the coefficients b and σ are Lipschitz continuous, and tthe fundamental solution of $Lu = 0$ satisfies Hypothesis **H**. Then (130) has a unique solution $u(t,x)$ which is continuous in L^2 and satisfies*

$$\sup_{(t,x) \in [0,T] \times \mathbf{R}^d} \mathrm{E}\left(|u(t,x)|^2 \right) < \infty, \tag{162}$$

for all $T > 0$.

5.2 Regularity of the Law

We will show that under suitable nondegeneracy conditions, the solution to Equation (146), at any point $(t,x) \in (0,T] \times \mathbf{R}^d$, is a random variable whose law admits a density with respect to Lebesgue measure on $\mathbf{R}$, and the density is smooth. We will make use of the techniques of the Malliavin calculus. Notice that the underlying Hilbert space here is $\mathscr{H}_T = L^2([0,T]; \mathscr{H})$.

The first step is to study the differentiability of $u(t,x)$, for all fixed $(t,x) \in (0,T] \times \mathbf{R}^d$. For any random variable F in $\mathbf{D}^{1,p}$, the derivative DF defines an $\mathscr{H}_T$-valued random variable, or an $\mathscr{H}$-valued stochastic process denoted by $\{D_s F, s \geq 0\}$. The proof of the following proposition is similar to that of Proposition 4.3 (see Theorem 2 in [16], and also Proposition 4.4 in [13]).

Proposition 5.5. *Assume that G satisfies Hypothesis **H**. Suppose also that the coefficients b and σ are C^1 functions with bounded Lipschitz continuous derivatives. Then, for any $(t,x) \in [0,T] \times \mathbf{R}^d$, $u(t,x)$ belongs to $\mathbf{D}^{1,p}$, for any $p \geq 2$. Moreover, the derivative $Du(t,x)$ satisfies the following linear stochastic differential equation:*

$$D_r u(t,x) = \sigma(u(r,\cdot))G(t-r,x-\cdot)$$

$$+ \int_r^t \int_{\mathbf{R}^d} G(t-s,x-y)\sigma'(u(s,y))D_r u(s,y)\,W(ds,dy)$$

$$+ \int_r^t \int_{\mathbf{R}^d} b'(u(s,x-y))D_r u(s,x-y)\,G(t-s,dy)\,ds, \quad (163)$$

for all $r \in [0,T]$. Moreover,

$$\sup_{(t,x)\in[0,T]\times\mathbf{R}^d} \mathrm{E}\left(\|D(t,x)\|_{\mathscr{H}_T}^p\right) < \infty. \quad (164)$$

Furthermore if the coefficients b and σ are infinitely differentiable with bounded derivatives of all orders, then $u(t,x)$ belongs to the space $\mathbf{D}^\infty$.

We can now show the regularity of the density of $u(t,x)$ under ellipticity conditions.

Theorem 5.6. *Assume that G satisfies Hypothesis $\mathbf{H}$, the coefficients σ and b are C^∞ functions with bounded derivatives of any order greater than or equal to one and $|\sigma(z)| \geq c > 0$, for all $z \in \mathbf{R}$. Moreover, suppose that there exist $\gamma > 0$ such that for all $\tau > 0$,*

$$\int_0^\tau \int_{\mathbf{R}^d} |\mathscr{F}G(s)(\xi)|^2 \,\mu(d\xi)\,ds \geq C\tau^\gamma, \quad (165)$$

for some positive constant C. Then, for all $(t,x) \in (0,T] \times \mathbf{R}^d$, the law of $u(t,x)$ has a C^∞ density with respect to Lebesgue measure on $\mathbf{R}$.

Proof. From the general criterion for smoothness of densities (see Theorem 3.5), it suffices to show that the inverse of the Malliavin matrix of $u(t,x)$ has moments of all order, that is

$$\mathrm{E}\left(\left|\int_0^T \|D_s u(t,x)\|_{\mathscr{H}}^2\,ds\right|^{-q}\right) < +\infty, \quad (166)$$

for all $q \geq 2$. Then, by Lemma 4.4 it is enough to show that for any $q \geq 2$ there exists an $\varepsilon_0(q) > 0$ such that for all $\varepsilon \leq \varepsilon_0$

$$\mathrm{P}\left(\int_0^t \|D_s u(t,x)\|_{\mathscr{H}}^2\,ds < \varepsilon\right) \leq C\varepsilon^q. \quad (167)$$

For $\delta > 0$ sufficiently small we have

$$\int_0^t \|D_s u(t,x)\|_{\mathscr{H}}^2\,ds \geq \frac{1}{2}\int_{t-\delta}^t \|G(t-s,x-\cdot)\sigma(u(s,\cdot))\|_{\mathscr{H}}^2\,ds - I_\delta, \quad (168)$$

where

$$I_\delta = \int_{t-\delta}^t \left\| \int_s^t \int_{\mathbf{R}^d} G(t-r, x-z)\sigma'(u(r, z))D_s u(r, z) W(dr, dz) \right.$$

$$\left. + \int_s^t \int_{\mathbf{R}^d} G(t-r, dz)b'(u(r, x-z))D_s u(r, x-z)dr \right\|_{\mathscr{H}}^2 ds. \tag{169}$$

In order to get a lower bound for the first term in the right-hand side of Equation (168), we regularize the measures G and $G(t, x - \cdot)\sigma(u(t-s, \cdot))$ with an approximation to the identity $\{\psi_n\}$:

$$G_n(t) = \psi_n * G(t),$$
$$J_n(t, s) = \psi_n * G(t, x - \cdot)\sigma(u(t-s, \cdot)). \tag{170}$$

Then, by the non-degeneracy assumption on σ, we have

$$\int_{t-\delta}^t \|G(t-s, x - \cdot)\sigma(u(s, \cdot))\|_{\mathscr{H}}^2 ds$$

$$= \int_0^\delta \|G(s, x - \cdot)\sigma(u(t-s, \cdot))\|_{\mathscr{H}}^2 ds$$

$$= \lim_{n\to\infty} \int_0^\delta \|J_n(t, s)\|_{\mathscr{H}}^2 ds$$

$$= \lim_{n\to\infty} \int_0^\delta \int_{\mathbf{R}^d} \int_{\mathbf{R}^d} J_n(t, s, y)f(y-z)J_n(t, s, z)\, dy\, dz\, ds \tag{171}$$

$$\geq c^2 \lim_{n\to\infty} \int_0^\delta \int_{\mathbf{R}^d} \int_{\mathbf{R}^d} G_n(s, x-y)f(y-z)G_n(s, x-z)\, dy\, dz\, ds$$

$$= c^2 \lim_{n\to\infty} \int_0^\delta \|G_n(s, x - \cdot)\|_{\mathscr{H}}^2 ds$$

$$= c^2 \int_0^\delta \|G(s, x - \cdot)\|_{\mathscr{H}}^2 ds$$

$$= c^2 g(\delta),$$

where

$$g(\delta) := \int_0^\delta \int_{\mathbf{R}^d} |\mathscr{F}G(s)(\xi)|^2 \mu(d\xi)\, ds. \tag{172}$$

Substituting (171) into (168) yields

$$\mathrm{P}\left(\int_0^t \|D_s u(t, x)\|_{\mathscr{H}}^2 ds < \varepsilon\right) \leq \mathrm{P}\left(I_\delta \geq \frac{c^2}{2}g(\delta) - \varepsilon\right)$$

$$\leq \left(\frac{c^2}{2}g(\delta) - \varepsilon\right)^{-p} \mathrm{E}(|I_\delta|^p), \tag{173}$$

for any $p > 0$.

The term I_δ can be bounded by $2(I_{\delta,1} + I_{\delta,2})$, with

$$I_{\delta,1} = \int_0^\delta \left\| \int_{t-s}^t \int_{\mathbf{R}^d} G(t-r, x-z)\sigma'(u(r,z))D_{t-s}u(r,z)\,W(dr,dz) \right\|_{\mathscr{H}}^2 ds,$$

$$I_{\delta,2} = \int_0^\delta \left\| \int_{t-s}^t \int_{\mathbf{R}^d} G(t-r, dz)b'(u(r,x-z))D_{t-s}u(r,x-z)\,dr \right\|_{\mathscr{H}}^2 ds.$$

We need to find upper bounds for $E(|I_{\delta,i}|)^p$, $i = 1, 2$. On one hand, owing to Hölder's inequality and (161) we get

$$E(|I_{\delta,1}|^p) = E\left(\int_0^\delta \|Z_{x,s,t}\|_{\mathscr{H}}^2\,ds \right)^p$$

$$\leq \delta^{p-1} E\left(\int_0^\delta \|Z_{x,s,t}\|_{\mathscr{H}}^{2p}\,ds \right) \tag{174}$$

$$\leq \delta^{p-1} (g(\delta))^p \sup_{(s,y)\in[0,T]\times\mathbf{R}^d} E\left(\int_0^s \|D_r u(s,y)\|_{\mathscr{H}}^{2p}\,dr \right),$$

where

$$Z_{x,s,t} = \int_{t-s}^t \int_{\mathbf{R}^d} G(t-r, x-z)\sigma'(u(r,z))D_{t-s}u(r,z)\,W(dr,dz). \tag{175}$$

The above estimate (174) together with (164) let us conclude that

$$E(|I_{\delta,1}|^p) \leq C\delta^{p-1}(g(\delta))^p. \tag{176}$$

On the other hand, using similar arguments but for the Hilbert-space-valued pathwise integral one proves that $E(|I_{\delta,2}|^p)$ may be bounded, up to some positive constant, by $\delta^{p-1}(g(\delta))^p$. Thus, we have proved that

$$P\left(\int_0^t \|D_s u(t,x)\|_{\mathscr{H}}^2\,ds < \epsilon \right) \leq C\left(\frac{c^2}{2}g(\delta) - \epsilon \right)^{-p} \delta^{p-1}(g(\delta))^p. \tag{177}$$

At this point, we choose $\delta = \delta(\epsilon)$ in such a way that $g(\delta) = 4\varepsilon/c^2$. By (165), this implies that $4\varepsilon/c^2 \geq C\delta^\gamma$, that is $\delta \leq C\varepsilon^{1/\gamma}$. Hence,

$$P\left(\int_0^t \|D_s u(t,x)\|_{\mathscr{H}}^2\,ds < \varepsilon \right) \leq C\varepsilon^{(p-1)/\gamma}, \tag{178}$$

and it suffices to take p sufficiently large such that $(p-1)/\gamma \geq q$. $\square$

Remark 5.7. Assume that G satisfies Hypothesis **H**. Suppose also that the coefficients b and σ are C^1 functions with bounded Lipschitz continuous derivatives and that $|\sigma(z)| \geq c > 0$, for all $z \in \mathbf{R}$ and some positive constant c. Then, for all $t > 0$ and $x \in \mathbf{R}^d$, the random variable $u(t,x)$ has an absolutely continuous law with respect to Lebesgue measure on $\mathbf{R}$. This result can be proved using the criterion for absolute continuity given in Theorem 3.3 (see Theorem 3 in the reference [16] and Theorem 5.2 in [13], in the case of the three dimensional wave equation).

Exercise 5.8. If G is the fundamental solution of the wave equation in $\mathbf{R}^d$, with $d = 1, 2, 3$, show that condition (165) is satisfied with $\gamma = 3$. On the other hand, if G is the fundamental solution of the heat equation on $\mathbf{R}^d$, $d \geq 1$, show that condition (165) is satisfied, with any $\gamma \geq 1$ (see Lemma 3.1 in [9]).

Exercise 5.9. Show that for the heat equation, condition (136) is always true when $d = 1$; for $d = 2$, (136) holds if and only if

$$\int_{|x| \leq 1} f(x) \log \left(\frac{1}{|x|} \right) dx < \infty, \tag{179}$$

and for $d \geq 3$, (136) holds if and only if

$$\int_{|x| \leq 1} \frac{f(x)}{|x|^{d-2}} dx < \infty. \tag{180}$$

References

[1] V. Bally, A. Millet, and M. Sanz-Solé (1995). Approximation and support theorem in Hšlder norm for parabolic stochastic partial differential equations, *Ann. Probab.* **23**, 178–222

[2] N. Bouleau and F. Hirsch (1991). *Dirichlet Forms and Analysis on Wiener Space*, Walter de Gruyter & Co., Berlin

[3] V. Bally and E. Pardoux (1998). Malliavin calculus for white noise driven Parabolic SPDEs, *Pot. Anal.* **9**, 27–64

[4] R. Cairoli and J. B. Walsh (1975). Stochastic integrals in the plane, *Acta Mathematica* **134**, 111–183

[5] G. Da Prato and J. Zabczyk (1992). *Stochastic Equations in Infinite Dimensions*, Cambridge University Press, Cambridge

[6] R. Dalang (1999). Extending the martingale measure stochastic integral with applications to spatially homogeneous s.p.d.e.'s, *Electron. J. Probab.* **4**, 29 pp

[7] K. Itô (1951). Multiple Wiener integral, *J. Math. Soc. Japan* **3**, 157–169

[8] P. Malliavin (1978). Stochastic calculus of variation and hypoelliptic operators, In: *Proceedings of the International Symposium on Stochastic Differential Equations* (Res. Inst. Math. Sci., Kyoto Univ., Kyoto, 1976), pp. 195–263, Wiley, New York-Chichester-Brisbane

[9] D. Màrquez-Carreras, M. Mellouk and M. Sarrà (2001). On stochastic partial differential equations with spatially correlated noise: smoothness of the law. *Stochastic Process. Appl.* **93**, 269–284

[10] A. Millet and M. Sanz-Solé (1999). A stochastic wave equation in two space dimension: Smoothness of the law. *Ann. Probab.* **27**, 803–844

[11] C. Mueller (1991). On the support of solutions to the heat equation with noise. *Stochastics Stochastics Rep.* **37**, 225–245

[12] D. Nualart (2006). *The Malliavin Calculus and Related Topics.* Second edition, Springer-Verlag, New York

[13] D. Nualart and Ll. Quer-Sardanyons (2007). Existence and smoothness of the density for spatially homogeneous SPDEs. *Potential Analysis* **27**, 281–299

[14] D. Nualart and M. Zakai (1988). Generalized multiple stochastic integrals and the representation of Wiener functionals. *Stochastics* **23**, 311–330

[15] E. Pardoux and T. Zhang (1993). Absolute continuity of the law of the solution of a parabolic SPDE. *J. Funct. Anal.* **112**, 447–458

[16] Ll. Quer-Sardanyons and M. Sanz-Solé (2004). Absolute continuity of the law of the solution to the 3-dimensional stochastic wave equation. *J. Funct. Anal.* **206**, 1–32

[17] Ll. Quer-Sardanyons and M. Sanz-Solé (2004). A stochastic wave equation in dimension 3: smoothness of the law. *Bernoulli* **10**, 165–186

[18] D. W. Stroock (1978). Homogeneous chaos revisited. In: Seminaire de Probabilités XXI, Lecture Notes in Mathematics **1247**, 1–8

[19] J. B. Walsh (1986). *An Introduction to Stochastic Partial Differential Equations.* In: Ecole d'Ete de Probabilites de Saint Flour XIV, Lecture Notes in Mathematics **1180**, 265–438

[20] S. Watanabe (1984). *Lectures on Stochastic Differential Equations and Malliavin Calculus.* Tata Institute of Fundamental Research, Springer-Verlag

Some Tools and Results for Parabolic Stochastic Partial Differential Equations

Carl Mueller

Summary. These notes give an informal introduction to parabolic stochastic partial differential equations. We emphasize material coming from particle systems, including duality and the Dawson–Watanabe superprocess. We also deal with large deviations and comparison theorems. Applications include blow-up, hitting theorems, and compact support.

1 Introduction

1.1 Outline of the notes

The goal of these notes is to explain a set of ideas in stochastic partial differential equations (SPDE) which I have found useful. The notes are meant for graduate students, so they are not written in a formal style. Sometimes I will explain an idea in a simple case, and leave it to the reader to develop the topic more broadly.

I will begin with some general thoughts on the field. SPDE, and perhaps PDE as well, find their primary motivations in science and engineering. It is best not to think of SPDE as objects in pure mathematics, but as models for physical phenomena. To study SPDE it is not necessary to have a Ph.D. in biology, for example, but it is often helpful to think of an SPDE in terms of a population of organisms. Similar helpful motivations come from other sciences as well. When thinking of the heat equation, with or without noise, it helps to visualize a physical object with varying temperature. It would be foolish to set forth a foundation for the entire field of SPDE, but my goal is to explain some tools which others may find useful.

Both ordinary differential equations (ODE) and partial differential equations (PDE) play a fundamental role in describing reality. However, any model of the real world must take into account uncertainty or random fluctuations. It is therefore surprising that while stochastic ODE were studied intensively throughout the twentieth century, SPDE only received attention much later.

D. Khoshnevisan and F. Rassoul-Agha (eds.) *A Minicourse on Stochastic Partial Differential Equations.*
Lecture Notes in Mathematics 1962.
© Springer-Verlag Berlin Heidelberg 2009

Some early work stemmed from the Zakai equation in filtering theory, see [32], and on the theoretical side there was the work of Pardoux [31] and Krylov and Rozovskii [17]. Much of this early work centered on foundational questions such as setting up the appropriate function spaces for studying solutions, or using such analytic tools as the method of monotonicity [31]. Later, Walsh [38] introduced the notion of martingale measures as an alternative framework. The diverse origins of SPDE have led to a lively interplay of viewpoints. Some people feel that SPDE should be based on such tools as Sobolev spaces, as is the case for PDE. Others, with a background in probability, feel that an SPDE describes a special kind of stochastic process. Applied mathematicians may feel that the study of SPDE should follow the ideas used in their domain.

By a historical accident, particle systems, which may be considered as discrete SPDE, were studied much earlier than SPDE. Such pioneers as Ted Harris and Frank Spitzer laid the groundwork for this theory. Their research was also influenced by results in percolation, by such mathematicians as Harry Kesten. Particle systems has changed its emphasis over the years, and some of this early work is being forgotten. However, I believe that the main methods of particle systems will always be relevant to SPDE. In particular, I will introduce the method of duality. Unfortunately, there was no time in the course to discuss percolation, which I also believe has fundamental importance for SPDE. Both duality and percolation, as well as many other ideas, are described in more detail in three classics: [21; 22] give detailed technical accounts of the field, and [9] provides a lively intuitive treatment.

Secondly, Watanabe and Dawson found that the scaling limit of critical branching Brownian motions give a fundamentally important model, called the Dawson–Watanabe process or superprocess. Because this model involves independently moving particles, there are powerful tools for studying its behavior, and many of these tools help in the study of SPDE. For example, the heat equation can be thought of as the density of a cloud of Brownian particles. Any SPDE which involves a density of particles can be studied via the Dawson–Watanabe process. There is a huge literature in this area, but two useful surveys are written by Dawson [6] and Perkins [33].

Thirdly, as one might expect, tools from PDE are useful for SPDE. Of course, Sobolev spaces and Hilbert spaces play a role, as in the work of Da Prato and Zabczyk [8] and Krylov [19]. But here I wish to concentrate on qualitative tools, such as the maximum principle. In particular, comparison theorems hold for many SPDE. Given two solutions, suppose that one is initially larger than the other. Then that relationship will continue to hold for later times.

Finally, tools from probability also find applications in SPDE. For example, the theory of large deviations of dynamical systems developed by Wentzell and Freidlin [13] also applies to SPDE. If the noise is small, we can estimate the probability that the solutions of the SPDE and corresponding PDE (without noise) differ by more than a given amount. Unfortunately, I had no time to discuss questions of coupling and invariant measures, which play a large role in the study of the stochastic Navier–Stokes equation.

After introducing these ideas, I give some applications to properties of parabolic SPDE.

1.2 Some Comments on the Current State of SPDEs

To finish this introduction, let me indicate some of the current directions in SPDE. The subject is expanding in many directions, so all I can do is give a personal view.

I believe that the foundations of SPDE are settled. One can use either Walsh's approach [38] using martingale measures, the Hilbert space approach of Da Prato and Zabczyk [8], or Krylov's L_p theory [19]. The goal should be to study properties of the equations.

Furthermore, the "naive" study of qualitative properties of SPDE, involving the superprocess or simple variants of the heat equation, is also largely finished. I do recommend learning the main ideas of this theory, as well as the foundational approaches mentioned earlier, since they can help us study of more complicated equations.

Luckily, finally, scientists are jumping into the field with a vengeance, and SPDE is expanding chaotically in all directions. I believe that the sciences will continue to provide important SPDE models and conjectures. For example, the reader can consult [7] for polymer models, and [3] for SPDEs in the physics of surfaces. There is broad interest in the stochastic Navier–Stokes equation [23]. Scientists seem to have finally grasped the importance of SPDE models, so the reader should stay alert for new developments.

2 Basic Framework

Expanding on Chapter 1, Section 6 of this book, we will mostly consider parabolic equations of the form

$$\frac{\partial u}{\partial t} = \Delta u + a(u) + b(u)\,\dot{W}(t,x),$$ (1)

$$u(0,x) = u_0(x).$$

where $\dot{W}(t,x)$ is $d+1$ parameter white noise and $x \in \mathbf{R}^d$. This set-up has been used for a long time, see [38] or [8]. If we expect our noise to arise from many small influences which are independent at different positions in space and time, then white noise is a good model. As with stochastic differential equations (SDE), the solution $u(t,x)$ is not differentiable in t or x, so (1) does not make sense as written. In Da Prato and Zabczyk's approach, we regard (1) as a stochastic differential equation (SDE) in function space. In Walsh's theory, which we will adopt here, we regard (1) as a shorthand for the following integral equation, which is often called the mild form of the equation. This is also explained in Chapter 1, Section 6. Of course, Da Prato and Zabczyk's theory can deal with the mild form as well.

$$u(t,x) = \int_{\mathbf{R}^d} G(t,x-y)u_0(y)\,dy + \int_0^t \int_{\mathbf{R}^d} G(t-s,x-y)a(u(s,y))\,dy\,ds$$

$$+ \int_0^t \int_{\mathbf{R}^d} G(t-s,x-y)b(u(s,y))\,W(dy\,ds). \tag{2}$$

Here $G(t,x)$ is the heat kernel,

$$G(t,x) = \frac{1}{(4\pi t)^{d/2}} \exp\left(-\frac{|x|^2}{4t}\right), \tag{3}$$

and the final integral in (2) is a stochastic integral in the sense of Walsh [38], see Chapter 1 of this book. Since there is a unique time direction t, such integrals can be constructed along the lines of Ito's theory, and their properties are mostly the same as for the Ito integral. Let $\mathscr{F}_t$ denote the σ-field generated by the noise W up to time t. That is, $\mathscr{F}_t$ is generated by the integral

$$\int_0^t \int_{\mathbf{R}^d} g(s,y)\,W(dy\,ds), \tag{4}$$

for deterministic function $g \in \mathbf{L}^2(dy\,ds)$. For convenience, when $f(s,y)$ is nonanticipating with respect to $\mathscr{F}_t$ and

$$E \int_0^t \int_{\mathbf{R}^d} G^2(t-s,x-y)f^2(s,y)\,dy\,ds < \infty, \tag{5}$$

we often write

$$N(t,x) = \int_0^t \int_{\mathbf{R}^d} G(t-s,x-y)f(s,y)\,W(dy\,ds), \tag{6}$$

so that if $f(s,y) = b(u(s,y))$, then $N(t,x)$ is the "noise term" in (2). Then, in particular,

$$E\left[N(t,x)^2\right] = \int_0^t \int_{\mathbf{R}^d} E\left[G^2(t-s,x-y)f(s,y)^2\right]\,dy\,ds. \tag{7}$$

Exercise 2.1. For $f(s,y) :=$, show that for any $t > 0$ and $x \in \mathbf{R}^d$,

$$E\left[N(t,x)^2\right] \quad \begin{cases} < \infty & \text{if } d = 1, \\ = \infty & \text{if } d > 1. \end{cases} \tag{8}$$

Thus, if $d \geq 2$, (1) is likely to have solutions which are generalized functions, but do not exist as functions in $\mathbf{L}^2(dP)$, for instance. But then nonlinear functions such as $a(u), b(u)$ are hard to define. Since we live in a 3-dimensional world, this situation gives rise to many difficulties. One common solutions is to smooth the noise, perhaps by convolving it with another function of the variable x. Such noise is called colored noise. Another possibility is to replace $\mathbf{R}^d$ by a lattice, and replace the Laplacian by the discrete Laplacian. It would be nice to deal with (1) more directly; maybe you have an idea of how to do this.

Actually, some authors deal with solutions to (1) which are generalized functions over Wiener space. Usually, generalized functions are defined in terms of integrals against a test function. Thus, we would have to define our solution in terms of an integral over all points in the probability space. But in the physical world, we are doomed to experience only a single point in the probability space. Maybe I'm being too pessimistic here, since repeated experiments sample different points in the probability space; readers can form their own conclusion.

Another point is that for x fixed, the process $t \to u(t, x)$ is Hölder continuous with parameter $1/4 - \varepsilon$ for every $\varepsilon > 0$, and therefore is not a semimartingale. Therefore there is no Ito's lemma in the usual sense, and this has caused a lot of problems for the theory. However, if $\phi(x)$ is an $\mathbf{L}^2$ function, then

$$X_t = \int_{\mathbf{R}^d} u(t, x)\phi(x) \, dx \qquad (9)$$

is a semimartingale with quadratic variation

$$\langle X \rangle_t = \int_0^t \int_{\mathbf{R}^d} \left[\int_{\mathbf{R}^d} \phi(x) G(t - s, x - y) \, dx \right]^2 b(u(s, y))^2 dy \, ds. \qquad (10)$$

Note that the inner integral smoothes out the singularity of $G(t - s, x - y)$.

Since it doesn't fit in elsewhere in this paper, let me mention a very nice survey by Ferrante and Sanz-Solé [12] which deals with SPDE driven by colored noise. For a colored noise $\dot{F}(t, x)$, the covariance in x is not the δ-function as in the case of white noise, so heuristically $E[\dot{F}(t, x)\dot{F}(s, y)] = \delta(t - s)R(x - y)$ for some covariance function R.

Finally two recent books, Chow [4] and Rockner [34], also give nice introductions to SPDE. Both deal with the functional analytic approach similar to Da Prato and Zabczyk [8].

3 Duality

3.1 Definitions

The reader might wonder why I am devoting so much space to duality, since there are only a few papers on SPDE with that word in the title. I firmly believe that many SPDE should be viewed as limits of particle systems, and duality is perhaps the leading method in particle systems. Secondly, the most important tool in superprocesses is the Laplace functional, which is a form of duality. Thirdly, duality is getting relatively little attention, but it remains a powerful tool.

Duality is a relationship between two stochastic processes which allows us to use information from one process in analyzing the other. There are at least two kinds of duality. One involves a duality function, and the other arises from studying the ancestry of particles in a particle system.

I will describe the functional form of duality first; details can be found in [10], pages 188–189. Two processes X_t, Y_t are said to be in duality with respect to a function $H(x, y)$ if, for all t in some interval $[0, T]$ we have

$$E\Big[H\left(X_t, Y_0\right)\Big] = E\Big[H\left(X_0, Y_t\right)\Big]. \tag{11}$$

All probabilists know simple examples of duality, although they may not realize it. Let $u(t, x)$ satisfy the heat equation

$$\frac{\partial u}{\partial t} = \frac{1}{2}\Delta u, \tag{12}$$

and let B_t be Brownian motion. It is well known that under the appropriate conditions,

$$u(t, x) = E_x\Big[u(0, B_t)\Big]. \tag{13}$$

Thus, the processes $u(t, \cdot)$ and B_t are in duality with respect to the function $H(u, B) = u(B)$. Note that deterministic processes such as $u(t, x)$ still count as processes.

3.2 Duality for Feller's Diffusion

Feller derived the following SDE as a limit for critical birth and death processes; a critical birth and death process has expected offspring size equal to 1, so that the expected number of particles does not change:

$$dX = \sqrt{X} \, dB. \tag{14}$$

Now consider a function $v(t)$ which satisfies

$$v' = -\frac{1}{2}v^2. \tag{15}$$

Explicit solutions $v(t)$ are easy to find. Next, Ito's lemma implies that for $0 < t < T$, if

$$M_t = \exp\left(-X_t v(T - t)\right), \tag{16}$$

then

$$\begin{aligned}
dM &= M \cdot (-\sqrt{X} \, dB) + \frac{1}{2}M \cdot Xv^2 \, dt - M \cdot \frac{1}{2}Xv^2 \, dt \\
&= M \cdot (-\sqrt{X} \, dB).
\end{aligned} \tag{17}$$

Thus M_t is a martingale, and

$$\exp\left(-X_0 v(T)\right) = M_0 = \mathrm{E}[M_T] = \mathrm{E}\left[\exp\left(-X_T v(0)\right)\right]. \qquad (18)$$

So, X_t and $v(t)$ are in duality with respect to the function $H(x,v) = \exp(-xv)$. In this case, duality gives us the Laplace transform of X_t. Duality implies that X_t is unique in law, provided X_0 is specified.

Exercise 3.1. Explicitly solve for $v(t)$ and hence find the Laplace transform of X_t.

3.3 Duality for the Wright–Fisher SDE

Next, let X_t satisfy the following SDE, named after Fisher and Wright. We can think of a population of constant size, consisting of 2 subpopulations whose percentages of the total are X and $1-X$. Due to competition, the population fluctuates randomly, and the variance of the fluctuations are proportional to the number of encounters between the two types. This leads to a standard deviation of $\sqrt{X(1-X)}$, and gives rise to the following SDE.

$$dX = \sqrt{X(1-X)}\, dB$$
$$X_0 = x_0 \in [0,1].$$

For n a nonnegative integer, Ito's formula gives

$$dX^n = nX^{n-1}\sqrt{X(1-X)}\, dB + \frac{n(n-1)}{2}X^{n-2}X(1-X)\, dt$$
$$= \mathrm{mart} - \binom{n}{2}(X^n - X^{n-1})\, dt, \qquad (19)$$

where "mart" indicates a martingale term whose expectation is 0 when integrated. The final term in (19) has the intuitive meaning that X^n is replaced by X^{n-1} at rate $\binom{n}{2}$, except that there is a negative sign.

Let N_t be a process taking values in the nonnegative integers, and independent of X_t. We think of N_t as the number of particles present at time t, and let each pair of particles coalesce at rate 1. In other words, N_t is a Markov process such that as $h \to 0$,

$$\mathrm{P}(N_{t+h} = N_t \mid N_t) = 1 - \binom{N_t}{2}h + o(h)$$
$$\mathrm{P}(N_{t+h} = N_t - 1 \mid N_t) = \binom{N_t}{2}h + o(h). \qquad (20)$$

Thus, for $x \in [0,1]$, one has

$$E\left[x^{N_{t+h}} - x^{N_t} \mid N_t\right]$$

$$= x^{N_t}\left(-\binom{N_t}{2}h + o(h)\right) + x^{N_t-1}\left(\binom{N_t}{2}h + o(h)\right) \qquad (21)$$

$$= -\binom{N_t}{2}\left(x^{N_t} - x^{N_t-1}\right)h + o(h).$$

Exercise 3.2. Prove (21).

Note that the last lines in (19) and (21) match. Then X_t, N_t are in duality with respect to the function $H(x, n) = x^n$. In other words,

$$E\left[X_t^{N_0}\right] = E\left[X_0^{N_t}\right]. \qquad (22)$$

This allows us to compute the moments of X_t in terms of X_0 and N_t.

Exercise 3.3. Prove (22).

We can regard duality relations of this kind as a means of calculating moment. Note that since N_t is a nonincreasing process, (22) implies that the nth moment of X_t only depends on X_0^k for $k \leq n$. Physicists and others often compute moments by finding systems of differential equations which they solve recursively. These equations are called closed if the derivative of the nth moment only depends on kth moments for $k \leq n$.

Exercise 3.4. Show that in the above example, the moment equations are closed. Modify this example so that the moment equations are not closed.

If the moment equations are not closed, we may still have a duality relationship, but the process N_t may not be nonincreasing. Thus, even if a system of moment equations is not closed, we may still be able to find a dual process and draw certain conclusions. One may view duality arguments as an attractive packaging for the idea of moment equations.

One important use of duality is the study of uniqueness. Suppose that processes X_t, Y_t are in duality with respect to the function $H(x, y)$. This relationship is often enough to show that X_t is unique in law, at least if $X_0 = x_0$ is specified. Indeed, let $X_t, \tilde{X}_t$ be two processes with the same initial value x_0, and assume that both $X_t, \tilde{X}_t$ are in dual to Y_t with respect to $H(x, y)$. Then

$$E\left[H(\tilde{X}_t, Y_0)\right] = E\left[H(\tilde{X}_0, Y_t)\right]$$

$$= E\left[H(X_0, Y_t)\right] \qquad (23)$$

$$= E\left[H(X_t, Y_0)\right].$$

If (23) is true for many initial values Y_0, we can often conclude that X_t and $\tilde{X}_t$ are equal in law. Using the Markov property (if X_t is Markov) and repeating this procedure gives us the uniqueness of the finite dimensional distributions. Then the Kolmogorov extension theorem shows that X_t is unique in law.

3.4 The Voter Model and the Contact Process

The voter model and contact process were some of the first models studied in particle systems. Although interest has largely shifted to other models, these processes give a useful introduction to duality. For more details, see [9] or [21; 22].

First we describe the voter model. In keeping with the style of these notes, our description is informal. The state space is the set of functions $F : \mathbf{Z}^d \to \{0,1\}$, with d fixed. We imagine that there is a voter at each site of $\mathbf{Z}^d$, whose opinion is either 0 (Republican) or 1 (Democrat). Let $\mathbf{E}^d$ denote the set of directed edges connecting nearest neighbors of $\mathbf{Z}^d$. In other words, each nearest neighbor pair $p, q \in \mathbf{Z}^d$ is associated with two edges, which either point from p to q or vice versa. If the edge e points from p to q, we set $e_0 = p$ and $e_1 = q$. To each edge $e \in \mathbf{E}^d$ we associate an independent rate-one Poisson process N_t^e. At the times τ_e of the Poisson process N_t^e, we change the opinion of e_1 to equal the opinion at e_0, that is, we redefine $F_{\tau_e}(e_1)$ to equal $F_{\tau_e}(e_0)$. Another common notation for the voter model involves the set ξ_t of sites $p \in \mathbf{Z}^d$ where voters have opinion 1, i.e., $F_t(p) = 1$. One needs to show that this procedure gives a consistent definition of a stochastic process, and this is done in [22] and other places.

The dual process of the voter model is given in terms of the ancestry of opinions. Fix a site $p \in \mathbf{Z}^d$ and a time $T > 0$. We wish to define a process X_t^p, which traces the history of the opinion found at position p at time T. We will let time t run backwards in the definition of X_t^p, so that t measures the amount of time before T. To be specific, let τ_e be the most recent Poisson time involving an edge e with $e_1 = p$. For $0 \le t < T - \tau_e$ let $X_t^p = p$, and let $X_{T-\tau_e} = e_0$. Now we repeat the construction for the site e_0 and time τ_e, until $t = T$.

Considering our definition of the voter model in terms of Poisson events, we see that X_t^p is a continuous-time nearest neighbor simple random walk on $\mathbf{Z}^d$, with transition rate $2d$. In fact, $\{X_t^p : p \in \mathbf{Z}^d\}$ is a collection of independent coalescing random walks. That is, they evolve independently, but each pair of random walks merge as soon as they land on the same site. One can also view this process in terms of the directed edges. X_t^p starts at p, and moves backwards through time starting at time T. Whenever it encounters an edge e, it moves from e_1 to e_0 if it finds itself at e_1. So, it moves along the edges in the reverse orientation.

Exercise 3.5. Convince yourself that our description of the dual for the voter model is correct.

This form of duality can also be expressed in terms of a function $H(A, \xi)$. Here, ξ_t is the set of sites with opinion 1 at time 0, and A_t is the set of sites where particles of the coalescing system of random walks are found at time t.

Exercise 3.6. What is the duality function H which matches our previous description? The answer is in [9, p. 23].

A beautiful application of duality for the voter model is the study of clustering. Clustering means that for a set of sites $S \subset \mathbf{Z}^d$, if time t is large, then with high probability all of the sites have the same opinion. This will certainly happen if opinions at the sites in S all come from a common ancestor. For simplicity, consider the case of 2 sites, $S = \{x, y\}$. Certainly $\xi_t(x) = \xi_t(y)$ if in the dual process, the random walks starting at x and y have coalesced by time t. But in dimensions 1 or 2, this coalescence occurs with probability 1. The probability is less than one in higher dimensions, and it is not hard to show that clustering does not occur in higher dimensions. We leave these details to the reader, who can also consult [9].

3.5 The Biased Voter Model

We might modify the voter model by assuming that one opinion is stronger than another. Using the same notation as for the unbiased voter model, for each directed edge e we construct 2 Poisson processes $N_t^{e,1}, N_t^{e,2}$ with rates λ_1, λ_2 respectively. Recall that the edge e points from the e_0 to e_1. Suppose that at time t there is an event of the Poisson process $N_t^{e,1}$. Then, at time t the point e_1 takes on the opinion at e_0. Secondly, suppose that at time t there is an event of the Poisson process $N_t^{e,2}$. In this case, if e_0 has opinion 1, then e_1 changes its opinion to 1 as well. On the other hand, if e_0 has opinion 0 at time t, then nothing happens. Thus, the opinion 1 is stronger than opinion 0. As of today (January 2006) we could say that 1 means the voter is a democrat.

Exercise 3.7. Verify that the ancestry of opinions is as described below, and construct an appropriate dual process.

The path of ancestry of the opinion at position x at time t should go backward in time, and if for an edge e pointing toward x the process $N_t^{e,1}$ has an event, it should follow that edge in the reverse direction to a new site. On the other hand, if $N_t^{e,2}$ has an event, then the ancestral path should split, with one branch staying at the same point, and another branch following the edge e backwards. If at time 0 the cloud of ancestral particles meets a 1, then the original site has opinion 1. Otherwise it has opinion 0. The various branches of the ancestry are meant to sample all possible occurrences of 1.

3.6 The Contact Process

In the contact process, there are particles at various sites in $\mathbf{Z}^d$. Particles die at exponential rate 1. At rate $\lambda > 0$, they give birth. When a particle at site x gives birth, the new particle chooses a site y at random among the nearest neighbors of x. If site y is already occupied, then the birth is aborted. Otherwise, the new particle moves to position y.

Exercise 3.8. Using the preceding ideas, verify that tracing the ancestry of particles gives us another contact process, and hence the contact process is self-dual.

3.7 The Dawson–Watanabe Superprocess

The Dawson–Watanabe superprocess, which we will simply call the superprocess, arose as a limit in the theory of spatial branching processes, more specifically in population biology. It is one of the few nonlinear SPDE with solutions in the space of generalized functions or Schwartz distributions. There are several good introductions to superprocesses; [20] is a classic of clear exposition. See also Etheridge's book [11] and the surveys of Dawson [6] and Perkins [33], to name a few. Since there are so many good sources, we will not systematically develop the theory of superprocesses, but rather describe the applications to SPDE.

Here is the intuition. Let μ be a given finite nonnegative measure on $\mathbf{R}^d$; the finiteness condition can be weakened. Fix a natural number m, and let $\{B_t^{(i)}\}_{i=1}^{N(t)}$ be a collection of critical branching Brownian motions taking values in $\mathbf{R}^d$. We assume that the Brownian particles are independent. $N(t)$ is the number of particles existing at time t, and critical branching means that each particle splits in two or dies with equal probability. We assume that the times between branching are independently distributed exponential variables with mean $1/m$. We define a measure-valued process by putting a delta function at the location of each particle, and then dividing by m:

$$X_t^{(m)}(A) = \frac{1}{m} \sum_{i=1}^{N(t)} \delta_{B_t^{(i)}}(A). \tag{24}$$

Here δ_x is the delta measure centered at x. Suppose that $X_0^{(m)}$ converges weakly to μ as $m \to \infty$. The main existence theorem for superprocesses asserts that in the appropriate topology, $X_t^{(m)}$ converges weakly to a limiting process X_t.

The limiting superprocess X_t has many fascinating properties. For example, in $\mathbf{R}^d$ with $d \geq 2$, with probability one X_t is a measure whose support has Hausdorff dimension 2. However, in $\mathbf{R}^2$ the measure X_t is singular with respect to Lebesgue measure. In fact, Perkins has determined the exact Hausdorff measure function, and we can loosely say that the Hausdorff dimension of the support is infinitesimally less than 2, meaning that the exact Hausdorff measure function is x^2 with extra logarithmic terms. These properties and more can be found in [6] and [33].

One important tool is the martingale problem formulation. For appropriate functions φ on $\mathbf{R}^d$, we denote

$$Z_t(\varphi) = X_t(\varphi) - \int_0^t \frac{1}{2} X_s(\Delta \varphi)\, ds. \tag{25}$$

Then $Z_t(\varphi)$ is a continuous martingale with quadratic variation

$$\langle Z \rangle_t = \int_0^t X_s(\varphi^2)\, ds. \tag{26}$$

If ν is a measure, we write $\nu(\varphi) = \int \varphi(x)\,\nu(dx)$. The martingale problem allows us to use Ito calculus. Indeed, since $Z_t(\varphi)$ is a continuous martingale with quadratic variation given by (26), we can use Ito calculus for $Z_t(\varphi)$. But (25) gives $X_t(\varphi)$ in terms of $Z_t(\varphi)$.

Even more important than the martingale problem is the Laplace functional. It is an expansion of the duality relation for Feller's diffusion explained in Subsection 3.2. Suppose that $v(t,x)$ satisfies

$$\frac{\partial v}{\partial t} = \frac{1}{2}\Delta v - \frac{1}{2}v^2 \tag{27}$$

We can solve (27) for a wide variety of initial functions $v(0,x)$, but suppose $v(0,x)$ is bounded and nonnegative, say $0 \le v(0,x) \le M$. Then we can replace (27) by

$$\frac{\partial v}{\partial t} = \frac{1}{2}\Delta v - \frac{M}{2}v \tag{28}$$

which is a linear equation and can be solved by standard PDE theory.

Exercise 3.9. Show that if $0 \le v(0,x) \le M$, then a solution of (28) also solves (27).

Under the appropriate conditions on $v(0,x)$ and X_0, for $0 \le t \le T$,

$$M_t = \exp\left(-X_t(v(T-t,\cdot))\right) \tag{29}$$

is a martingale. For example, we could assume that $v(0,x)$ is nonnegative and bounded.

Exercise 3.10. Under the assumption that $v(0,x)$ is nonnegative and bounded, show that M_t is a martingale.

Therefore

$$\mathrm{E}\left[\exp\left(-X_T(v(0,\cdot))\right)\right] = \exp\left(-X_0(v(T,\cdot))\right). \tag{30}$$

In other words, X_t and $v(t,\cdot)$ are in duality with respect to the function

$$H(X,v) = \exp\left(-X(v)\right). = \exp\left(-\int v(x)\,X(dx)\right). \tag{31}$$

Exercise 3.11. Verify the above duality relation.

This duality relation gives very useful information about the Laplace transform of X_t, and also proves uniqueness in law for the superprocess.

For $x \in \mathbf{R}^1$, the superprocess has a density $X_t(dx) = u(t,x)dx$ which satisfies

$$\frac{\partial u}{\partial t} = \frac{1}{2}\Delta u + \sqrt{u}\,\dot{W}(t,x). \tag{32}$$

Thus, we have uniqueness in law for this equation. Almost sure uniqueness is an unsolved problem which has attracted the attention of many of the best probabilists, and I have heard at least two announcements of false proofs. The lack of Ito's lemma hurts us here.

3.8 Branching Brownian Motion and a Population Equation

Consider a population with two types of genes, and let $u(t, x)$ be the population density of one type of individual at time t at position x. We assume that individuals perform independent Brownian motions, so the population density might be modeled by the heat equation. Due to mating between individuals, there might be a random contribution to the population density. The variance of this random contribution should be proportional to the product of the two population densities, namely $u(1 - u)$. Therefore, its standard deviation should be $\sqrt{u(1 - u)}$. This leads us to the following model on $t \geq 0$, $x \in \mathbf{R}$.

$$\frac{\partial u}{\partial t} = \Delta u + \sqrt{u(1 - u)}\, \dot{W}(t, x)$$

$$u(0, x) = u_0(x),$$

(33)

where $0 \leq u_0(x) \leq 1$. If there were no dependence on x, this equation would be identical to (19). Using the duality we derived for (19), it is not hard to guess that $u(t, x)$ will be dual to a system of Brownian motions $\{B_i(t)\}_{i=1}^{N(t)}$, where each pair of particles $B_i(t), B_j(t)$ coalesce at exponential rate 1, measured with respect to the local time at 0 for the process $B_i(t) - B_j(t)$. To be specific, let τ be an independent exponential variable with parameter 1, and let $\ell_{i,j}(t)$ be the local time at 0 of $B_i(t) - B_j(t)$. If there were no other particles, then the particles $B_i(t), B_j(t)$ would coalesce at time t for which $\ell_{i,j}(t) = \tau$. One has the duality relation

$$H\left(u, \{b_i\}_{i=1}^N\right) = \prod_{i=1}^N (1 - u(b_i)),$$

(34)

so that

$$\mathrm{E}\left[\prod_{i=1}^{N(0)} \left(1 - u(t, B_i(0))\right)\right] = \mathrm{E}\left[\prod_{i=1}^{N(t)} \left(1 - u(0, B_i(t))\right)\right].$$

(35)

See [37] for details. Since we can choose $\{B_i(0)\}_{i=1}^{N(0)}$, this gives us a formula for the moments of $u(t, x)$. Notice that the moment equations are closed, since there cannot be more particles at time t than at time 0.

Exercise 3.12. Compute the moments of u.

Also, if $u(0, x) \approx 1$, then the right side of (35) is close to 0 if there are any particles near x. This gives us a way of relating the size of $u(t, x)$ to the probabilities of the Brownian particles.

Among other things, this duality relation gives uniqueness in law for (37). In [27], this duality was used to study the width $D(t)$ of the interface where $0 < u(t, x) < 1$, assuming that $u(0, x) = \mathbf{1}_{(-\infty, 0]}(x)$. This interface was proved to have stochastically finite width, that is

$$\lim_{\lambda \to \infty} \sup_{t \geq 0} P(D(t) > \lambda) = 0. \tag{36}$$

At about the same time, and independently, Cox and Durrett [5] proved a similar result for the long-range voter model. They also used duality.

3.9 Branching Coalescing Brownian Motion and the KPP Equation with Noise

The KPP equation is one of the simplest equations exhibiting traveling wave solutions.

$$\frac{\partial u}{\partial t} = \Delta u + u(1 - u),$$
$$u(0, x) = u_0(x). \tag{37}$$

Often, one takes $u_0(x) = \mathbf{1}_{(-\infty,0]}(x)$. One can prove that there is a function $h(x)$ with $\lim_{x \to \infty} h(x) = 0$ and $\lim_{x \to -\infty} h(x) = 1$, and a function $v(t)$ satisfying $\lim_{t \to \infty} v(t)/t = 2$ for which

$$\lim_{t \to \infty} \sup_{x \in \mathbf{R}} |u(t, x) - h(x - v(t))| = 0. \tag{38}$$

Detailed properties of this equation have been derived by Bramson [1; 2] using the Feynman-Kac formula. Suppose that $\{B_i(t)\}_{i=1}^{\infty}$ is a collection of Brownian motions. At exponential rate 1, each particle splits in two. Then $u(t, x)$ and $\{B_i(t)\}_{i=1}^{\infty}$ are in duality, and the duality function $H(\cdot, \cdot)$ is the same as in (34). As a thought exercise, the reader may wish to verify that the cloud of particles spread at about the same rate as $u(t, x)$. Hint: fix $x_0 > 0$, and let T be the first time t that $u(t, x) = 1/2$. Start a single particle at position x_0 at time 0, and write down the duality equation.

If we consider the population model in the previous section, but suppose that one type kills the other, there might be a drift term proportional to the frequency of interactions, which is proportional to $u(1 - u)$. This would give us the equation

$$\frac{\partial u}{\partial t} = \Delta u + u(1 - u) + \sqrt{u(1 - u)}\, \dot{W}(t, x),$$
$$u(0, x) = u_0(x). \tag{39}$$

Combining the two previous types of duality, we might conjecture that $u(t, x)$ is dual to a system of Brownian particles, in which pairs of particles $B_i(t), B_j(t)$ coalesce at rate 1 according to the local time where $B_i(t) - B_j(t) = 0$, and each particle splits in two at an exponential rate with parameter 1.

This kind of duality is not easy to work with, but it does prove uniqueness for (39). In [26], the traveling wave behavior for (39) was studied.

4 Large Deviations for SPDEs

Roughly speaking, large deviations measures how far the solution of an SPDE can get from the solution of the corresponding PDE. Either noise or time is taken to be small. From another point of view, we might wish to see how large the solution of an SPDE can be. If we know how large the solution of the corresponding PDE is, then large deviations can give an upper bound for the SPDE.

There are many excellent books on large deviations; our goal here is to give an overview with emphasis on intuition. To start at the basic level, we give a tail estimate for a $N(0,1)$ random variable Z.

$$P(Z > \lambda) = \int_\lambda^\infty \frac{1}{\sqrt{2\pi}} \exp\left(-\frac{x^2}{2}\right) dx. \tag{40}$$

Note that for $a > 0$ fixed,

$$\lim_{x \to \infty} \frac{\exp\left(-(x+a)^2/2\right)}{\exp\left(-x^2/2\right)} = 0. \tag{41}$$

So it seems reasonable that for large λ, the integral in (40) will be dominated by values of x very close to λ, since the other values of the integrand are very small in comparison. Then as $\lambda \to \infty$,

$$-\log P(Z > \lambda) \sim -\log\left(\frac{1}{\sqrt{2\pi}} \exp\left(-\frac{\lambda^2}{2}\right)\right) \sim \frac{\lambda^2}{2}, \tag{42}$$

in the sense that the ratio of the two sides tends to 1.

Exercise 4.1. Prove (42).

The large deviations philosophy states that as some parameter tends to 0 or ∞, the probability in question will be determined by a single point in the probability space, along with a small neighborhood around it. In our example, the point would be where $Z = \lambda$.

For SPDE, consider the following setup. Let

$$N(t,x) = \int_0^t \int_{\mathbf{R}} G(t-s, x-y) f(s,y) W(dy\,ds), \tag{43}$$

where $f(s,y)$ is a predictable random function, with the almost sure bound

$$\sup_{s,y} |f(s,y)| \le K. \tag{44}$$

Here is our large deviations theorem.

Theorem 4.2. *Let $M > 0$. There exist constants $C_1, C_2 > 0$ such that for all $T, K, \lambda > 0$,*

$$P\left(\sup_{0 \le t \le T} \sup_{|x| \le M} |N(t,x)| > \lambda\right) \le C_1 \exp\left(-\frac{C_2 \lambda^2}{T^{1/2} K^2}\right). \tag{45}$$

Proof (Theorem 4.2). The reader should go back to Chapter 1, Chapter 4.2 of this book to see the similarities between Theorem 4.2 and Kolmogorov's continuity theorem. Theorem 4.2 can be proved from the Garsia–Rodemich–Rumsey lemma [14], and the proof has much in common with similar ideas from Gaussian processes. We prefer to give a proof from first principles, which duplicates part of the proof of the Garsia–Rodemich–Rumsey lemma.

We need the fact that $N(t,x)$ is continuous with probability 1. Actually, this can be deduced from our proof below, but for simplicity, we refer the reader to [38].

Next, observe that by scaling we can assume that $K = 1$. By cutting up the x-axis into intervals of size 1 and adding the corresponding estimates, we can reduce to the case where the supremum over x is taken on the interval $[0,1]$.

Furthermore, G and W have the following scaling for $x \in \mathbf{R}$.

$$aG(a^2 t, ax) = G(t,x),$$
$$W(d(ax), d(a^2 t)) \overset{\mathscr{D}}{=} a^{3/2} W(dx, dt). \tag{46}$$

By taking $a = t^{-1/2}$ and using the above scaling and setting $t = T$, the reader can verify that we need only prove the theorem for $T = 1$.

To summarize, we must show that if

$$\sup_{t,x} |f(t,x)| \le 1 \quad \text{almost surely,} \tag{47}$$

then there exist constants C_1, C_2 such that for all $\lambda > 0$ we have

$$P\left(\sup_{0 \le t \le T} \sup_{0 \le x \le 1} |N(t,x)| > \lambda\right) \le C_1 \exp\left(-C_2 \lambda^2\right). \tag{48}$$

Recall that we have reduced to $0 \le x \le 1$ by chopping up the interval $[-M, M]$.

To prove (48), we need the following estimates.

Lemma 4.3. *There exists a constant C such that for all $0 < s < t < 1$ and $x, y \in [-1, 1]$ we have*

$$\int_0^t \int_{\mathbf{R}} [G_t(x - z) - G_t(y - z)]^2 \, dz \, ds \le C|x - y|,$$

$$\int_s^t G_{t-r}(z)^2 \, dz \, dr \le C|t - s|^{1/2}, \tag{49}$$

$$\int_0^s [G_{t-r}(z) - G_{s-r}(z)]^2 \, dz \, dr \le C|t - s|^{1/2}.$$

The proof of Lemma 4.3 is an exercise in calculus or perhaps real analysis.

Exercise 4.4. Verify Lemma 4.3.

The details can also be found in [38]. Observe that Lemma 4.3 has the following corollary.

Corollary 4.5. *Assume that* $|f(t,x)| \leq 1$ *almost surely for all* t, x. *Then there exist constants* C_1, C_2 *such that for all* $0 < s < t < 1$, $x, y \in [-1, 1]$, *and* $\lambda > 0$,

$$P\left(|N(t,x) - N(t,y)| > \lambda\right) \leq C_1 \exp\left(-\frac{C_2\lambda^2}{|x - y|}\right),$$

$$P\left(|N(t,x) - N(s,x)| > \lambda\right) \leq C_1 \exp\left(-\frac{C_2\lambda^2}{|t - s|^{1/2}}\right). \tag{50}$$

Proof. We prove only the first assertion of Corollary 4.5, leaving the second assertion to the reader. Let

$$\bar{N}_t(s,x) = \int_0^s \int_{\mathbf{R}} G(t - r, x - y) f(r, y) W(dy\, dr), \tag{51}$$

and note that $\bar{N}_t(t, x) = N(t, x)$. In other words, we have frozen the variable t which occurs inside G, in order for the stochastic integral to be a martingale. Let

$$M_s = \bar{N}_t(s, x) - \bar{N}_t(s, y) \tag{52}$$

and note that $M_t = N(t, x) - N(t, y)$. Thus M_s is a continuous martingale and hence a time-changed Brownian motion, see [36]. By Lemma 4.3, we have

$$\langle M \rangle_t \leq C|x - y|. \tag{53}$$

Readers should convince themselves that the time scale of the time changed Brownian motion is given by $\langle M \rangle_t$. In other words, there is a Brownian motion B_t such that $M_t = B_{\langle M \rangle_t}$. Hence by the reflection principle for Brownian motion,

$$P(N(t,x) - N(t,y) > \lambda) \leq P(B_{C|x-y|} > \lambda)$$

$$\leq C_1 \exp\left(-\frac{C_2\lambda^2}{|x - y|}\right). \tag{54}$$

The assertion in Corollary 4.5 then follows by making a similar estimate for $P(-N(t,x) + N(t,y) > \lambda)$. □

Continuing with the proof of Theorem 4.2, we define the grid

$$\mathcal{G}_n = \left\{\left(\frac{j}{2^{2n}}, \frac{k}{2^n}\right) : 0 \leq j \leq 2^{2n}, \ 0 \leq k \leq 2^n\right\}. \tag{55}$$

We write

$$\left(t_j^{(n)}, x_k^{(n)}\right) = \left(\frac{j}{2^{2n}}, \frac{k}{2^n}\right).$$ (56)

Two points $(t_{j_i}^{(n)}, x_{k_i}^{(n)}) : i = 1, 2$ are nearest neighbors if either

1. $j_1 = j_2$ and $|k_1 - k_2| = 1$, or
2. $|j_1 - j_2| = 1$ and $k_1 = k_2$

Claim 1 *Let $p \in \mathscr{G}_n$ for some n. There exists a path $(0,0) = p_0, p_1, \ldots, p_N = p$ of points in $\mathscr{G}_n$ such that each pair p_{i-1}, p_i are nearest neighbors in some grid $\mathscr{G}_m, m \leq n$. Furthermore, at most 4 such pairs are nearest neighbors in any given grid $\mathscr{G}_m$.*

We will prove a one-dimensional version of the claim for $x \in [0, 1]$, and leave the rest to the reader. Let $\mathscr{H}_n = \{k/2^n : k = 0, \ldots, 2^n\}$ denote the dyadic rational numbers of order n lying in $[0, 1]$.

Claim 2 *Let $x \in \mathscr{H}_n$. There exists a path $0 = p_0, p_1, \ldots, p_N = x$ of points in $\mathscr{H}_n$ such that each pair p_{i-1}, p_i are nearest neighbors in some grid $\mathscr{H}_m, m \leq n$. Furthermore, at most one such pair consists of points which are nearest neighbors in any given grid $\mathscr{H}_m$.*

Proof (Claim 2). Let

$$x = 0.x_1 x_2 \cdots x_n$$ (57)

denote the binary expansion of x, that is, its expansion in base 2, and let

$$p_m = 0.x_1 x_2 \cdots x_m.$$ (58)

Then Claim 2 follows. Claim 1 is proved using a similar argument, where we write $p = (t, x)$, take the binary expansion of x, and the base 4 expansion of t. □

Exercise 4.6. Prove Claim 1.

Next, let $K, \alpha > 0$, and let $A(n, \lambda)$ be the event that for all nearest neighbors $p, q \in \mathscr{G}_n$ we have

$$|N(p) - N(q)| \leq \lambda K 2^{-(2-\alpha)n}.$$ (59)

By Corollary 4.5, for each pair of nearest neighbors $p, q \in \mathscr{G}_n$, we have

$$P\left(|N(p) - N(q)| > \lambda K 2^{-(2-\alpha)n}\right) \leq C_1 \exp\left(-C_2 \lambda^2 2^{(2-\alpha)n}\right).$$ (60)

Since there are 2^{3n} nearest neighbors in $\mathscr{G}_n$, we have

$$\begin{aligned}
P(A^c(n, \lambda)) &\leq C_1 2^{3n} \exp\left(-C_2 \lambda^2 2^{(2-\alpha)n}\right) \\
&\leq C_3 \exp\left(-C_4 \lambda^2 2^{(2-\alpha)n}\right),
\end{aligned}$$ (61)

for appropriate constants C_3, C_4. Let $A(\lambda) = \cup_{n=0}^{\infty} A(n, \lambda)$. Summing the previous estimates over n, we get

$$P(A(\lambda)^c) \leq C_1 \exp\left(-C_2\lambda^2\right), \tag{62}$$

where C_1, C_2 might be different than before. Let $p = (t, x)$. If $A(\lambda)$ holds, then using the path $0 = p_0, \ldots, p_N = (t, x)$, we have

$$|N(t, x)| = |N(t, x) - N(0, 0)| \leq \sum_{j=1}^{N} |N(p_{j-1}) - N(p_k)|$$

$$\leq 4\sum_{j=1}^{\infty} \lambda K 2^{-(2-\alpha)j} \leq CK\lambda \leq \lambda, \tag{63}$$

with the appropriate choice of K. This proves (48) and hence finishes the proof of Theorem 4.2. $\square$

Exercise 4.7. Modify the proof of Theorem 4.2 to get the following modulus of continuity for $u(t, x)$: For all $\varepsilon, T, M > 0$ there exists $K = K(\omega, \varepsilon, T, M)$ such that for all $0 \leq s < t \leq T$ and $-M \leq x < y \leq M$, we have

$$|u(t, x) - u(s, x)| \leq K|t - s|^{1/4-\varepsilon}$$
$$|u(t, x) - u(t, y)| \leq K|x - y|^{1/2-\varepsilon}$$

5 A Comparison Theorem

The maximum principle is a powerful method for studying elliptic and parabolic equations, and it can be used to prove comparison theorems of the following type. For simplicity, let $\mathscr{C}$ be the circle $[0, 1]$ with endpoints identified. Consider two solutions $u_1(t, x)$ and $u_2(t, x)$, $t \geq 0$, $x \in \mathscr{C}$, of the heat equation

$$\frac{\partial u}{\partial t} = \Delta u. \tag{64}$$

Here we assume that $u(t, x)$ has periodic boundary conditions on $[0, 1]$, that is $u(t, 0) = u(t, 1)$ and $u_x(t, 0) = u_x(t, 1)$. Suppose that $u_1(0, x) \leq u_2(0, x)$. Then for every $t > 0$ and $x \in \mathscr{C}$, $u_1(t, x) \leq u_2(t, x)$. Indeed, let $v(t, x) = u_1(t, x) - u_2(t, x)$. Then $v(0, x)$ is nonpositive, and the maximum principle states that since v satisfies the heat equation, its maximum must be attained on the boundary of the domain, namely at $t = 0$. Thus, $v(t, x) \leq 0$ for all $t \geq 0$ and $x \in \mathscr{C}$, so $u_1(t, x) \leq u_2(t, x)$. This argument can be extended to many semilinear heat equations, see [35].

For stochastic equations, comparison principles for finite dimensional diffusions are known, see [16, Theorem 1.1, Chapter VI]. For example, suppose that $a(x), b(x)$ are Lipschitz functions, and that $x_0 \leq y_0$. Suppose that $X(t), Y(t)$ satisfy

$$dX = a(X)\, dt + b(X)\, dB,$$
$$dY = a(Y)\, dt + b(Y)\, dB,$$
$$X_0 = x_0,$$
$$Y_0 = y_0.$$
$$(65)$$

Then with probability 1, for all $t \geq 0$ we have $X_t \leq Y_t$.

Theorem 5.1. Let $a(u), b(u)$ be Lipschitz functions on **R**, and consider solutions $u_1(t,x), u_2(t,x)$, $t \geq 0$, $x \in \mathscr{C}$ to the SPDE

$$\frac{\partial u}{\partial t} = \Delta u + a(u) + b(u)\dot{W}, \qquad (66)$$

with $u_1(0,x) \leq u_2(0,x)$. Then with probability 1, for all $t \geq 0$, $x \in \mathscr{C}$, we have

$$u_1(t,x) \leq u_2(t,x). \qquad (67)$$

Proof. We will only give an outline of the proof, and only for fixed t. A special case is treated in [28, Section 3], but the proof carries over to our situation. For other approaches, see [30] and [37].

All such proofs follow the same strategy: discretize the SPDE and then use the comparison result for diffusions. Fix $N > 0$ and for $k = 1, \ldots, N$ let

$$u_{i,k,N}(0) = N \int_{k/N}^{(k+1)/N} u_i(0,x)\, dx, \qquad (68)$$

where the interval $[k/N, (k+1)/N]$ is taken to be a subset of $\mathscr{C}$. Also, let

$$B_k(t) = N^{1/2} \int_0^t \int_{k/N}^{(k+1)/N} W(dx\, ds), \qquad (69)$$

and note that $B_k(t)$ is a standard Brownian motion. Define the operator $\Delta^{(N)}$ by

$$\Delta^{(N)} u_{i,k,N} = N^2 \left(u_{i,k+1,N} - 2u_{i,k,N} + u_{i,k-1,N} \right). \qquad (70)$$

In other words, $\Delta^{(N)}$ is the discrete Laplacian. Because of our periodic boundary conditions, we let $u_{i,N,N} = u_{i,0,N}$.

We will construct $u_{i,k,N}(t)$ in stages. For $j/N^2 < t < (j+1)/N^2$ and $j \geq 0$, let $u_{i,k,N}(t)$ satisfy

$$du_{i,k,N} = a(u_{i,k,N})\, dt + N^{1/2} b(u_{i,k,N})\, dB_k. \qquad (71)$$

For $t = j/N^2$, let

$$u_{i,k,N}(t) = u_{i,k,N}(t-) + \frac{1}{N^2} \Delta^{(N)} u_{i,k,N}(t-) \qquad (72)$$

where the operator $\Delta^{(N)}$ acts on k. Finally, for

$$\left| x - \frac{k}{N^2} \right| < \frac{1}{2N^2} \tag{73}$$

let $v_{i,N}(t,x) = u_{i,k,N}(t)$. It can be shown that for any $T > 0$,

$$\lim_{N\to\infty} \mathrm{E}\left[\int_0^T \int_{\mathscr{C}} |u_i(t,x) - v_{i,N}(t,x)|^2 \, dx \, dt \right] = 0. \tag{74}$$

Exercise 5.2. Verify (74).

Furthermore, the comparison theorem for diffusions from [16, Theorem 1.1, Chapter VI], and the positivity of the operator $\Delta^{(N)}$ shows that with probability 1, $u_{1,k,N}(t) \le u_{2,k,N}(t)$. See [28, Section 3] for details.

Modulo the missing details, this gives us our comparison theorem.   □

6 Applications

We give several applications of the preceding ideas.

6.1 Blow-up

Blow-up in finite time is a well-studied property for PDE, arising in such applications as flame propagation and the shrinkage of an elastic string to a point. In this section we will show that a certain SPDE does not blow up in finite time. The basic idea is to show that if $\sup_x u(t,x)$ is large, then it is more likely to decrease than increase. This intuitive idea is implemented by comparison with a random walk whose steps have negative expectation. Probabilities are controlled using large deviations. We express the solution as a sum of two terms, bound one by large deviations, and bound the other by elementary heat kernel estimates.

Consider the following SPDE on the circle $\mathscr{C}$, which we take to be $[0,1]$ with endpoints identified. We impose periodic boundary conditions. Let $\gamma \ge 1$. Assume that $u_0(x)$ is a continuous and nonnegative function on $\mathscr{C}$.

$$\frac{\partial u}{\partial t} = \Delta u + u^\gamma \dot{W},$$
$$u(0,x) = u_0(x) \ge 0. \tag{75}$$

For this subsection, let $G(t,x,y)$ denote the heat kernel on the circle. Equivalently, we could consider $G(t,x,y)$ to be the heat kernel on $[0,1]$ with periodic boundary conditions.

Exercise 6.1. Show that

$$G(t,x,y) = \sum_{n\in\mathbf{Z}} \Big[G(t,x-y+2n) + G(t,x+y+2n) \Big], \tag{76}$$

where $G(t,x)$ is the heat kernel on $\mathbf{R}$.

As earlier, the rigorous meaning of (75) is given in terms of an integral equation. Since we are concerned about blowup, we will truncate u^γ. Let

$$
\begin{aligned}
u_N(t,x) = &\int_{\mathscr{C}} G(t,x,y)u_0(y)\,dy \\
&+ \int_0^t \int_{\mathscr{C}} G(t-s,x,y)(u_N \wedge N)^\gamma(s,y)\,W(dy\,ds).
\end{aligned}
\tag{77}
$$

Since $(u_N \wedge N)^\gamma$ is a Lipschitz function of N, the usual theory implies existence and uniqueness for solutions u_N to (77). See [38], Chapter 3, for example. Let τ_N be the first time $t \geq 0$ that $u_N(t,x) \geq N$ for some $x \in [0,1]$. Then we can construct $u_N : N = 1, 2, \dots$ on the same probability space, and the limit $\tau_N \uparrow \tau$ exists almost surely. Note that almost surely, $u_m(t,x) = u_n(t,x)$ as long as $t < \tau_m \wedge \tau_n$. Then we may define $u(t,x)$ for $t < \tau$ by setting $u(t,x) = u_n(t,x)$ whenever $t < \tau_n$ for some n, and we see that $\tau \in [0,\infty]$ is the time at which $\sup_x u(t,x) = \infty$, and $\tau = \infty$ if $u(t,x)$ never reaches ∞.

Definition 6.2. *We say that $u(t,x)$ blows up in finite time if $\tau < \infty$.*

Then we have

Theorem 6.3. *If $\gamma < 3/2$ then with probability 1, $u(t,x)$ does not blow up in finite time.*

We remark that Krylov [18] has a different proof of Theorem 6.3, including generalizations, based on his L_p theory of SPDE.

Proof (Theorem 6.3). First, we claim that

$$
U(t) := \int_{\mathscr{C}} u(t,x)\,dx
\tag{78}
$$

is a local martingale. It suffices to show that $U_N(t) := \int_0^1 u_N(t,x)\,dx$ is a martingale for each N. Of course

$$
\int_{\mathscr{C}} G(t,x,y)\,dx = 1.
\tag{79}
$$

Integrating (77) over x and using the stochastic Fubini's lemma [38], we get

$$
U_N(t) = \int_{\mathscr{C}} u_0(y)\,dy + \int_0^t \int_{\mathscr{C}} (u_N \wedge N)^\gamma(s,y)\,W(dy\,ds),
\tag{80}
$$

which is a martingale.

Since $U(t)$ is a nonnegative continuous local martingale, it must be bounded, and so

$$
U(t) \leq K = K(\omega).
\tag{81}
$$

for some $K(\omega) < \infty$ almost surely. We would like to study the maximum of $u(t, x)$ over x,

$$M(t) = \sup_{x \in \mathscr{C}} u(t, x). \tag{82}$$

Our goal is to show that $M(t)$ does not reach ∞ in finite time, with probability 1. To that end, we define a sequence of stopping times $\tau_n : n \geq 0$ as follows. Let $\tau_0 = 0$, and for simplicity assume that $M(0) = 1$. Given τ_n, let τ_{n+1} be the first time $t > \tau_n$ such that $M(t)$ equals either $2M(\tau_n)$ or $(1/2)M(\tau_n)$. The reader can easily verify that τ_n is well defined for all values of n. Next, we wish to show that if $M(\tau_n)$ is large enough, say $M(\tau_n) > 2^{N_0}$ for some $N_0 > 0$, then

$$P\left(M(\tau_{n+1}) = 2M(\tau_n) \,\Big|\, \mathscr{F}_n\right) < \frac{1}{3}. \tag{83}$$

Assuming (83), we can compare $Z_n = \log_2[M(\tau_n) - N_0]$ to a nearest-neighbor random walk R_n, with $P(R_{n+1} = R_n + 1) = 1/3$. In particular we claim that Z_n, R_n can be constructed on the same probability space such that the following comparison holds. Let σ_n be the first time n such that $Z_n \leq N_0$. Then almost surely, for all $n \leq \sigma_{N_0}$,

$$Z_n \leq R_n.$$

Such a random walk R_n always returns to 0 if $R_0 > 0$.

Exercise 6.4. Fill in the details of the above comparison.

Therefore, Z_n either visits the level 2^{N_0} infinitely often, or tends to 0. If $Z_n \to 0$, then clearly $u(t, x)$ does not blow up, either in finite or infinite time. On the other hand, suppose Z_n visits the level 2^{N_0} infinitely often, at the times $\tau_{n_k} : k = 1, 2, \ldots$. We claim that there exists a constant $C > 0$ such that $E[\tau_{n_k+1} - \tau_{n_k} \,|\, \mathscr{F}_n] > C$ for all k. This follows from the strong Markov property of solutions, namely, if $\mathscr{F}_t$ is the σ-field generated by the noise $\dot{W}(s, x)$ for time $s \leq t$, then our solution $u(t, x)$ is a Markov process. It is intuitively clear that the heat equation starts afresh at any given time, with the current solution as new initial value.

Exercise 6.5. Prove the Markov property of $u(t, x)$ from (77).

Exercise 6.6. Give a full proof of the existence of the constant C mentioned above.

Therefore $\lim_{t \to \infty} \tau_{n_k} = \infty$, and so with probability 1, $u(t, x)$ does not blow up in finite time.

Now we turn to the proof of (83). It suffices to consider $u_{2^m+1}(t, x)$. Suppose that $M(\tau_n) = 2^m$ and let $v(t, x) = u_{2^m+1}(\tau_n + t, x)$. Using the strong Markov property for $u(t, x)$, and the fact that $v(t, x) \leq 2^{m+1}$, we have

$$v(t,x) = \int_{\mathscr{C}} G(t,x,y)v(0,y)\,dx$$

$$+ \int_0^t \int_{\mathscr{C}} G(t-s,x,y)v(s,y)^\gamma W(y,s)\,dy\,ds \tag{84}$$

$$=: V_1(t,x) + V_2(t,x).$$

Note that by the maximum principle, $\sup_x V_1(t,x)$ is nonincreasing. Our goal is to choose a nonrandom time T such that for m large enough,

$$P\left(\sup_{0 \le t \le T} \sup_{x \in \mathscr{C}} V_2(t,x) > 2^{m/4}\right) < \frac{1}{3} \tag{85}$$

and

$$\sup_{x \in \mathscr{C}} V_1(t,x) \le 2^{m/4}. \tag{86}$$

Exercise 6.7. Check that (85) and (86) imply (83).

Now for the heat kernel $G(t,x)$ on $x \in \mathbf{R}$,

$$G(t,x-y) = \frac{1}{\sqrt{4\pi t}} \exp\left(-\frac{(x-y)^2}{4t}\right) \le Ct^{-1/2} \tag{87}$$

for $C = (4\pi)^{-1/2}$.

Exercise 6.8. Use (76) and (87) to show that for some constant $C > 0$, the heat kernel $G(t,x,y)$ on $\mathscr{C}$ satisfies

$$G(t,x,y) \le Ct^{-1/2} \tag{88}$$

for all $t > 0$, $x,y \in \mathscr{C}$.

So for all $x \in \mathscr{C}$ we have

$$V_1(t,x) \le Ct^{-1/2} \int_{\mathscr{C}} v(t,y)\,dy \le CKt^{-1/2}, \tag{89}$$

where $K = K(\omega)$ was our upper bound for the integral of $u(t,x)$ over x.
Now choose K_0 such that if $\mathscr{A} = \{K(\omega) > K_0\}$ then

$$P(\mathscr{A}) < \frac{1}{6}. \tag{90}$$

Also choose

$$T = C^2 K_0^2 2^{-m/2-6}. \tag{91}$$

This choice of T gives us (86).
Next we prove (85). Note that

$$u_{2^{m+1}}(t,x)^\gamma \le C_3 2^{\gamma m} \tag{92}$$

We ask the reader to believe that Theorem 4.2 also holds for equations on $\mathscr{C}$. If this is granted, we have

$$\mathrm{P}\left(\sup_{0 \leq t \leq T} \sup_{x \in \mathscr{C}} V_2(t,x) > 2^{m/4}\right)$$

$$\leq \mathrm{P}(\mathscr{A}^c) + C_0 \exp\left(-\frac{C_1(2^{m/4})^2}{T^{1/2}(C_3 2^{\gamma m})^{1/2}}\right) \qquad (93)$$

$$\leq \frac{1}{6} + C_0 \exp\left(-C_4 2^{(m/2)(1-\gamma+(1/2))}\right)$$

But if $\gamma < 3/2$ then $1 - \gamma + (1/2) > 0$, and the above probability is less than $1/6$ for $m > m_0$ and m_0 large enough.

This verifies (86), and finishes the proof of Theorem 6.3. □

6.2 Hitting Zero

Next we determine the critical drift for an SPDE to hit 0, given that the initial function is strictly positive and bounded away from 0. The argument is similar to the proof of Theorem 6.3. We show that if $\inf_x u(t,x)$ is small, then it is more likely to increase than decrease. We express the solution as a sum of two terms, bound one by large deviations, and bound the other by elementary heat kernel estimates. For $t > 0, x \in \mathscr{C}$, we consider solutions $u(t,x)$ to

$$\frac{\partial u}{\partial t} = \Delta u + u^{-\alpha} + \dot{W}(t,x)$$

$$u(0,x) = u_0(x) \qquad (94)$$

where $0 < c < u_0(x) < C$ for some constants c, C. The term $u^{-\alpha}$ is singular at $u = 0$, so once again we must restrict to $0 < t < \tau$ where τ is the first time t such that $u(t,x) = 0$. If there is no such time, we let $\tau = \infty$.

Theorem 6.9. *Let $\mathscr{A}$ be the event that $\tau < \infty$, that is, u hits 0 in finite time.*

1. If $\alpha > 3$ then $\mathrm{P}(\mathscr{A}) = 0$.
2. If $\alpha < 3$ then $\mathrm{P}(\mathscr{A}) > 0$.

Proof. We will only prove assertion (1). For (2) the reader can consult [25].

As in the previous section, our strategy is to compare $\inf_x \log_2 u(t,x)$ to a nearest-neighbor random walk. To verify assertion (1), we need to show that for u small enough, the random walk has a higher probability of moving up than down.

Let $I(t) = \inf_{x \in \mathscr{C}} u(t,x)$. We construct a sequence of stopping times τ_n : $n \geq 0$. Let $\tau_0 = 0$. Given τ_n, let τ_{n+1} be the first time $t > \tau_n$ such that $I(t) = 2I(\tau_n)$ or $I(t) = (1/2)I(\tau_n)$. Let $Z_n = \log_2 I(\tau_n)$. Fix n, and let $v(t,x) = u(\tau_n + t, x)$. Let $\mathscr{A}_n$ be the event that $Z_{n+1} = Z_n - 1$. We claim that for Z_n small enough,

$$P\left(\mathscr{A}_n^c \mid \mathscr{F}_{\tau_n}\right) < \frac{1}{3}. \tag{95}$$

First, by the comparison principle of Theorem 5.1, it is enough to show (95) for $v(0,x) = I(\tau_n)$. Let $I(\tau_n) = 2^{-m}$. Using the strong Markov property, we can write

$$\begin{aligned}
v(t,x) = &\int_{\mathscr{C}} G(t,x,y)v(0,y)\,dx \\
&+ \int_0^t \int_{\mathscr{C}} G(t-s,x,y)v(s,y)^{-\alpha}\,dy\,ds \\
&+ \int_0^t \int_{\mathscr{C}} G(t-s,x,y)W(dy\,ds)\,dy\,ds \\
=: &\, 2^{-m} + V_1(t,x) + V_2(t,x).
\end{aligned} \tag{96}$$

Fix $\delta > 0$, and let

$$T = 2^{-4m-2\delta}. \tag{97}$$

Let $\mathscr{B} = \mathscr{B}(T,\tau_n)$ be the event that

$$\sup_{0 \le t \le T}\, \sup_{x \in \mathscr{C}} |V_2(t,x)| \le 2^{-m-1}. \tag{98}$$

By Theorem 4.2 (for $\mathscr{C}$), we have

$$\begin{aligned}
P\left(\mathscr{B}^c\right) &\le C_0 \exp\left(-\frac{C_1(2^{-m-1})^2}{T^{1/2}}\right) \\
&\le C_0 \exp\left(-C_2 2^{\delta m}\right) \\
&< \frac{1}{3},
\end{aligned} \tag{99}$$

if m is large enough.

We claim that on $\mathscr{B}$,

$$\sup_{0 < t < T}\, \sup_{x \in \mathscr{C}} |V_1(t,x)| \le 2^{m-1}. \tag{100}$$

Indeed, let $v_m(t,x)$ satisfy

$$\begin{aligned}
v_m(t,x) = &\int_{\mathscr{C}} G(t,x,y)v(0,y)\,dx \\
&+ \int_0^t \int_{\mathscr{C}} G(t-s,x,y)\left(v_m(s,y) \vee 2^{-m-1}\right)^{-\alpha}\,dy\,ds \\
&+ \int_0^t \int_{\mathscr{C}} G(t-s,x,y)\,W(dy\,ds).
\end{aligned} \tag{101}$$

Then $v_m(t,x) = v(t,x)$ for all $x \in \mathscr{C}$ and for all $0 < t < \sigma_m$, where σ_m is the first time s that

$$\inf_{x \in \mathscr{C}} v_m(s\,,x) \le 2^{-m-1}. \tag{102}$$

However, on the event $\mathscr{B}$, since $v(0\,,x) \equiv 2^{-m}$, we see that

$$\inf_{0<t<T} \inf_{x \in \mathscr{C}} v_m(t\,,x) \ge 2^{-m} - \sup_{0<t<T} \sup_{x \in \mathscr{C}} |V_1(t\,,x)|$$
$$\ge 2^{-m-1}, \tag{103}$$

by the definition of $\mathscr{B}$. Therefore, on $\mathscr{B}$, $v_m(t\,,x) = v(t\,,x)$ for $0 < t < T$ and $x \in \mathscr{C}$. It follows that on $\mathscr{B}$, $v(t\,,x) \ge 2^{-m-1}$ for $0 < t < T$ and $x \in \mathscr{C}$. It follows that on $\mathscr{B}$ and for $0 < t < T$ and $x \in \mathscr{C}$,

$$v(t\,,x) \le \int_{\mathscr{C}} G(t\,,x\,,y) v(0\,,y)\,dx$$
$$+ \int_0^t \int_{\mathscr{C}} G(t-s\,,x\,,y) \left(2^{-m-1}\right)^{-\alpha} dy\,ds$$
$$+ \int_0^t \int_{\mathscr{C}} G(t-s\,,x\,,y)\,W(dy\,ds) \tag{104}$$
$$\le 2^{-m} + CT 2^{m\alpha} + 2^{-m-1}$$
$$< 2^{-m+1}.$$

for m large enough. Thus, if m is large enough,

$$\mathrm{P}\left(Z_{n+1} = Z_n + 1 \,\Big|\, \mathscr{F}_n\right) \le \mathrm{P}(\mathscr{B}) < \frac{1}{3}. \tag{105}$$

This proves the comparison of Z_n with a random walk, and finishes the proof of Theorem 6.9. □

Remark 6.10. Recently, Zambotti [39; 40] has found remarkable connections between the problem of hitting 0 and the random string, which is a vector-valued solution of the heat equation with noise.

6.3 Compact Support

We all know that nonnegative solutions to the heat equation which are not identically zero have support on all of $\mathbf{R}^d$. It is therefore of great interest that certain stochastic heat equations have solutions with compact support. Intuitively, at the edge of the support the noise term dominates, so it has a chance to push the solution to 0.

Consider solutions $u(t\,,x)$, $0 \le t < \infty$, $x \in \mathbf{R}$, to

$$\frac{\partial u}{\partial t} = \Delta u + u^\gamma \dot{W}(t\,,x)$$
$$u(0\,,x) = u_0(x), \tag{106}$$

where $u_0(x)$ is a continuous nonnegative function of compact support, and

$$\frac{1}{2} < \gamma < 1. \tag{107}$$

By an ingenious argument using duality, Mytnik [29] has proved uniqueness in law. Actually, approximate duality is used, not exact duality. Roughly speaking, the dual process $v(t,x)$ satisfies

$$\frac{\partial v}{\partial t} = \Delta v + v^{1/\gamma} \dot{L}(t,x)$$
$$v(0,x) = v_0(x) \geq 0, \tag{108}$$

where $\dot{L}(t,x)$ is a one-sided Lévy noise with positive jumps, of index 2γ. For details, see [29]. Mytnik's duality relation is

$$H(u,v) = \exp\left(-\int_{\mathbf{R}} u(x)v(x)\,dx\right), \tag{109}$$

which is one of the standard duality functions. (At least this last part is easy to guess).

Theorem 6.11. *With probability 1, $u(t,x)$ has compact support in x for all $t \geq 0$.*

Proof. We will mainly discuss the proof for $\gamma = 1/2$, which is much easier. This argument is essentially due to Iscoe [15]. Let

$$v(x) = \begin{cases} 12(x+R)^{-2} & \text{for } x > -R \\ \infty & \text{for } x \leq -R. \end{cases} \tag{110}$$

and note that for $x > -R$,

$$\Delta v(x) = \frac{1}{2}v(x)^2. \tag{111}$$

Define $0 \cdot \infty = 0$, and let

$$M_t = \exp\left(-\int_{\mathbf{R}} v(x)u(t,x)\,dx\right). \tag{112}$$

By Ito's lemma (see (9) and (10)), we have

$$dM_t = M_t\left(-\int_{x\in\mathbf{R}} \left[v(x)\Delta u(t,x) + \frac{1}{2}v(x)^2 u(t,x)\right]dx\right)dt$$
$$- M_t \int_{x\in\mathbf{R}} v(x)u(t,x)^{1/2}\, W(dx\,dt) \tag{113}$$
$$= -M_t \int_{x\in\mathbf{R}} v(x)u(t,x)^{1/2}\, W(dx\,dt).$$

Here we have used the martingale problem formulation to do integration by parts, ie to replace $v\Delta u$ by $u\Delta v$. Then we used (111) to substitute for

Δv. Actually, to justify this calculation, we must truncate $v(x)$ and let the truncation level tend to ∞. We leave these details to the reader.

Thus M_t is a local martingale. Fix $T > 0$, and let τ be the infimum of times $t \leq T$ for which

$$\int_{-\infty}^{R} u(t, x) \, dx > 0 \tag{114}$$

and let $\tau = T$ if there is no such time. If $\tau < T$, we say that u charges $(-\infty, R]$ before time T. Since $0 \leq M_t \leq 1$, we can apply the optional sampling theorem to conclude that

$$M_0 = EM_\tau. \tag{115}$$

Let $\mathscr{A}$ be the event that u does not charge $(-\infty, R]$ before time T. Note that on $\mathscr{A}^c$ we have $M_\tau = 0$, while on $\mathscr{A}$ we have $M_\tau \leq 1$. Therefore,

$$P(\mathscr{A}) \geq EM_\tau = M_0 = \exp\left(-\int_{\mathbf{R}} v(x) u(0, x) \, dx\right) \tag{116}$$

From (111) we see that $v(x) = v(R, x) \to 0$ uniformly on compact intervals. Since $u(0, x)$ is continuous with compact support, it follows that the right hand side tends to 0 as $R \to \infty$. $\qquad \square$

For $1/2 < \gamma < 1$, Mueller and Perkins [24] gave a more complicated argument proving compact support in this situation, but one which gave information which has proved useful for other problems.

Note 6.12. Compact support can also occur in deterministic heat equations such as

$$\frac{\partial u}{\partial t} = \Delta u - u^\rho \tag{117}$$

when $\rho < 1$. Assume that $u(t, x)$ is nonnegative. For small values of u, we have $u^\rho \gg u$, so the final term can push the equation to 0. More complicated equations of this type appear in chemical engineering, and the region where $u(t, x) = 0$ is called a dead zone. Chemical engineers try to minimize the dead zone, since no reactions take place there, and this leads to inefficient use of the reactor vessel.

6.4 Noncompact Support

On the other hand, some stochastic heat equations have solutions whose support is all of $\mathbf{R}^d$. Note that in the following equation, no matter what the size of u, the noise term is comparable to the other terms in the equation.

Theorem 6.13. *Suppose that $u(t, x)$, $t \geq 0$, $x \in \mathbf{R}$ satisfies*

$$\frac{\partial u}{\partial t} = \Delta u + u\dot{W}(t, x)$$
$$u(0, x) = u_0(x), \tag{118}$$

for $u_0(x)$ continuous, nonnegative, and not identically 0. Then with probability 1, $u(t,x) > 0$ for all $t > 0, x \in \mathbf{R}$.

The assertion seems intuitively obvious, since solutions to the heat equation have support on all of $\mathbf{R}$. However, the previous section shows that certain heat equations can have solutions of compact support.

Proof. Working with $x \in \mathbf{R}$ gives rise to technical complications, so let us suppose that $u(t,x) : x \in [-2R, 2R]$ satisfies

$$\frac{\partial u}{\partial t} = \Delta u + u\dot{W}(t,x),$$
$$u(t,-2R) = u(t,2R) = 0, \tag{119}$$
$$u(0,x) = u_0(x).$$

As before, we would give rigorous meaning to this equation in terms of the following integral equation.

$$u(t,x) = \int_{-2R}^{2R} G_D(t,x,y)u_0(y)\,dy$$
$$+ \int_0^t \int_{-2R}^{2R} G_D(t-s,x,y)u(s,y)\,W(dy\,ds), \tag{120}$$

where $G_D(t,x,y)$ is the fundamental solution to the heat equation on $[-2R, 2R]$ with Dirichlet boundary conditions. It is a standard fact that

$$G_D(t,x,y) \leq G(t,x-y). \tag{121}$$

It is not hard to modify the proof of the comparison theorem, Theorem 5.1, to show that the solution to (119) is less than or equal to the solution of (118). From now on, let $u(t,x)$ denote the solution to (119). It is enough to show that with probability 1, $u(t,x)$ satisfies

$$\operatorname{supp}(u(t,\cdot)) \supset [-R,R] \tag{122}$$

for all $t > 0$.

Note that the equation is linear, so $v(t,x) = cu(t,x)$ satisfies the same equation, with different initial conditions, of course. We will subdivide time into stages, and show that with high probability, at each stage the support expands a little more.

Translating $u(0,x)$ if necessary, we may assume that $u(0,x) \geq \delta\mathbf{1}_{[-a,a]}$ for some $\delta, a > 0$. For simplicity, assume that $a = 1$. Let

$$\mathscr{S}_t = \operatorname{supp}(u(t,\cdot)) \tag{123}$$

It suffices to show that for $T, R, \delta > 0$,

$$[-R,R] \subset \mathscr{S}_T. \tag{124}$$

For simplicity we will present the proof for $T = 1$ and $R > 2$. Fix $N > 0$ and let $t_k = Tk/N$. Let $\mathscr{A}_k$ be the event that

$$u(t_k, x) \geq \delta I_k(x) \tag{125}$$

for some δ, where

$$I_k(x) = \mathbf{1}\left(-1 - \frac{Rk}{N} \leq x \leq 1 + \frac{Rk}{N}\right). \tag{126}$$

Note that

$$I_N(x) \geq \mathbf{1}_{[-R,R]}(x). \tag{127}$$

We wish to show that for all $\varepsilon > 0$, we can choose N so large that for all $k = 1, \ldots, N$

$$\mathrm{P}\left(\mathscr{A}_{k+1}^c \,\middle|\, \mathscr{A}_1 \cap \cdots \cap \mathscr{A}_k\right) < \frac{\varepsilon}{N}. \tag{128}$$

Note that $\mathscr{A}_0$ occurs by assumption. It would then follow that

$$\mathrm{P}\left(\mathscr{A}_N^c\right) \leq \sum_{k=0}^{N-1} \mathrm{P}\left(\mathscr{A}_{k+1}^c \,\middle|\, \mathscr{A}_1 \cap \cdots \cap \mathscr{A}_k\right) \tag{129}$$
$$\leq \varepsilon.$$

But

$$\mathrm{P}\left(\mathscr{S}_1 \subset [-R, R]\right) \geq \mathrm{P}\left(\mathscr{A}_N\right) \geq 1 - \varepsilon. \tag{130}$$

and since ε is arbitrary, we would be done.

Now we turn to the proof of (128). Assuming that $\mathscr{A}_k$ occurs, we have $u(t_k, x) \geq \delta I_k(x)$. By the comparison theorem, Theorem 5.1, it is enough to show (128) assuming that $u(t_k, x) = \delta I_k(x)$. Now let

$$v_k(t, x) = (1/\delta)u(t_k + t, x), \tag{131}$$

so that $v_k(t, x)$ satisfies (118), and $v_k(0, x) = I_k(x)$. Let

$$\eta(N, k) = \int_{-2R}^{2R} G(t, 1 + (k+1)/N - y)v_k(0, y)\, dy, \tag{132}$$

and let

$$\eta(N) = \inf_{0 \leq k \leq N} \eta(N, k). \tag{133}$$

We leave it to the reader to show that $\eta(N, k)$ is increasing in k, so that $\eta(N) = \eta(N, 0)$ and that

$$\eta := \inf_N \eta(N) > 0. \tag{134}$$

Hint: roughly speaking, the heat kernel spreads a distance $1/N^{1/2}$ in time $1/N$, so it must be close to 1 at distance $1/N$ from the edge of the support of the indicator function.

Next, let $w_k(t, x)$ satisfy

$$w_k(t, x) = \int_{-2R}^{2R} G_D(t, x, y) I_k(y) \, dy$$

$$+ \int_0^t \int_{-2R}^{2R} G_D(t - s, x, y) u(s, y) \, W(dy \, ds), \tag{135}$$

and let

$$N_w(t, x) = \int_0^t \int_{-2R}^{2R} G_D(t - s, x, y) u(s, y) \, W(dy \, ds). \tag{136}$$

In the same way as for Theorem 4.2, it is not hard to prove that if $0 < M < R$, then there exist constants $C_1, C_2 > 0$ such that for all $T, K, \lambda > 0$,

$$P \left(\sup_{0 \le t \le T} \sup_{|x| \le M} |N_w(t, x)| > \lambda \right) \le C_1 \exp \left(-\frac{C_2 \lambda^2}{T^{1/2} K^2} \right). \tag{137}$$

Now let $\lambda = \eta/2$, and let $T = 1/N$. Thus, given ε, we can choose N so large that the right hand side of (137) is less than ε.

This proves (128), and so finishes the proof of Theorem 6.13.   □

References

[1] Maury D. Bramson (1978). Maximal displacement of branching Brownian motion, *Comm. Pure Appl. Math.* **31**, 531–581

[2] Maury Bramson (1983). Convergence of solutions of the Kolmogorov equation to travelling waves, *Mem. Amer. Math. Soc.* **44**(285)

[3] Albert-László Barabási and H. Eugene Stanley (1995). *Fractal Concepts in Surface Growth*, Cambridge University Press, Cambridge

[4] Pao-Liu Chow (2007). *Stochastic partial differential equations.* Chapman & Hall/CRC Applied Mathematics and Nonlinear Science Series. Chapman & Hall/CRC, Boca Raton, FL.

[5] J. T. Cox and R. Durrett (1995). Hybrid zones and voter model interfaces. *Bernoulli*, **1**(4), 343–370

[6] D. A. Dawson (1993). *Measure-valued Markov processes*, In: Lecture Notes in Mathematics **1180**, 1–260, Springer-Verlag, Berlin, Heidelberg, New York

[7] M. Doi and S.F. Edwards (1988). *The Theory of Polymer Dynamics*, Oxford University Press, Oxford

[8] Giuseppe Da Prato and Jerzy Zabczyk (1992). *Stochastic Equations in Infinite Dimensions*, Cambridge University Press, Cambridge

[9] R. Durrett (1988). *Lecture Notes on Particle Systems and Percolation*, Wadsworth and Brooks/Cole, Pacific Grove

[10] S. Ethier and T. Kurtz (1986). *Markov Processes: Characterization and Convergence*, John Wiley & Sons Inc., New York

[11] Alison M. Etheridge (2000). *An Introduction to Superprocesses*, Amer. Math. Soc., Providence

[12] Marco Ferrante and Marta Sanz-Solé (2006). SPDEs with coloured noise: Analytic and stochastic approaches, *ESAIM Probab. Stat.* **10**, 380–405 (electronic)

[13] M. I. Freidlin and A. D. Wentzell (1998). *Random Perturbations of Dynamical Systems*, Springer-Verlag, New York, second edition. Translated from the 1979 Russian original by Joseph Szücs

[14] A. M. Garsia, E. Rodemich, and H. Rumsey, Jr. (1970/1971). A real variable lemma and the continuity of paths of some Gaussian processes. *Indiana Univ. Math. J.* **20**, 565–578

[15] I. Iscoe (1988). On the supports of measure-valued critical branching Brownian motion, *Probab. Th. Rel. Fields* **16**, 200–221

[16] N. Ikeda and S. Watanabe (1989). *Stochastic Differential Equations and Diffusion Processes*, North Holland/Kodansha, Amsterdam, Oxford, New York

[17] N. V. Krylov and B. L. Rozovskii (1982). Stochastic partial differential equations and diffusion processes, *Russian Math. Surveys* **37**(6), 81–105. Originally in Uspekhi Mat. Nauk, **37**(6), 75–95

[18] N. V. Krylov (1996). On L_p-theory of stochastic partial differential equations in the whole space, *SIAM J. Math Anal.* **27**(2), 313–340

[19] Nicolai V. Krylov (2006). On the foundation of the L_p-theory of stochastic partial differential equations, In: *Stochastic Partial Differential Equations and Applications–VII, Lect. Notes Pure Appl. Math.* **245**, 179–191, Chapman & Hall/CRC, Boca Raton

[20] Jean-François Le Gall (1999). *Spatial Branching Processes, Random Snakes and Partial Differential Equations*, Lectures in Mathematics ETH Zürich, Birkhäuser-Verlag, Basel

[21] Thomas M. Liggett (1999). *Stochastic Interacting Systems: Contact, Voter and Exclusion Processes*, Springer-Verlag, Berlin

[22] Thomas M. Liggett (2005). *Interacting Particle Systems*, Springer-Verlag, Berlin. Reprint of the 1985 original

[23] Jonathan C. Mattingly (2003). On recent progress for the stochastic Navier Stokes equations, In: *Journées "Équations aux Dérivées Partielles"*, pages Exp. No. XI, 52. Univ. Nantes, Nantes

[24] C. Mueller and E. Perkins (1992). The compact support property for solutions to the heat equation with noise, *Probab. Th. Rel. Fields* **93**, 325–358

[25] Carl Mueller and Etienne Pardoux (1999). The critical exponent for a stochastic PDE to hit zero, In: *Stochastic Analysis, Control, Optimization and Applications*, 325–338, Birkhäuser Boston, Boston

[26] C. Mueller and R. Sowers (1995). Random traveling waves for the KPP equation with noise, *J. Funct. Anal.* **128**, 439–498

[27] C. Mueller and R. Tribe (1997). Finite width for a random stationary interface, *Electronic J. Prob.* **2**, Paper no. 7, 1–27

[28] C. Mueller (1991). On the support of the heat equation with noise, *Stochastics* **37**(4), 225–246

[29] L. Mytnik (1998). Weak uniqueness for the heat equation with noise, *Ann. Probab.*, **26**(3), 968–984

[30] D. Nualart and É. Pardoux (1992). White noise driven quasilinear SPDEs with reflection, *Probab. Theory Related Fields* **93**(1), 77–89

[31] Étienne Pardoux (1972). Sur des équations aux dérivées partielles stochastiques monotones, *C. R. Acad. Sci. Paris Sér. A-B* **275**, A101–A103

[32] Étienne Pardoux (1991). Filtrage non linéaire et équations aux dérivées partielles stochastiques associées, In: *Lecture Notes in Math.* **1464**, 67–163, Springer, Berlin

[33] Edwin Perkins (2002). Dawson–Watanabe superprocesses and measure-valued diffusions, In: *Lecture Notes in Math.* **1781**, 125–324, Springer, Berlin

[34] Claudia Prévôt and Michael Röckner (2007). *A concise course on stochastic partial differential equations*, volume 1905 of *Lecture Notes in Mathematics*. Springer, Berlin

[35] Murray H. Protter and Hans F. Weinberger (1984). *Maximum Principles in Differential Equations*, Springer-Verlag, New York

[36] Daniel Revuz and Marc Yor (1999). *Continuous Martingales and Brownian Motion*, Springer-Verlag, Berlin, third edition

[37] T. Shiga (1994). Two contrasting properties of solutions for one-dimensional stochastic partial differential equations, *Can. J. Math* **46**(2), 415–437

[38] J. B. Walsh (1986). *An Introduction to Stochastic Partial Differential Equations*, In: Lecture Notes in Mathematics **1180**, 265–439, Springer-Verlag, Berlin, Heidelberg, New York

[39] Lorenzo Zambotti (2003). Integration by parts on δ-Bessel bridges, $\delta \geq 3$ and related SPDEs, *Ann. Probab.* **31**(1), 323–348

[40] Lorenzo Zambotti (2005). Integration by parts on the law of the reflecting Brownian motion, *J. Funct. Anal.* **223**(1), 147–178

Sample Path Properties of Anisotropic Gaussian Random Fields

Yimin Xiao

Summary. Anisotropic Gaussian random fields arise in probability theory and in various applications. Typical examples are fractional Brownian sheets, operator-scaling Gaussian fields with stationary increments, and the solution to the stochastic heat equation.

This paper is concerned with sample path properties of anisotropic Gaussian random fields in general. Let $X = \{X(t), t \in \mathbf{R}^N\}$ be a Gaussian random field with values in $\mathbf{R}^d$ and with parameters $H_1, \ldots, H_N$. Our goal is to characterize the anisotropic nature of X in terms of its parameters explicitly.

Under some general conditions, we establish results on the modulus of continuity, small ball probabilities, fractal dimensions, hitting probabilities and local times of anisotropic Gaussian random fields. An important tool for our study is the various forms of strong local nondeterminism.

1 Introduction

Gaussian random fields have been extensively studied in probability theory and applied in a wide range of scientific areas including physics, engineering, hydrology, biology, economics and finance. Two of the most important Gaussian random fields are respectively the Brownian sheet and fractional Brownian motion.

The Brownian sheet $W = \{W(t), t \in \mathbf{R}_+^N\}$, which was first introduced by a statistician J. Kitagawa in 1951, is a centered Gaussian random field with values in $\mathbf{R}^d$ and covariance function given by

$$\mathrm{E}\big[W_i(s)W_j(t)\big] = \delta_{ij} \prod_{k=1}^{N} (s_k \wedge t_k), \qquad \forall s, t \in \mathbf{R}_+^N, \tag{1}$$

D. Khoshnevisan and F. Rassoul-Agha (eds.) *A Minicourse on Stochastic Partial Differential Equations.*
Lecture Notes in Mathematics 1962.
© Springer-Verlag Berlin Heidelberg 2009

where $\delta_{ij} = 1$ if $i = j$ and 0 if $i \neq j$. When $N = 1$, W is the ordinary Brownian motion in $\mathbf{R}^d$. For $N \geq 2$, W has independent increments over disjoint intervals in $\mathbf{R}_+^N$ and such increments are stationary. We refer to [1; 49] for systematic accounts on properties of the Brownian sheet and to [94] and the articles in this volume for its important roles in stochastic partial differential equations.

For a fixed constant $0 < \alpha < 1$, an (N, d)-fractional Brownian motion with index α is a centered Gaussian random field $X^\alpha = \{X^\alpha(t), t \in \mathbf{R}^N\}$ with values in $\mathbf{R}^d$ and covariance function given by

$$ \mathrm{E}\big[X_i^\alpha(s)X_j^\alpha(t)\big] = \frac{1}{2}\delta_{ij}\Big(|s|^{2\alpha} + |t|^{2\alpha} - |s - t|^{2\alpha}\Big), \quad \forall s, t \in \mathbf{R}^N, \qquad (2) $$

where $|\cdot|$ denotes the Euclidean norm in $\mathbf{R}^N$. The existence of X^α follows from the positive semi-definiteness of the kernel on the right hand side of (2); see [82] for a proof. When $N = 1$ and $\alpha = 1/2$, X^α is again the Brownian motion in $\mathbf{R}^d$; when $N > 1$, $\alpha = 1/2$ and $d = 1$, it is the multiparameter Brownian motion introduced by P. Lévy; see [46; 82] for more historical information, probabilistic and statistical properties of fractional Brownian motion.

By using (2) one can verify that X^α is self-similar with exponent α, i.e. for every constant $c > 0$,

$$ \big\{X^\alpha(ct), t \in \mathbf{R}^N\big\} \overset{\mathrm{d}}{=} \big\{c^\alpha X^\alpha(t), t \in \mathbf{R}^N\big\}, \qquad (3) $$

where $\overset{\mathrm{d}}{=}$ means equality in finite dimensional distributions. Moreover, X^α has stationary increments in the *strong* sense; see Section 8.1 of [82]. In particular, X is isotropic in the sense that the distribution of $X(s) - X(t)$ depends only on the Euclidean distance $|s - t|$. Fractional Brownian motion is naturally related to long range dependence which makes it important for modelling phenomena with self-similarity and/or long memory properties. In the last decade the literature on statistical analysis and applications of fractional Brownian motion has grown rapidly [30].

On the other hand, many data sets from various areas such as image processing, hydrology, geostatistics and spatial statistics have anisotropic nature in the sense that they have different geometric and probabilistic characteristics along different directions, hence fractional Brownian motion is not adequate for modelling such phenomena. Many people have proposed to apply anisotropic Gaussian random fields as more realistic models [11; 18; 29].

Several classes of anisotropic Gaussian random fields have been introduced for theoretical and application purposes. For example, Kamont [47] introduced fractional Brownian sheets [see the definition in Section 2.1] and studied some of their regularity properties. Benassi et al. [10] and Bonami and Estrade [18] considered some anisotropic Gaussian random fields with stationary increments. Biermé et al. [17] constructed a large class of operator self-similar Gaussian or stable random fields with stationary increments. Anisotropic Gaussian random fields also arise naturally in stochastic partial differential equations [20; 73; 74; 75]; and in studying the most visited sites of symmetric Markov processes [35]. Hence it is of importance in both theory and applications to investigate the probabilistic and statistical properties of anisotropic random fields.

This paper is concerned with sample path properties of anisotropic Gaussian random fields in general. From the recent works on fractional Brownian sheets [see Section 2.1 for a list of references] it is known that the behavior of anisotropic Gaussian random fields may differ significantly from those of the Brownian sheet and fractional Brownian motion. Our objective is to gather and develop some general methods for studying the analytic and geometric properties of anisotropic Gaussian fields. In particular our results are applicable to the solutions of stochastic partial differential equations including the stochastic heat and wave equations. In a similar vein, Pitt and Robeva [78], Robeva and Pitt [79], Balan and Kim [8] have proposed to study the Markov properties of (anisotropic) Gaussian random fields and the solutions to the stochastic heat equations.

The rest of this paper is organized as follows. Section 2 contains definitions and basic properties of several classes of anisotropic Gaussian random fields including fractional Brownian sheets, Gaussian random fields with stationary increments and solutions to stochastic partial differential equations. We also provide the general conditions [i.e., Conditions (C1), (C2), (C3) and (C3′)] for the Gaussian random fields that will be studied in this paper.

An important technical tool in this paper is the properties of strong local nondeterminism for anisotropic Gaussian random fields, extending the concept of *local nondeterminism* first introduced by Berman [14] for Gaussian processes. In Section 3, we recall the recent result of Wu and Xiao [97] on the property of sectorial local nondeterminism for fractional Brownian sheets; and we prove a sufficient condition for an anisotropic Gaussian field with stationary increments to be strongly locally nondeterministic (with respect to an appropriate metric).

Section 4 is concerned with analytic and asymptotic properties of the sample functions of anisotropic Gaussian fields. We summarize three methods for deriving a sharp modulus of continuity for any anisotropic Gaussian random field satisfying Condition (C1). The first method is to use an extension, due to Arnold and Imkeller [2], Funaki, Kikuchi and Potthoff [39], Dalang, Khoshnevisan and Nualart [22], of the powerful Garsia-Rodemich-Rumsey continuity lemma; the second is the "minorizing metric" method of Kwapień and Risiński [60]; and the third is based on the Gaussian isoperimetric inequality. While the first two methods have wider applicability, the third method produces more precise results for Gaussian random fields.

Section 5 provides an application of strong local nondeterminism in studying small ball probabilities of anisotropic Gaussian fields.

In Section 6, we consider the Hausdorff and packing dimensions of the range $X([0,1]^N) = \{X(t) : t \in [0,1]^N\}$ and graph $\mathrm{Gr}X([0,1]^N) = \{(t, X(t)) : t \in [0,1]^N\}$ of X. Due to anisotropy, these results are different from the corresponding results for fractional Brownian motion and the Brownian sheet. We also establish an explicit formula for the Hausdorff dimension of the image $X(E)$ in terms of the generalized Hausdorff dimension of E (with respect to an appropriate metric) and the Hurst index H. Moreover, when $H = (\alpha, \dots, \alpha) \in (0,1)^N$, we prove the following uniform Hausdorff dimension result for the images of X: If $N \leq \alpha d$, then with probability one,

$$\dim_{\mathscr{H}} X(E) = \frac{1}{\alpha} \dim_{\mathscr{H}} E \quad \text{for all Borel sets } E \subseteq (0,\infty)^N. \tag{4}$$

This extends the previous results of [51; 70; 72] for fractional Brownian motion and the Brownian sheet, respectively, and is another application of the strong local nondeterminism.

In Section 7, we determine the Hausdorff and packing dimensions of the level sets, and establish estimates on the hitting probabilities of Gaussian random fields X satisfying Conditions (C1) and (C2).

In Section 8, we study the existence and joint continuity of local times of anisotropic Gaussian random fields under Conditions (C3) and (C3'). Moreover, we discuss local and uniform Hölder conditions of the local times in the set variable and show their applications in evaluating the exact Hausdorff measure of the level sets of X.

We end the Introduction with some notation. Throughout this paper, the underlying parameter space is $\mathbf{R}^N$ or $\mathbf{R}_+^N = [0,\infty)^N$. We use $|\cdot|$ to denote the Euclidean norm in $\mathbf{R}^N$. The inner product and Lebesgue measure in $\mathbf{R}^N$ are denoted by $\langle \cdot, \cdot \rangle$ and λ_N, respectively. A typical parameter, $t \in \mathbf{R}^N$ is written as $t = (t_1, \dots, t_N)$, or as $\langle c \rangle$ if $t_1 = \dots = t_N = c$. For any $s, t \in \mathbf{R}^N$ such that $s_j < t_j$ $(j = 1, \dots, N)$, $[s, t] = \prod_{j=1}^N [s_j, t_j]$ is called a closed interval (or a rectangle). We will let $\mathscr{A}$ denote the class of all closed intervals in $\mathbf{R}^N$. For two functions f and g, the notation $f(t) \asymp g(t)$ for $t \in T$ means that the function $f(t)/g(t)$ is bounded from below and above by positive constants that do not depend on $t \in T$.

We will use c to denote an unspecified positive and finite constant which may not be the same in each occurrence. More specific constants in Section i are numbered as $c_{i,1}, c_{i,2}, \dots$.

2 Examples and General Assumptions

In this section, we give some important examples of anisotropic Gaussian random fields, among them, fractional Brownian sheets are the most studied. We will show that the methods for studying fractional Brownian sheets can be modified to investigate sample path properties of anisotropic Gaussian random fields in general. In §2.4, we provide the general conditions for the Gaussian random fields that will be studied in this paper.

Even though anisotropic random fields generally do not satisfy the ordinary self-similarity (3), they may have certain *operator-scaling* properties. Following the terminology of Biermé et al. [17], we say that a random field $X = \{X(t), t \in \mathbf{R}^N\}$ is *operator-self-similar* [or *operator-scaling*] in the *time variable* if there exist a linear operator A on $\mathbf{R}^N$ with positive real parts of the eigenvalues and some constant $\beta > 0$ such that

$$\{X(c^A t), t \in \mathbf{R}^N\} \overset{d}{=} \{c^\beta X(t), t \in \mathbf{R}^N\} \qquad \forall c > 0. \tag{5}$$

In the above, c^A is the linear operator defined by $c^A = \sum_{n=0}^\infty (\ln c)^n A^n / n!$. The linear operator A is called a self-similarity exponent [which may not be unique].

There is also a notion of operator-self-similarity in *the space variable* [65; 98]. We will not discuss this topic in this paper.

2.1 Fractional Brownian Sheets

Fractional Brownian sheets were first introduced by Kamont [47], who also studied some of their regularity properties. For $H = (H_1, \ldots, H_N) \in (0,1)^N$, an $(N,1)$-fractional Brownian sheet $B_0^H = \{B_0^H(t), t \in \mathbf{R}^N\}$ with Hurst index H is a real-valued, centered Gaussian random field with covariance function given by

$$\mathrm{E}\big[B_0^H(s)B_0^H(t)\big] = \prod_{j=1}^N \frac{1}{2}\Big(|s_j|^{2H_j} + |t_j|^{2H_j} - |s_j - t_j|^{2H_j}\Big), \qquad s,t \in \mathbf{R}^N. \tag{6}$$

It follows from (6) that $B_0^H(t) = 0$ a.s. for every $t \in \mathbf{R}^N$ with at least one zero coordinate.

Note that if $N = 1$, then B_0^H is a fractional Brownian motion in $\mathbf{R}$ with Hurst index $H_1 \in (0,1)$; if $N > 1$ and $H = \langle 1/2 \rangle$, then B^H is the Brownian sheet in $\mathbf{R}$. Hence B_0^H can be regarded as a natural generalization of one parameter fractional Brownian motion as well as a generalization of the Brownian sheet.

It follows from (6) that B_0^H has the following operator-scaling property: For all constants $c > 0$,

$$\{B_0^H(c^A t), t \in \mathbf{R}^N\} \overset{d}{=} \{c^N B_0^H(t), t \in \mathbf{R}^N\}, \tag{7}$$

where $A = (a_{ij})$ is the $N \times N$ diagonal matrix with $a_{ii} = 1/H_i$ for all $1 \le i \le N$ and $a_{ij} = 0$ if $i \ne j$. Thus, B_0^H is operator-self-similar with exponent A and $\beta = N$.

The covariance structure of B_0^H is more complicated than those of fractional Brownian motion and the Brownian sheet. The following stochastic integral representations are useful. They were established by Ayache et al. [4]

and Herbin [43], respectively, and can be verified by checking the covariance functions.

• **Moving average representation:**

$$B_0^H(t) = \kappa_H^{-1} \int_{-\infty}^{t_1} \cdots \int_{-\infty}^{t_N} g(t,s)\, W(ds), \tag{8}$$

where $W = \{W(s),\, s \in \mathbf{R}^N\}$ is a standard real-valued Brownian sheet and

$$g(t,s) = \prod_{j=1}^{N} \left[\left((t_j - s_j)_+\right)^{H_j - 1/2} - \left((-s_j)_+\right)^{H_j - 1/2} \right]$$

with $s_+ = \max\{s,0\}$, and where $\kappa_H > 0$ is a normalization constant.

To give a harmonizable representation for B_0^H, let us recall briefly the definition of a complex-valued Gaussian measure. Let $(E, \mathscr{E}, \Delta)$ be a measure space and let $\mathscr{A} = \{A \in \mathscr{E} : \Delta(A) < \infty\}$. We say that $\widetilde{\mathscr{M}}$ is a centered complex-valued Gaussian measure on $(E, \mathscr{E}, \Delta)$ if $\{\widetilde{\mathscr{M}}(A), A \in \mathscr{A}\}$ is a centered complex-valued Gaussian process satisfying

$$\mathrm{E}\left(\widetilde{\mathscr{M}}(A)\overline{\widetilde{\mathscr{M}}(B)} \right) = \Delta(A \cap B) \quad \text{and} \quad \widetilde{\mathscr{M}}(-A) = \overline{\widetilde{\mathscr{M}}(A)}, \tag{9}$$

for all $A,\, B \in \mathscr{A}$. The measure Δ is called the *control measure* of $\widetilde{\mathscr{M}}$. For any complex valued function $\widetilde{f} \in L^2(E, \mathscr{E}, \Delta)$, the stochastic integral $\int_E \widetilde{f}(\xi)\, \widetilde{\mathscr{M}}(d\xi)$ can be defined; see, e.g., Section 7.2.2 of [82]. With this notion, we give the following:

• **Harmonizable representation:**

$$B_0^H(t) = K_H^{-1} \int_{\mathbf{R}^N} \psi_t(\lambda)\, \widetilde{W}(d\lambda), \tag{10}$$

where $\widetilde{W}$ is a centered complex-valued Gaussian random measure in $\mathbf{R}^N$ with Lebesgue control measure and

$$\psi_t(\lambda) = \prod_{j=1}^{N} \frac{e^{it_j \lambda_j} - 1}{|\lambda_j|^{H_j + \frac{1}{2}}}, \tag{11}$$

where $K_H > 0$ is a constant. Recently, Wang [95] gives another stochastic integral representation for B_0^H.

Let $B_1^H, \ldots, B_d^H$ be d independent copies of B_0^H. Then the (N,d)-fractional Brownian sheet with Hurst index $H = (H_1, \ldots, H_N)$ is the Gaussian random field $B^H = \{B^H(t) : t \in \mathbf{R}^N\}$ with values in $\mathbf{R}^d$ defined by

$$B^H(t) = \left(B_1^H(t), \ldots, B_d^H(t) \right), \quad t \in \mathbf{R}^N. \tag{12}$$

Several authors have studied various properties of fractional Brownian sheets. For example, Ayache et al. [4] provided the moving average representation (8) for B_0^H and studied its sample path continuity as well as its continuity in H. Dunker [32], Mason and Shi [66], Belinski and Linde [9], Kühn and Linde [59] studied the small ball probabilities of B_0^H. Mason and Shi [66] also computed the Hausdorff dimension of some exceptional sets related to the oscillation of the sample paths of B_0^H. Ayache and Taqqu [5] derived an optimal wavelet series expansion for fractional Brownian sheet B_0^H; see also [33; 59] for other optimal series expansions for B_0^H. Øksendal and Zhang [75], and Hu, Øksendal and Zhang [45] studied stochastic partial differential equations driven by fractional Brownian sheets.

For fractal properties, Kamont [47] and Ayache [3] studied the box and Hausdorff dimensions of the graph set of an $(N, 1)$-fractional Brownian sheet. Ayache and Xiao [7] investigated the uniform and local asymptotic properties of B^H by using wavelet methods, and determined the Hausdorff dimensions of the image $B^H([0, 1]^N)$, the graph $\mathrm{Gr}B^H([0, 1]^N)$ and the level set $L_x = \{t \in (0, \infty)^N : B^H(t) = x\}$, where $x \in \mathbf{R}^d$. Further results on the geometric and Fourier analytic properties of the images of B^H can be found in Wu and Xio [97].

Xiao and Zhang [110] studied the existence of local times of an (N, d)-fractional Brownian sheet B^H and proved a sufficient condition for the joint continuity of the local times. Ayache, Wu and Xiao [6] established the joint continuity of the local times under the optimal condition and studied the local and uniform Hölder conditions for the maximum local times. Related to the above results, we mention that Tudor and Xiao [93] have obtained results on Hausdorff dimensions of the sample paths, local times and their chaos expansion for (N, d)-bifractional Brownian sheets.

2.2 Anisotropic Gaussian Random Fields with Stationary Increments

Let $X = \{X(t), t \in \mathbf{R}^N\}$ be a real-valued, centered Gaussian random field with $X(0) = 0$. We assume that X has stationary increments and continuous covariance function $R(s, t) = \mathrm{E}[X(s)X(t)]$. According to Yaglom [111], $R(s, t)$ can be represented as

$$R(s, t) = \int_{\mathbf{R}^N} (e^{i\langle s, \xi \rangle} - 1)(e^{-i\langle t, \xi \rangle} - 1)\Delta(d\xi) + \langle s, \Theta t \rangle, \qquad (13)$$

where $\langle x, y \rangle$ is the ordinary inner product in $\mathbf{R}^N$, Θ is an $N \times N$ non-negative definite matrix and $\Delta(d\xi)$ is a nonnegative symmetric measure on $\mathbf{R}^N \backslash \{0\}$ that satisfies

$$\int_{\mathbf{R}^N} \frac{|\xi|^2}{1 + |\xi|^2}\, \Delta(d\xi) < \infty. \qquad (14)$$

The measure Δ and its density (if it exists) $f(\xi)$ are called the *spectral measure* and *spectral density* of X, respectively.

It follows from (13) that X has the following stochastic integral representation:

$$X(t) = \int_{\mathbf{R}^N} \left(e^{i\langle t,\xi\rangle} - 1\right) \widetilde{\mathscr{M}}(d\xi) + \langle \mathbf{Y}, t\rangle, \qquad (15)$$

where $\mathbf{Y}$ is an N-dimensional Gaussian random vector with mean 0 and covariance matrix Θ, and where $\widetilde{\mathscr{M}}$ is a centered complex-valued Gaussian random measure in $\mathbf{R}^N$ with control measure Δ, which is independent of $\mathbf{Y}$. Since the linear term $\langle \mathbf{Y}, t\rangle$ in (15) will not have any effect on the problems considered in this paper, we will from now on assume $\mathbf{Y} = 0$. This is equivalent to assuming $\Theta = 0$ in (13). Consequently, we have

$$\sigma^2(h) = \mathrm{E}\left[\left(X(t+h) - X(t)\right)^2\right] = 2\int_{\mathbf{R}^N} \left(1 - \cos\langle h, \xi\rangle\right) \Delta(d\xi). \qquad (16)$$

It is important to note that $\sigma^2(h)$ is a negative definite function [12] and, by the Lévy-Khintchine formula, can be viewed as the characteristic exponent of a symmetric infinitely divisible distribution.

If the function $\sigma^2(h)$ depends only on $|h|$, then X is called an *isotropic* random field. We say that a Gaussian random field X is *approximately isotropic* if $\sigma^2(h) \asymp \phi(|h|)$ in a neighborhood of $h = 0$ for some nonnegative function ϕ. Sample path properties of such Gaussian random fields have been studied widely. See [83; 102; 108] and the references therein for more information. The results in [7; 97] on fractional Brownian sheets indicate that the properties of anisotropic Gaussian random fields can be very different and often more difficult to be established.

Many Gaussian random fields can be constructed by choosing the spectral measures appropriately. For example, if we consider the spectral density

$$f(\xi) = \frac{1}{\left(\sum_{j=1}^{N} |\xi_j|^{H_j}\right)^{2+Q}} \qquad \forall \xi \in \mathbf{R}^N \setminus \{0\}, \qquad (17)$$

where the constants $H_j \in (0, 1)$ for $j = 1, \ldots, N$ and $Q = \sum_{j=1}^{N} H_j^{-1}$, then the corresponding Gaussian random field X has stationary increments and is operator-self-similar with exponent $A = (a_{ij})$, where $a_{ii} = H_i^{-1}$ and $a_{ij} = 0$ if $i \neq j$ and $\beta = 1$. This Gaussian random field is similar to that in Example 3 of [18].

The following class of Gaussian random fields constructed by Biermé, Meerschaert and Scheffler [17, Section 4] is more general.

Theorem 2.1. *Let A be a real $N \times N$ matrix with the real parts of the eigenvalues $1 < a_1 \leq a_2 \leq \cdots \leq a_N$ and let $Q = \mathrm{trace}(A)$. If $\psi : \mathbf{R}^N \to [0, \infty)$ is a continuous, A'-homogeneous function [i.e., $\psi(c^{A'}\xi) = c\psi(\xi)$ for all $c > 0$ and $\xi \in \mathbf{R}^N$. Here A' denotes the transpose of A] such that $\psi(\xi) > 0$ for $\xi \neq 0$. Then the Gaussian random field*

$$X_\psi(t) = \mathrm{Re} \int_{\mathbf{R}^N} \left(e^{i\langle t,\xi\rangle} - 1\right) \frac{\widetilde{W}(d\xi)}{\psi(\xi)^{1+Q/2}}, \qquad x \in \mathbf{R}^N, \qquad (18)$$

where $\widetilde{\mathscr{W}}$ is a centered complex-valued Gaussian random measure in $\mathbf{R}^N$ with Lebesgue control measure, has stationary increments and is operator-self-similar in the sense of (5) with exponent A and $\beta = 1$.

Compared with (15), we see that the spectral measure of X_ψ is $\Delta(d\xi) = \psi(\xi)^{-(2+Q)} d\xi$. As the results of this paper will suggest, the sample functions of X_ψ share many properties with fractional Brownian sheets and many of them can be described in terms of the real parts of the eigenvalues of A. See [15] for more details.

2.3 Solutions to Stochastic Partial Differential Equations

Gaussian random fields arise naturally as solutions to stochastic partial differential equations. In the following we list as examples the solutions to the stochastic heat equation and stochastic wave equation, and discuss possible ways to study their sample path properties using general methods for Gaussian random fields. We refer to [20; 21; 22; 23; 24; 27; 28; 73; 94], and the articles in this volume for more information.

2.3.1 The Stochastic Heat Equation

Funaki's model for random string in $\mathbf{R}$ is specified by the following stochastic heat equation:

$$\frac{\partial u(t,x)}{\partial t} = \frac{\partial^2 u(t,x)}{\partial x^2} + \dot{W}, \tag{19}$$

where $\dot{W}(x,t)$ is an $\mathbf{R}$-valued space-time white noise, which is assumed to be adapted with respect to a filtered probability space $(\Omega, \mathscr{F}, \mathscr{F}_t, \mathrm{P})$, where $\mathscr{F}$ is complete and the filtration $\{\mathscr{F}_t, t \geq 0\}$ is right continuous [38; 73].

Recall from [73] that a solution of (19) is defined as an $\mathscr{F}_t$-adapted, continuous random field $\{u(t,x), t \in \mathbf{R}_+, x \in \mathbf{R}\}$ with values in $\mathbf{R}$ satisfying the following properties:

(i) $u(0,\cdot) \in \mathscr{E}_{\exp}$ almost surely and is adapted to $\mathscr{F}_0$, where $\mathscr{E}_{\exp} = \cup_{\lambda>0}\mathscr{E}_\lambda$ and
$$\mathscr{E}_\lambda = \left\{ f \in C(\mathbf{R}) : |f(x)| e^{-\lambda|x|} \to 0 \text{ as } |x| \to \infty \right\};$$

(ii) For every $t > 0$, there exists $\lambda > 0$ such that $u(s,\cdot) \in \mathscr{E}_\lambda$ for all $s \leq t$, almost surely;

(iii) For every $t > 0$ and $x \in \mathbf{R}$, the following Green's function representation holds

$$u(t,x) = \int_{\mathbf{R}} G(t,x-y)u(0,y)\,dy + \int_0^t \int_{\mathbf{R}} G(t-r,x-y)\,W(dy\,dr), \tag{20}$$

where $G(t,x) = (4\pi t)^{-1/2} \exp\{-x^2/(4t)\}$ is the fundamental solution of the heat equation.

We call each solution $\{u(t,x),\ t \in \mathbf{R}_+,\ x \in \mathbf{R}\}$ of (19) a random string process with values in $\mathbf{R}$, or simply a random string as in [73]. Note that, in general, a random string may not be Gaussian, a powerful step in the proofs of [73] is to reduce the problems about a general random string process to those of the stationary pinned string $U_0 = \{U_0(t,x), t \in \mathbf{R}_+,\ x \in \mathbf{R}\}$, obtained by taking the initial function $u(0,\cdot)$ in (20) to be defined by

$$u(0,x) = \int_0^\infty \int_{\mathbf{R}} (G(r,x-z) - G(r,z))\,\widehat{W}(dz\,dr), \qquad (21)$$

where $\widehat{W}$ is a space-time white noise independent of the white noise $\dot{W}$. Consequently, the stationary pinned string is a continuous version of the following Gaussian field

$$\begin{aligned}
U_0(t,x) = &\int_0^\infty \int_{\mathbf{R}} \Big(G(t+r,x-z) - G(t+r,z)\Big)\,\widehat{W}(dz\,dr) \\
&+ \int_0^t \int_{\mathbf{R}} G(r,x-z)\,W(dz\,dr),
\end{aligned} \qquad (22)$$

Mueller and Tribe [73] proved that the Gaussian field $U_0 = \{U_0(t,x), t \in \mathbf{R}_+, x \in \mathbf{R}\}$ has stationary increments and satisfies the Conditions (C1) and (C2) in Section 2.4. Let $U_1,\ldots,U_d$ be d independent copies of U_0, and consider the Gaussian random field $U = \{U(t,x), t \in \mathbf{R}_+, x \in \mathbf{R}\}$ with values in $\mathbf{R}^d$ defined by $U(t,x) = (U_1(t,x),\ldots,U_d(t,x))$. Mueller and Tribe [73] found necessary and sufficient conditions [in terms of the dimension d] for U to hit points or to have double points of various types. They also studied the question of recurrence and transience for $\{U(t,x),\ t \in \mathbf{R}_+,\ x \in \mathbf{R}\}$. Continuing the work of Mueller and Tribe [73], Wu and Xiao [96] studied the fractal properties of various random sets generated by the random string processes. Further results on hitting probabilities of non-linear stochastic heat equations can be found in [22; 23].

On the other hand, Robeva and Pitt [79, Proposition 3] showed that the Gaussian random field

$$u_0(t,x) = \frac{1}{2\pi} \int_{\mathbf{R}^2} \frac{e^{i(\xi_1 t + \xi_2 x)} - 1}{i\xi_1 + \xi_2^2}\,\widetilde{W}(d\xi_1\,d\xi_2), \qquad \forall t \in \mathbf{R}_+, x \in \mathbf{R} \quad (23)$$

is another solution to (19) satisfying $u_0(0,0) = 0$. Here, $\widetilde{W}$ is a centered complex Gaussian random measure in $\mathbf{R}^2$ with Lebesgue control measure. This Gaussian random field has stationary increments with spectral density

$$f(\xi) = \frac{1}{\xi_1^2 + \xi_2^4}. \qquad (24)$$

This density function is comparable to (17) with $H_1 = 1/4$, $H_2 = 1/2$ and $Q = 6$. Hence, it follows from Theorem 3.2 that the Gaussian field $\{u_0(t,x),\ t \in \mathbf{R}_+, x \in \mathbf{R}\}$ satisfies the Conditions (C1) and (C3$'$) in §2.4.

If we define a $(2, d)$-Gaussian random field $\{u(t, x), t \in \mathbf{R}_+, x \in \mathbf{R}\}$ by $u(t, x) = (u_1(t, x), \ldots, u_d(t, x))$, where $u_1, \ldots, u_d$ are independent copies of u_0, then many of its sample path properties follow from the results in later sections of this paper.

If $x \in \mathbf{R}^N$ and $N \geq 2$, the stochastic heat equation (19) has no process solution [the solution is a random Schwartz distribution]. It might be helpful to remark that our random field notation is different from that in the references on s.p.d.e.'s: now the parameter $(t, x) \in \mathbf{R}^{N+1}$ and $\mathbf{R}^d$ is reserved for the state space of random fields.

The approach of Dalang [20] is to replace $\dot{W}$ by a Gaussian noise $\dot{F}$ which is white in time and has spatial covariance induced by a kernel f [not to be confused with the spectral density above], which is the Fourier transform of a tempered measure μ in $\mathbf{R}^N$. The covariance of F is of the form

$$\mathrm{E}\big(F(dt\,dx)F(ds\,dy)\big) = \delta(t - s)f(x - y), \tag{25}$$

where $\delta(\cdot)$ is the Dirac delta function. The case $f(r) = \delta(r)$ would correspond to the case of space-time white noise. More precisely, let $\mathscr{D}(\mathbf{R}^{N+1})$ be the topological space of functions $\phi \in C_0^\infty(\mathbf{R}^{N+1})$ with the topology that corresponds to the following notion of convergence: $\phi_n \to \phi$ if and only if the following two conditions hold:

(i) There exists a compact set $K \subseteq \mathbf{R}^{N+1}$ such that $\mathrm{supp}(\phi_n - \phi) \subseteq K$ for all $n \geq 1$, and
(ii) $\lim_{n \to \infty} D^a \phi_n = D^a \phi$ uniformly on K for every multi-index a.

Let $F = \{F(\phi), \phi \in \mathscr{D}(\mathbf{R}^{N+1})\}$ be an $L^2(\Omega, \mathscr{F}, \mathrm{P})$-valued, centered Gaussian process with covariance of the form $(\phi, \psi) \mapsto \mathrm{E}\big(F(\phi)F(\psi)\big) = J(\phi, \psi)$, where

$$J(\phi, \psi) = \int_{\mathbf{R}_+} dt \int_{\mathbf{R}^N} dx \int_{\mathbf{R}^N} \phi(t, x)f(x - y)\psi(t, y)\,dy. \tag{26}$$

As shown by Dalang [20], $\phi \mapsto F(\phi)$ can be extended to a worthy martingale measure $(t, A) \mapsto M_t(A)$ in the sense of Walsh [94, pp. 289–290], with covariance measure

$$\begin{aligned}
Q([0, t] \times A \times B) &= \langle M(A); M(B)\rangle_t \\
&= t \int_{\mathbf{R}^N} dx \int_{\mathbf{R}^N} \mathbf{1}_A(x)f(x - y)\mathbf{1}_B(y)\,dy,
\end{aligned} \tag{27}$$

and dominating measure $K \equiv Q$ such that

$$F(\phi) = \int_{\mathbf{R}_+} \int_{\mathbf{R}^N} \phi(t, x)\,M(dt\,dx), \qquad \forall \phi \in \mathscr{D}(\mathbf{R}^{N+1}). \tag{28}$$

Moreover, Dalang [20] constructed generalized stochastic integrals with respect to the martingale measure M.

Now we consider the stochastic heat equation with vanishing initial conditions, written formally as

$$\frac{\partial u(t,x)}{\partial t} = \Delta u(t,x) + \dot{F}, \qquad \forall (t,x) \in (0,T) \times \mathbf{R}^N \qquad (29)$$

and $u(0,\cdot) \equiv 0$. Here $T > 0$ is any fixed constant and $\dot{F}$ is the Gaussian noise defined above.

Dalang [20] proved that (29) has a process solution if and only if

$$\int_{\mathbf{R}^N} \frac{1}{1+|\xi|^2} \mu(d\xi) < \infty. \qquad (30)$$

Under this condition, the mean zero Gaussian field $u = \{u(t,x); t \in [0,T], x \in \mathbf{R}^N\}$ defined by

$$u(t,x) = \int_0^T \int_{\mathbf{R}^N} G(t-s,x-y) M(ds\,dy) \qquad (31)$$

is the *process solution* of the stochastic heat equation (29) with vanishing initial condition. In the above, $G(r,x) = (4\pi r)^{-N/2} \exp(-|x|^2/(4r))$ ($r > 0, x \in \mathbf{R}^N$) is the fundamental solution of the heat equation.

Many interesting examples can be constructed by choosing $\mu(d\xi)$ suitably [8; 20]. As we mentioned in the Introduction, [8; 79] studied the Markov property of the solution of stochastic heat equation (29). In view of the results in this paper, it would be interesting to see when the solutions of (29) satisfy Conditions (C3) or (C3′) in §2.4.

2.3.2   The Stochastic Wave Equation

The stochastic wave equation in one spatial dimension [i.e., $N = 1$]

$$\frac{\partial^2 u(t,x)}{\partial^2 t} - \frac{\partial^2 u(t,x)}{\partial x^2} = \dot{W}(t,x), \qquad t > 0, \ x \in \mathbf{R}, \qquad (32)$$

driven by the white noise was considered by Walsh [94] and many other authors [21; 24]. In spatial dimension two or higher, however, the stochastic wave equation driven by the white noise has no solution in the space of real valued measurable processes [94].

For $N = 2$, Dalang and Frangos [21] considered the stochastic wave equation driven by the Gaussian noise $\dot{F}$ with covariance (25):

$$\begin{cases} \dfrac{\partial^2 u(t,x)}{\partial t^2} = \Delta u(t,x) + \dot{F}, \\ u(0,x) = 0, \qquad\qquad\qquad \forall (t,x) \in (0,\infty) \times \mathbf{R}^2. \qquad (33) \\ \dfrac{\partial u}{\partial t}(0,x) = 0, \end{cases}$$

They proved that (33) has a process solution $u = \{u(t,x) : t \geq 0, x \in \mathbf{R}^2\}$ if and only if

$$\int_{0+} f(r) r \, \log\left(\frac{1}{r}\right) dr < \infty, \tag{34}$$

where f is the kernel in (25). Under the latter condition, $u = \{u(t,x) : t \geq 0, x \in \mathbf{R}^2\}$ can be represented as

$$u(t,x) = \int_0^t \int_{\mathbf{R}^2} S(t-s, x-y) \, M(ds\,dy), \tag{35}$$

where $S(t,x) = (2\pi)^{-1}(t^2 - |x|^2)^{-1/2} \mathbf{1}_{\{|x|<t\}}$. Sample path regularity of the solution $\{u(t,x) : t \geq 0, x \in \mathbf{R}^2\}$ has been investigated by Dalang and Frangos [21] and Dalang and Sanz-Solé [27].

For the stochastic wave equation with spatial dimension three, we refer to [24; 28] for information on the existence of a process solution and its sample path regularities. It seems that, in all the cases considered so far, the questions on fractal properties, existence and regularity of the local times of the solutions remain to be investigated.

2.4 General Assumptions

Let $X = \{X(t), t \in \mathbf{R}^N\}$ be a Gaussian random field in $\mathbf{R}^d$ defined on some probability space $(\Omega, \mathscr{F}, \mathrm{P})$ by

$$X(t) = \big(X_1(t), \ldots, X_d(t)\big), \qquad t \in \mathbf{R}^N, \tag{36}$$

where $X_1, \ldots, X_d$ are independent copies of X_0. We assume that X_0 is a mean zero Gaussian random field with $X_0(0) = 0$ a.s.

Let $(H_1, \ldots, H_N) \in (0,1)^N$ be a fixed vector. In order to study anisotropic Gaussian fields, we have found the following metric ρ on $\mathbf{R}^N$ is often more convenient than the Euclidean metric:

$$\rho(s,t) = \sum_{j=1}^N |s_j - t_j|^{H_j}, \qquad \forall\, s,t \in \mathbf{R}^N. \tag{37}$$

For any $r > 0$ and $t \in \mathbf{R}^N$, we denote by $B_\rho(t,r) = \{s \in \mathbf{R}^N : \rho(s,t) \leq r\}$ the closed (or open) ball in the metric ρ.

Let $I \in \mathscr{A}$ be a fixed closed interval, and we will consider various sample path properties of $X(t)$ when $t \in I$. For simplicity we will mostly assume $I = [\varepsilon, 1]^N$, where $\varepsilon \in (0,1)$ is fixed. Typically, the assumption for I to be away from the axis is only needed for Gaussian fields similar to fractional Brownian sheets. Even in these later cases, many results such as those on Hausdorff and packing dimensions remain to be true for $I = [0,1]^N$.

Many sample path properties of X can be determined by the following function:

$$\sigma^2(s,t) = \mathrm{E}\left(|X_0(s) - X_0(t)|^2\right), \qquad \forall s,t \in \mathbf{R}^N. \tag{38}$$

In this paper, we will make use of the following general conditions on X_0:

(C1) There exist positive constants $c_{2,1}, \ldots, c_{2,4}$ such that

$$c_{2,1} \leq \sigma^2(t) := \sigma^2(0, T) \leq c_{2,2} \qquad \forall t \in I \qquad (39)$$

and

$$c_{2,3} \sum_{j=1}^{N} |s_j - t_j|^{2H_j} \leq \sigma^2(s, t) \leq c_{2,4} \sum_{j=1}^{N} |s_j - t_j|^{2H_j}, \qquad (40)$$

for all $s, t \in I$. It may be helpful to note that (40) is in terms of $\rho(s, t)^2$.

(C2) There exists a constant $c_{2,5} > 0$ such that for all $s, t \in I$,

$$\text{Var}\big(X_0(t) \,\big|\, X_0(s)\big) \geq c_{2,5}\, \rho(s, t)^2.$$

Here and in the sequel, $\text{Var}(Y \,|\, Z)$ denotes the conditional variance of Y given Z.

(C3) There exists a constant $c_{2,6} > 0$ such that for all integers $n \geq 1$ and all $u, t^1, \ldots, t^n \in I$,

$$\text{Var}\big(X_0(u) \,\big|\, X_0(t^1), \ldots, X_0(t^n)\big) \geq c_{2,6} \sum_{j=1}^{N} \min_{0 \leq k \leq n} \big|u_j - t_j^k\big|^{2H_j},$$

where $t_j^0 = 0$ for every $j = 1, \ldots, N$.

(C3′) There exists a constant $c_{2,7} > 0$ such that for all integers $n \geq 1$ and all $u, t^1, \ldots, t^n \in I$,

$$\text{Var}\big(X_0(u) \,\big|\, X_0(t^1), \ldots, X_0(t^n)\big) \geq c_{2,7} \min_{0 \leq k \leq n} \rho(u, t^k)^2,$$

where $t^0 = 0$.

Remark 2.2. The following are some remarks about the above conditions.

- Conditions (C1)–(C3) can be significantly weakened and/or modified in various parts of the paper to obtain more general results. The present formulation of these conditions has the advantage that it is more convenient and produces cleaner results.
- Condition (39) assumes that X is non-degenerate on I. If (40) holds for $s = 0$ as well, then (39) is true automatically.
- Under condition (C1), X has a version which has continuous sample functions on I almost surely. Henceforth we will assume without loss of generality that the Gaussian random field X has continuous sample paths.
- Conditions (C1) and (C2) are related. It is easy to see that (C1) implies that $\text{Var}(X_0(t) \,|\, X_0(s)) \leq c_{2,4} \sum_{j=1}^{N} |s_j - t_j|^{2H_j}$ for all $s, t \in I$ and, on the other hand, (C2) implies $\sigma^2(s, t) \geq c_{2,5} \sum_{j=1}^{N} |s_j - t_j|^{2H_j}$. Moreover, if the function $\sigma(0, t)$ satisfies certain smoothness condition, say, it has continuous first order derivatives on I, then one can show that (C1) implies

(C2) by using the following fact [which can be easily verified]: If (U, V) is a Gaussian vector, then

$$\text{Var}(U \mid V) = \frac{\left(\rho_{U,V}^2 - (\sigma_U - \sigma_V)^2\right)\left((\sigma_U + \sigma_V)^2 - \rho_{U,V}^2\right)}{4\sigma_V^2}, \qquad (41)$$

where $\rho_{U,V}^2 = \text{E}[(U - V)^2]$, $\sigma_U^2 = \text{E}(U^2)$ and $\sigma_V^2 = \text{E}(V^2)$.

- Pitt [77] proved that fractional Brownian motion X^α satisfies Condition (C3′) for all $I \in \mathscr{A}$ with $H = \langle \alpha \rangle$; Khoshnevisan and Xiao [55] proved that the Brownian sheet satisfies the property (C3) with $H = \langle 1/2 \rangle$ for all $I \in \mathscr{A}$ which are away from the boundary of $\mathbf{R}_+^N$. It has been proved in [7; 97] that, for every $\varepsilon \in (0, 1)$, fractional Brownian sheets satisfy Conditions (C1), (C2) and (C3) for all $I \subseteq [\varepsilon, \infty)^N$.

- Let X be a Gaussian random field with stationary increments and spectral density comparable to (17). Then one can verify that X satisfies Condition (C1). In the next section, we will prove that X satisfies Condition (C3′) [thus it also satisfies (C2)]. Therefore, all the results in this paper are applicable to such Gaussian random fields.

- Note that Condition (C3′) implies (C3). It can be verified that the converse does not even hold for the Brownian sheet [this is an exercise]. Roughly speaking, when $H = \langle \alpha \rangle$, the behavior of a Gaussian random field X satisfying conditions (C1) and (C3′) is comparable to that of a fractional Brownian motion of index α; while the behavior of a Gaussian random field X satisfying conditions (C1) and (C3) [but not (C3′)] is comparable to that of a fractional Brownian sheet. Hence, in analogy to the terminology respectively for fractional Brownian motion and the Brownian sheet, Condition (C3′) will be called *the strong local nondeterminism* [in metric ρ] and Condition (C3) will be called *the sectorial local nondeterminism*.

- It is well-known that there is a close relation between Gaussian processes and operators in Hilbert spaces [62]. Recently Linde [64] has extended the notion of strong local nondeterminism to a linear operator $u : \mathscr{H} \to C(T)$, where $\mathscr{H}$ is a real Hilbert space and $C(T)$ is the Banach space of continuous functions on the compact metric space T, and applied this property to derive a lower bound for the entropy number of u. As examples, Linde [64] showed that the integral operators related to fractional Brownian motion and fractional Brownian sheets are strongly locally nondeterministic in his sense. Following this line, it would be interesting to further study the properties of strong local nondeterminism analogous to (C3) and (C3′) for linear operators related to anisotropic Gaussian random fields such as the solutions to the stochastic heat and wave equations.

3 Properties of Strong Local Nondeterminism

One of the main difficulties in studying sample path properties of anisotropic Gaussian random fields such as fractional Brownian sheets is the complexity of their dependence structure. For example, unlike fractional Brownian motion which is locally nondeterministic [77] or the Brownian sheet which has independent increments, a fractional Brownian sheet has neither of these properties. The same is true for anisotropic Gaussian random fields in general. The main technical tool which we will apply to study anisotropic Gaussian random fields is the properties of strong local nondeterminism [SLND] and sectorial local nondeterminism.

Recall that the concept of local nondeterminism was first introduced by Berman [14] to unify and extend his methods for studying local times of real-valued Gaussian processes, and then extended by Pitt [77] to Gaussian random fields. The notion of strong local nondeterminism was later developed to investigate the regularity of local times, small ball probabilities and other sample path properties of Gaussian processes and Gaussian random fields. We refer to [107; 108] for more information on the history and applications of the properties of local nondeterminism.

For Gaussian random fields, the aforementioned properties of local nondeterminism can only be satisfied by those with approximate isotropy. It is well-known that the Brownian sheet does not satisfy the properties of local nondeterminism in the senses of Berman or Pitt. Because of this, many problems for fractional Brownian motion and the Brownian sheet have to be investigated using different methods.

Khoshnevisan and Xiao [55] have recently proved that the Brownian sheet satisfies the sectorial local nondeterminism [i.e., (C3) with $H = \langle 1/2 \rangle$] and applied this property to study various analytic and geometric properties of the Brownian sheet; see also [51].

Wu and Xiao [97] extended the result of [55] and proved that fractional Brownian sheet B_0^H satisfies Condition (C3).

Theorem 3.1. Let $B_0^H = \{B_0^H(t), t \in \mathbf{R}^N\}$ be an $(N,1)$-fractional Brownian sheet with index $H = (H_1, \ldots, H_N) \in (0,1)^N$. For any fixed number $\varepsilon \in (0,1)$, there exists a positive constant $c_{3,1}$, depending on ε, H and N only, such that for all positive integers $n \geq 1$, and all $u, t^1, \ldots, t^n \in [\varepsilon, \infty)^N$, we have

$$\mathrm{Var}\left(B_0^H(u) \mid B_0^H(t^1), \ldots, B_0^H(t^n) \right) \geq c_{3,1} \sum_{j=1}^{N} \min_{0 \leq k \leq n} \left| u_j - t_j^k \right|^{2H_j}, \qquad (42)$$

where $t_j^0 = 0$ for $j = 1, \ldots, N$.

Proof. While the argument of [55] relies on the property of independent increments of the Brownian sheet and its connection to Brownian motion, the proof for B_0^H is based on a Fourier analytic argument in [46, Chapter 18] and the harmonizable representation (10) of B_0^H. We refer to [97] for details.   □

Now we prove a sufficient condition for an anisotropic Gaussian random field with stationary increments to satisfy Condition (C3').

Theorem 3.2. *Let $X = \{X(t), t \in \mathbf{R}^N\}$ be a centered Gaussian random field in $\mathbf{R}$ with stationary increments and spectral density $f(\lambda)$. Assume that there is a vector $H = (H_1, \ldots, H_N) \in (0,1)^N$ such that*

$$f(\lambda) \asymp \frac{1}{\left(\sum_{j=1}^N |\lambda_j|^{H_j}\right)^{2+Q}} \qquad \forall \lambda \in \mathbf{R}^N \backslash \{0\}, \tag{43}$$

where $Q = \sum_{j=1}^N \frac{1}{H_j}$. Then there exists a constant $c_{3,2} > 0$ such that for all $n \geq 1$, and all $u, t^1, \ldots, t^n \in \mathbf{R}^N$,

$$\mathrm{Var}\left(X(u) \mid X(t^1), \ldots, X(t^n)\right) \geq c_{3,2} \min_{0 \leq k \leq n} \rho(u, t^k)^2, \tag{44}$$

where $t^0 = 0$.

Remark 3.3. The following are some comments about Theorem 3.2.

(i) When $H_1 = \cdots = H_N$, (44) is of the same form as the SLND of fractional Brownian motion [77]. As shown by Xiao [102; 108] and Shieh and Xiao [83], many sample path properties of such Gaussian random fields are similar to those of fractional Brownian motion.

(ii) Condition (43) can be significantly weakened. In particular, one can prove that similar results hold for Gaussian random fields with stationary increments and discrete spectrum measures; see [109] for details.

(iii) It would be interesting to study under which conditions the solutions to the stochastic heat and wave equations (29) and (33) are strongly local nondeterministic.

Proof of Theorem 3.2. Denote $r \equiv \min_{0 \leq k \leq n} \rho(u, t^k)$. Since the conditional variance in (44) is the square of the $L^2(\mathrm{P})$-distance of $X(u)$ from the subspace generated by $\{X(t^1), \ldots, X(t^n)\}$, it is sufficient to prove that for all $a_k \in \mathbf{R}$ ($1 \leq k \leq n$),

$$\mathrm{E}\left(\left| X(u) - \sum_{k=1}^n a_k X(t^k)\right|^2\right) \geq c_{3,2} r^2, \tag{45}$$

and $c_{3,2} > 0$ is a constant which may only depend on H and N.

By the stochastic integral representation (15) of X, the left hand side of (45) can be written as

$$E\left(\left|X(u) - \sum_{k=1}^{n} a_k X(t^k)\right|^2\right)$$

$$= \int_{\mathbf{R}^N} \left|e^{i\langle u, \lambda\rangle} - 1 - \sum_{k=1}^{n} a_k \left(e^{i\langle t^k, \lambda\rangle} - 1\right)\right|^2 f(\lambda)\, d\lambda. \tag{46}$$

Hence, we need to only show that

$$\int_{\mathbf{R}^N} \left|e^{i\langle u, \lambda\rangle} - \sum_{k=0}^{n} a_k\, e^{i\langle t^k, \lambda\rangle}\right|^2 f(\lambda)\, d\lambda \geq c_{3,2}\, r^2, \tag{47}$$

where $t^0 = 0$ and $a_0 = -1 + \sum_{k=1}^{n} a_k$.

Let $\delta(\cdot) : \mathbf{R}^N \to [0,1]$ be a function in $C^\infty(\mathbf{R}^N)$ such that $\delta(0) = 1$ and it vanishes outside the open ball $B_\rho(0,1)$ in the metric ρ. Denote by $\widehat{\delta}$ the Fourier transform of δ. Then $\widehat{\delta}(\cdot) \in C^\infty(\mathbf{R}^N)$ as well and $\widehat{\delta}(\lambda)$ decays rapidly as $|\lambda| \to \infty$.

Let E be the diagonal matrix with $H_1^{-1}, \ldots, H_N^{-1}$ on its diagonal and let $\delta_r(t) = r^{-Q}\delta(r^{-E}t)$. Then the inverse Fourier transform and a change of variables yield

$$\delta_r(t) = (2\pi)^{-N} \int_{\mathbf{R}^N} e^{-i\langle t, \lambda\rangle}\, \widehat{\delta}(r^E \lambda)\, d\lambda. \tag{48}$$

Since $\min\{\rho(u, t^k) : 0 \leq k \leq n\} \geq r$, we have $\delta_r(u - t^k) = 0$ for $k = 0, 1, \ldots, n$. This and (48) together imply that

$$\begin{aligned}
J &:= \int_{\mathbf{R}^N} \left(e^{i\langle u, \lambda\rangle} - \sum_{k=0}^{n} a_k\, e^{i\langle t^k, \lambda\rangle}\right) e^{-i\langle u, \lambda\rangle}\, \widehat{\delta}(r^E \lambda)\, d\lambda \\
&= (2\pi)^N \left(\delta_r(0) - \sum_{k=0}^{n} a_k\, \delta_r(u - t^k)\right) \\
&= (2\pi)^N\, r^{-Q}.
\end{aligned} \tag{49}$$

On the other hand, by the Cauchy–Schwarz inequality and (46), we have

$$\begin{aligned}
J^2 &\leq \int_{\mathbf{R}^N} \left|e^{i\langle u, \lambda\rangle} - \sum_{k=0}^{n} a_k\, e^{i\langle t^k, \lambda\rangle}\right|^2 f(\lambda)\, d\lambda \cdot \int_{\mathbf{R}^N} \frac{1}{f(\lambda)} \left|\widehat{\delta}(r^E \lambda)\right|^2 d\lambda \\
&\leq E\left(\left|X(u) - \sum_{k=1}^{n} a_k X(t^k)\right|^2\right) \cdot r^{-Q} \int_{\mathbf{R}^N} \frac{1}{f(r^{-E}\lambda)} \left|\widehat{\delta}(\lambda)\right|^2 d\lambda \\
&\leq c\, E\left(\left|X(u) - \sum_{k=1}^{n} a_k X(t^k)\right|^2\right) \cdot r^{-2Q-2},
\end{aligned} \tag{50}$$

where $c > 0$ is a constant which may only depend on H and N.

We square both sides of (49) and use (50) to obtain

$$(2\pi)^{2N} r^{-2Q} \leq c\, r^{-2Q-2}\, \mathrm{E}\left(\left|X(u) - \sum_{k=1}^{n} a_k X(t^k)\right|^2\right). \tag{51}$$

Hence (47) holds. This finishes the proof of the theorem.

Given a Gaussian vector $(Z_1, \ldots, Z_n)$, we denote the determinant of its covariance matrix by $\det \mathrm{Cov}(Z_1, \ldots, Z_n)$. If $\det \mathrm{Cov}(Z_1, \ldots, Z_n) > 0$, then we have the identity

$$\frac{(2\pi)^{n/2}}{\det \mathrm{Cov}(Z_1, \ldots, Z_n)} = \int_{\mathbf{R}^n} \mathrm{E}\exp\left(-i\sum_{k=1}^{n} u_k Z_k\right) du_1 \cdots du_n. \tag{52}$$

By using the fact that, for every k, the conditional distribution of Z_k given $Z_1, \ldots, Z_{k-1}$ is still Gaussian with mean $\mathrm{E}(Z_k \,|\, Z_1, \ldots, Z_{k-1})$ and variance $\mathrm{Var}(Z_k \,|\, Z_1, \ldots, Z_{k-1})$, one can evaluate the integral in the right-hand side of (52) and thus verify the following formula:

$$\det \mathrm{Cov}(Z_1, \ldots, Z_n) = \mathrm{Var}(Z_1) \prod_{k=2}^{n} \mathrm{Var}(Z_k \,|\, Z_1, \ldots, Z_{k-1}). \tag{53}$$

A little thought reveals that (53) still holds when $\det \mathrm{Cov}(Z_1, \ldots, Z_n) = 0$. Note that the left-hand side of (53) is permutation invariant for $Z_1, \ldots, Z_n$, one can represent $\det \mathrm{Cov}(Z_1, \ldots, Z_n)$ in terms of the conditional variances in $n!$ different ways.

Combined with (42) or (44), the identity (53) can be applied to estimate the joint distribution of the Gaussian random variables $X(t^1), \ldots, X(t^n)$, where $t^1, \ldots, t^n \in \mathbf{R}^N$. This is why the properties of strong local nondeterminism are not only essential in this paper, but will also be useful in studying self-intersection local times, exact Hausdorff measure of the sample paths and other related problems [68; 69].

The following simple result will be needed in Section 8.

Lemma 3.4. *Let X be a Gaussian random field satisfying Condition (C3')* *[resp., (C3)]. Then for all integers $n \geq 1$ and for all distinct points $t^1, \ldots, t^n \in$* *$[\varepsilon, 1]^N$ [resp., all points $t^1, \ldots, t^n \in [\varepsilon, 1]^N$ with distinct coordinates, i.e.,* *$t_i^k \neq t_j^l$ when $(i, k) \neq (j, l)$], the Gaussian random variables $X(t^1), \ldots, X(t^n)$* *are linearly independent.*

Proof. We assume Condition (C3') holds and let $t^1, \ldots, t^n \in [\varepsilon, 1]^N$ be n distinct points. Then it follows from (53) that $\det \mathrm{Cov}(X(t^1), \ldots, X(t^n)) > 0$. This proves the lemma. Similar conclusion holds when Condition (C3) is satisfied. $\square$

4 Modulus of Continuity

It is sufficient to consider real-valued Gaussian random fields. Ayache and Xiao [7] established a sharp modulus of continuity (i.e., including the logarithmic correction) for fractional Brownian sheets as a consequence of their wavelet expansion for B_0^H. Since the wavelet method depends on the stochastic integral representation (8), it can not be easily applied to Gaussian random fields in general. In this section, we describe several ways to establish sharp modulus of continuity for all anisotropic Gaussian random fields satisfying Condition (C1). The first two methods, i.e., the extended Garsia-Rodemich-Rumsey continuity lemma and the minorization metric method of Kwapień and Rosiński [60], can be applied to random fields which are not necessarily Gaussian. Hence they can be more convenient when applied to solutions of stochastic partial differential equations. The third method, which is based on Dudley's entropy theorem and the Gaussian isoperimetric inequality, provides a stronger result in the sense that the upper bound is a constant instead of a random variable [cf. (69)].

Theorem 4.1 is an extension of the well-known Garsia-Rodemich-Rumsey continuity lemma [40]. It follows from Theorem 2.1 of [39], which is slightly more general [because of its freedom in choosing the function p] than an analogous result of [2]. A similar result can also be found in [22].

For our purpose, we have formulated Theorem 4.1 in terms of the metric ρ defined in (37). Let $T \subseteq \mathbf{R}^N$ be a fixed closed interval. For any $r > 0$ and $s \in T$, recall that $B_\rho(s, r) = \{t \in T : \rho(t, s) \leq r\}$ denotes the closed ball (in T) with center s and radius r in the metric ρ.

Theorem 4.1. *Suppose that $Y : T \to \mathbf{R}$ is a continuous mapping. If there exist two strictly increasing functions Ψ and p on $\mathbf{R}_+$ with $\Psi(0) = p(0) = 0$ and $\lim_{u \to \infty} \Psi(u) = \infty$ such that*

$$K := \int_T \int_T \Psi\left(\frac{|Y(s) - Y(t)|}{p(\rho(s, t))}\right) ds\, dt < \infty. \tag{54}$$

Then for all $s, t \in T$, we have

$$|Y(s) - Y(t)| \leq 8 \max_{z \in \{s, t\}} \int_0^{\rho(s, t)} \Psi^{-1}\left(\frac{4K}{\lambda_N\left(B_\rho(z, u)\right)^2}\right) \tilde{p}(du), \tag{55}$$

where $\tilde{p}(u) = p(4u)$ for all $u \in \mathbf{R}_+$.

Applying Theorem 4.1, we prove the following theorem on the modulus of continuity of an anisotropic Gaussian random field.

Theorem 4.2. *Let $X = \{X(t), t \in \mathbf{R}^N\}$ be a centered Gaussian field in $\mathbf{R}$ satisfying Condition (C1). Then, almost surely, there exists a random variable A depending on N and $(H_1, \ldots, H_N)$ only such that A has finite moments of all orders and for all $s, t \in I$,*

$$|X(s) - X(t)| \le A\,\rho(s,t)\sqrt{\log\left(1 + \rho(s,t)^{-1}\right)}. \tag{56}$$

Proof. In Theorem 4.1, let $T = I$ and we choose the functions $\Psi(x) = \exp\left(\frac{x^2}{4c_{2,3}}\right) - 1$ and $p(x) = x$, where $c_{2,3} > 0$ is the constant in (40). It follows from Condition (C1) that the random variable K in (54) has finite moments of all orders and

$$\begin{aligned}
\mathrm{E}(K) &= \mathrm{E}\int_I\int_I \Psi\left(\frac{|X(s) - X(t)|}{\rho(s,t)}\right) ds\,dt \\
&\le \int_I\int_I \mathrm{E}\Psi\left(c\,|\xi|\right) ds\,dt = c_{4,1} < \infty.
\end{aligned} \tag{57}$$

In the above ξ is a standard normal random variable. Note that $\Psi^{-1}(u) = \sqrt{4c_{2,3}\log(1+u)}$ and $\lambda_N\left(B_\rho(z,u)\right) \asymp u^Q$ is independent of z. Hence by Theorem 4.1 we have

$$\begin{aligned}
|X(s) - X(t)| &\le c\int_0^{\rho(s,t)} \sqrt{\log\left(1 + \frac{4K}{u^Q}\right)}\,du \\
&\le A\,\rho(s,t)\sqrt{\log(1 + \rho(s,t)^{-1})},
\end{aligned} \tag{58}$$

where A is a random variable depending on K and we can choose it so that $A \le c\max\{1, \log K\}$. Thus all moments of A are finite. This finishes the proof of Theorem 4.2. □

Let $X = \{X(t), t \in T\}$ be a stochastic process defined on a separable metric space (T, d) and let ψ be a Young function [that is, ψ is strictly increasing, convex and $\psi(0) = 0$]. Recently, Kwapień and Rosiński [60] investigated the following problem: When can one find an appropriate metric τ on T such that the implication

$$\sup_{s,t\in T} \mathrm{E}\psi\left(\frac{|X(s) - X(t)|}{d(s,t)}\right) < \infty \quad\Rightarrow\quad \sup_{s,t\in T} \frac{|X(s) - X(t)|}{\tau(s,t)} < \infty, \quad \text{a.s.} \tag{59}$$

holds? Their results can be applied to derive sharp modulus of continuity for a large class of stochastic processes including Gaussian random fields [but not stable random fields].

Recall from [60] that a probability measure m on T is called a weakly majorizing measure relative to ψ and the metric d if for all $s, t \in T$,

$$\int_0^{d(s,t)} \psi^{-1}\left(\frac{1}{m(B_d(s,r))}\right) dr < \infty, \tag{60}$$

where ψ^{-1} denotes the inverse function of ψ and $B_d(s,r) = \{t \in T : d(t,s) \le r\}$. For every weakly majorizing measure m, the "minorizing metric" $\tau = \tau_{\psi,d,m}$ on T relative to ψ, d and m is defined as

$$\tau(s,t) = \max\left\{\int_0^{d(s,t)} \psi^{-1}\left(\frac{1}{m(B_d(s,r))}\right)dr, \int_0^{d(t,s)} \psi^{-1}\left(\frac{1}{m(B_d(t,r))}\right)dr\right\}.$$
(61)

The following theorem of [60] gives a sufficient condition for (59) to hold.

Theorem 4.3. *Let ψ be a Young function satisfying the following growth condition:*

$$\psi(x)\psi(y) \le \psi(c_{4,2}(x+y)) \quad \text{for all } x, y \ge 0,$$
(62)

where $c_{4,2} > 0$ is a constant. Let m be a weakly majorizing measure relative to ψ and d on T. Then there exists a positive constant $c_{4,3}$ depending only on ψ such that for every stochastic process $X = \{X(t), t \in T\}$,

$$\mathrm{E}\psi\left(c_{4,3} \sup_{s,t\in T} \frac{|X(s) - X(t)|}{\tau(s,t)}\right) \le 1 + \sup_{s,t\in T} \mathrm{E}\psi\left(\frac{|X(s) - X(t)|}{d(s,t)}\right),$$
(63)

where τ is the minorizing metric relative to ψ, d and m.

Note that, for any $\alpha > 0$, the function $\psi(x) = x^\alpha$ does not satisfy the growth condition (62), hence Theorem 4.3 is not applicable to stable random fields.

By applying Theorem 4.3 to the metric space (I, ρ) in our setting, we can provide more information about the random variable A in Theorem 4.2.

Corollary 4.4. *Let $X = \{X(t), t \in \mathbf{R}^N\}$ be a centered Gaussian field in $\mathbf{R}$ satisfying Condition (C1). Then there exists a constant $c_{4,4} > 0$ such that*

$$\mathrm{E}\exp\left(c_{4,4} \sup_{s,t\in I} \frac{|X(s) - X(t)|^2}{\rho^2(s,t)\log\left(1 + \rho(s,t)^{-1}\right)}\right) < \infty.$$
(64)

Proof. This can be verified by showing that the Lebesgue measure on I is a weakly majorizing measure relative to the Young function $\psi(x) = e^{x^2} - 1$ and the metric ρ; and the corresponding minorizing metric $\tau(s,t)$ satisfies

$$c_{4,5}\,\rho(s,t)\sqrt{\log\left(1 + \rho(s,t)^{-1}\right)} \le \tau(s,t) \le c_{4,6}\,\rho(s,t)\sqrt{\log\left(1 + \rho(s,t)^{-1}\right)},$$
(65)

for all $s, t \in I$. We leave the details to an interested reader. □

As a third method, we mention that it is also possible to obtain a uniform modulus of continuity for a Gaussian random field satisfying Condition (C1) by using the Gaussian isoperimetric inequality [85, Lemma 2.1]. To this end, we introduce an auxiliary Gaussian random field $Y = \{Y(s,t) : t \in I, s \in [0, h]\}$ defined by $Y(t,s) = X(t+s) - X(t)$, where $h \in (0,1)^N$. Then the canonical metric d on $T := I \times [0, h]$ associated with Y satisfies the following inequality:

$$d\big((t,s),(t',s')\big) \le c \min \big\{\rho(0,s) + \rho(0,s'),\ \rho(s,s') + \rho(t,t')\big\}. \tag{66}$$

Denote the d-diameter of T by D. It follows from (66) that $D \le c_{4,7} \sum_{j=1}^{N} h_j^{H_j} = c_{4,7}\, \rho(0,h)$, and the d-covering number of T satisfies

$$N_d(T,\varepsilon) \le c \left(\frac{1}{\varepsilon}\right)^{Q} \prod_{j=1}^{N} \left(\frac{h_j}{\varepsilon^{1/H_j}}\right) \le c_{4,8}\, \varepsilon^{-2Q}.$$

One can verify that

$$\int_0^D \sqrt{\log N_d(T,\varepsilon)}\, d\varepsilon \le c_{4,9}\, \rho(0,h)\, \sqrt{\log\big(1 + \rho(0,h)^{-1}\big)}. \tag{67}$$

It follows from Lemma 2.1 in [85] that for all $u \ge 2c_{4,9}\, \rho(0,h)\sqrt{\log\big(1 + \rho(0,h)^{-1}\big)}$,

$$\mathrm{P}\left\{ \sup_{(t,s)\in T} \big|X(t+s) - X(t)\big| \ge u \right\} \le \exp\left(-\frac{u^2}{D^2}\right). \tag{68}$$

By using (68) and a standard Borel-Cantelli argument, we can prove that

$$\limsup_{|h|\to 0} \frac{\sup_{t\in I, s\in[0,h]} |X(t+s) - X(t)|}{\rho(0,h)\sqrt{\log(1 + \rho(0,h)^{-1})}} \le c_{4,10}, \tag{69}$$

where $c_{4,10} > 0$ is a finite constant depending on $c_{2,4}$, I and H only.

We believe that, for Gaussian random fields satisfying (C1), the rate function in (56) is sharp. This has been partly verified by Meerschaert, Wang and Xiao [68] who proved that, if a Gaussian field X satisfies Conditions (C1) and (C3), then

$$c_{4,11} \le \limsup_{|h|\to 0} \frac{\sup_{t\in I, s\in[0,h]} |X(s) - X(t)|}{\rho(0,h)\sqrt{\log(1 + \rho(0,h)^{-1})}} \le c_{4,12}, \tag{70}$$

where $c_{4,11}$ and $c_{4,12}$ are positive constants depending on $c_{2,3}$, $c_{2,4}$, I and H only.

On the other hand, we can also use the above metric entropy method to prove that, for all $t_0 \in I$ and $u > 0$ large enough,

$$\mathrm{P}\left\{ \sup_{s\in[0,h]} \big|X(t_0 + s) - X(t_0)\big| \ge \rho(0,h)\, u \right\} \le \exp\big(-c_{4,13}\, u^2\big), \tag{71}$$

where $c_{4,13}$ is a positive constant depending on $c_{2,4}$, I and H only.

By using (71) and the Borel-Cantelli lemma, we derive the following local modulus of continuity for Gaussian random fields satisfying (C1): There exists a positive constant $c_{4,14}$ such that for every $t_0 \in I$,

$$\limsup_{|h|\to 0} \frac{\sup_{s\in[0,h]} |X(t_0 + s) - X(t_0)|}{\rho(0,h)\sqrt{\log\log(1 + \rho(0,h)^{-1})}} \le c_{4,14}, \qquad \text{a.s.} \tag{72}$$

Under certain mild conditions, it can be shown that (72) is sharp. For example, Meerschaert, Wang and Xiao [68] proved that, if X is a Gaussian random field with stationary increments and satisfies (C1), then for every $t_0 \in I$,

$$\limsup_{|h| \to 0} \frac{\sup_{s \in [0,h]} |X(t_0 + s) - X(t_0)|}{\rho(0,h) \sqrt{\log \log(1 + \rho(0,h)^{-1})}} = c_{4,15}, \qquad \text{a.s.,} \qquad (73)$$

where $c_{4,15}$ is a positive constant.

We should mention that one can also study the uniform and local moduli of continuity in terms of the increments of X over intervals. Related results of this type for fractional Brownian sheets have been obtained by Wang [95].

In the special case when X is a direct sum of independent fractional Brownian motions of indices $H_1, \ldots, H_N$, that is,

$$X(t) = X_1(t_1) + \cdots + X_N(t_N), \qquad \forall t = (t_1, \ldots, t_N) \in \mathbf{R}^N, \qquad (74)$$

where $X_1, \ldots, X_N$ are independent fractional Brownian motions in $\mathbf{R}$ of indices $H_1, \ldots, H_N$, respectively, Kôno [58] established integral tests for the uniform and local upper and lower classes. It is natural to ask whether his results hold for more general anisotropic Gaussian random fields.

5 Small Ball Probabilities

In recent years, there has been much interest in studying the small ball probabilities of Gaussian processes [62; 63]. Small ball properties of fractional Brownian sheets have been considered by Dunker [32], Mason and Shi [66], Belinski and Linde [9].

The small ball behavior of operator-scaling Gaussian random fields with stationary increments and the solution to the stochastic heat equation is different, as shown by the following general result.

Theorem 5.1. Let $X = \{X(t), t \in \mathbf{R}^N\}$ be a centered Gaussian field in $\mathbf{R}$ satisfying Conditions (C1) and (C3') on $I = [0,1]^N$. Then there exist positive constants $c_{5,1}$ and $c_{5,2}$ such that for all $\varepsilon > 0$,

$$\exp\left(-\frac{c_{5,1}}{\varepsilon^Q}\right) \leq \mathrm{P}\left\{\max_{t \in [0,1]^N} |X(t)| \leq \varepsilon\right\} \leq \exp\left(-\frac{c_{5,2}}{\varepsilon^Q}\right), \qquad (75)$$

where $Q = \sum_{j=1}^{N} \frac{1}{H_j}$.

In order to prove the lower bound in (75), we will make use of the following general result of [84], see also [61, p. 257].

Lemma 5.2. *Let* $Y = \{Y(t), t \in T\}$ *be a real-valued Gaussian process with mean zero and let d be the canonical metric on T defined by*

$$d(s,t) = \left\{ \mathrm{E}\left(|Y(s) - Y(t)|^2\right)\right\}^{1/2}, \quad s,t \in T, \tag{76}$$

and denote by $N_d(T, \varepsilon)$ *the smallest number of d-balls of radius* $\varepsilon > 0$ *needed to cover T. Assume that there is a nonnegative function* ψ *on* $\mathbf{R}_+$ *such that* $N_d(T, \varepsilon) \le \psi(\varepsilon)$ *for* $\varepsilon > 0$ *and such that*

$$c_{5,3}\psi(\varepsilon) \le \psi\left(\frac{\varepsilon}{2}\right) \le c_{5,4}\psi(\varepsilon) \tag{77}$$

for some constants $1 < c_{5,3} \le c_{5,4} < \infty$ *and all* $\varepsilon > 0$. *Then there is a constant* $c_{5,5} > 0$ *such that*

$$\mathrm{P}\left\{ \sup_{t,s \in T} |Y(t) - Y(s)| \le \varepsilon \right\} \ge \exp\left(-c_{5,5}\psi(\varepsilon)\right). \tag{78}$$

Proof of Theorem 5.1. It follows from (C1) that for all $\varepsilon \in (0,1)$,

$$N_\rho(I, \varepsilon) \le c\varepsilon^{-Q} := \psi(\varepsilon). \tag{79}$$

Clearly $\psi(\varepsilon)$ satisfies the condition (77). Hence the lower bound in (75) follows from Lemma 5.2.

The proof of the upper bound in (75) is based on Condition (C3′) and a conditioning argument in [71]. For any integer $n \ge 2$, we divide $[0,1]^N$ into n^Q rectangles of side-lengths n^{-1/H_j} $(j = 1, \ldots, N)$. We denote the lower-left vertices of these rectangles (in any order) by $t_{n,k}$ $(k = 1, 2, \ldots, n^Q)$. Then

$$\mathrm{P}\left\{ \max_{t \in [0,1]^N} |X(t)| \le \varepsilon \right\} \le \mathrm{P}\left\{ \max_{1 \le k \le n^Q} |X(t_{n,k})| \le \varepsilon \right\}. \tag{80}$$

It follows from Condition (C3′) that for every $1 \le k \le n^Q$,

$$\mathrm{Var}\left(X(t_{n,k}) \mid X(t_{n,i}), 1 \le i \le k - 1\right) \ge cn^{-1}. \tag{81}$$

This and Anderson's inequality for Gaussian measures imply the following upper bound for the conditional probabilities

$$\mathrm{P}\left\{ |X(t_{n,k})| \le \varepsilon \mid X(t_{n,j}), \ 1 \le j \le k - 1 \right\} \le \Phi(c\varepsilon n), \tag{82}$$

where $\Phi(x)$ is the distribution function of a standard normal random variable. It follows from (80) and (82) that

$$\mathrm{P}\left\{ \max_{t \in [0,1]^N} |X(t)| \le \varepsilon \right\} \le [\Phi(c\varepsilon n)]^{n^Q}. \tag{83}$$

By taking $n = (c\varepsilon)^{-1}$, we obtain the upper bound in (75).

Remark 5.3. If $H_1 = H_2 = \cdots = H_N$, then (75) is of the same form as the small ball probability estimates for multiparameter fractional Brownian motion [62; 85].

Among many applications, Theorem 5.1 can be applied to establish Chung-type laws of the iterated logarithm for anisotropic Gaussian random fields. Moreover, it would also be interesting to investigate the small ball probabilities of X in other norms such as the L^2 or Hölder norms; see [62; 63] for information on Gaussian processes.

6 Hausdorff and Packing Dimensions of the Range and Graph

In this section, we study the Hausdorff and packing dimensions of the range $X([0,1]^N) = \{X(t) : t \in [0,1]^N\}$ and the graph $\mathrm{Gr}X([0,1]^N) = \{(t, X(t)) : t \in [0,1]^N\}$ of a Gaussian random field X satisfying Condition (C1) on $[0,1]^N$.

Hausdorff dimension and Hausdorff measure have been extensively used in describing thin sets and fractals. For any set $E \subseteq \mathbf{R}^d$ and $\gamma > 0$, we will denote the Hausdorff dimension and the γ-dimensional Hausdorff measure of E by $\dim_{\mathscr{H}} E$ and $\mathscr{H}^\gamma(E)$, respectively [36; 46; 67]. More generally, for any nondecreasing, right continuous function $\varphi : [0,1] \to [0,\infty)$ with $\varphi(0) = 0$, one can define the Hausdorff measure of E with respect to φ and denoted it by $\mathscr{H}^\varphi(E)$. We say that a function φ is an exact Hausdorff measure function for E if $0 < \mathscr{H}^\varphi(E) < \infty$.

Now we recall briefly the definition of capacity and its connection to Hausdorff dimension. A kernel κ is a measurable function $\kappa : \mathbf{R}^d \times \mathbf{R}^d \to [0,\infty]$. For a Borel measure μ on $\mathbf{R}^d$, the energy of μ with respect to the kernel κ (in short, κ energy of μ) is defined by

$$\mathscr{E}_\kappa(\mu) = \iint \kappa(x,y)\,\mu(dx)\,\mu(dy). \tag{84}$$

For any Borel set $E \subseteq \mathbf{R}^d$, the *capacity of E with respect to κ*, denoted by $\mathscr{C}_\kappa(E)$, is defined by

$$\mathscr{C}_\kappa(E) = \left[\inf_{\mu \in \mathscr{P}(E)} \mathscr{E}_\kappa(\mu) \right]^{-1}, \tag{85}$$

where $\mathscr{P}(E)$ is the family of probability measures carried by E, and, by convention, $\infty^{-1} = 0$. Note that $\mathscr{C}_\kappa(E) > 0$ if and only if there is a probability measure μ on E with finite κ-energy. We will mostly consider the case when $\kappa(x,y) = f(|x-y|)$, where f is a non-negative and non-increasing function on $\mathbf{R}_+$. In particular, if

$$f(r) = \begin{cases} r^{-\alpha} & \text{if } \alpha > 0, \\ \log\left(\dfrac{e}{r \wedge 1}\right) & \text{if } \alpha = 0, \end{cases} \tag{86}$$

then the corresponding $\mathscr{E}_\kappa(\mu)$ and $\mathscr{C}_\kappa(E)$ will be denoted by $\mathscr{E}_\alpha(\mu)$ and $\mathscr{C}_\alpha(E)$ respectively; and the former will be called the α-energy of μ and the latter will be called the *Bessel–Riesz capacity of E of order α*. The capacity dimension of E is defined by

$$\dim_c(E) = \sup\{\alpha > 0 : \mathscr{C}_\alpha(E) > 0\}. \tag{87}$$

The well-known Frostman's theorem (see [46, p. 133] or [49]) states that $\dim_{\mathscr{H}} E = \dim_c(E)$ for every compact set E in $\mathbf{R}^d$. This result provides a very useful analytic way for the lower bound calculation of Hausdorff dimension. That is, for $E \subseteq \mathbf{R}^d$ in order to show $\dim_{\mathscr{H}} E \geq \alpha$, one only needs to find a measure μ on E such that the α-energy of μ is finite. For many deterministic and random sets such as self-similar sets or the range of a stochastic process, there are natural choices of μ. This argument is usually referred to as the *capacity argument*.

Packing dimension and packing measure were introduced by Tricot [92] and Taylor and Tricot [89] as dual concepts to Hausdorff dimension and Hausdorff measure. We only recall briefly a definition of packing dimension, which will be denoted by $\dim_{\mathscr{P}}$. For any $\varepsilon > 0$ and any bounded set $F \subseteq \mathbf{R}^d$, let $N(F, \varepsilon)$ be the smallest number of balls of radius ε needed to cover F. Then the *upper box-counting dimension* of F is defined as

$$\overline{\dim}_{\mathscr{B}} F = \limsup_{\varepsilon \to 0} \frac{\log N(F, \varepsilon)}{-\log \varepsilon}. \tag{88}$$

The packing dimension of F can be defined by

$$\dim_{\mathscr{P}} F = \inf \left\{ \sup_n \overline{\dim}_{\mathscr{B}} F_n : \; F \subseteq \bigcup_{n=1}^\infty F_n \right\}. \tag{89}$$

It is known that for any bounded set $F \subseteq \mathbf{R}^d$,

$$\dim_{\mathscr{H}} F \leq \dim_{\mathscr{P}} F \leq \overline{\dim}_{\mathscr{B}} F \leq d. \tag{90}$$

Further information on packing dimension and packing measure can be found in [36; 67]. We mention that various tools from fractal geometry have been applied to studying sample path properties of stochastic processes since 1950's. The survey papers [88] and [106] summarize various fractal properties of random sets related to sample paths of Markov processes.

Throughout the rest of this paper, we will assume that

$$0 < H_1 \leq \ldots \leq H_N < 1. \tag{91}$$

Theorem 6.1. *Let $X = \{X(t), t \in \mathbf{R}^N\}$ be an (N, d)-Gaussian field satisfying Condition (C1) on $I = [0, 1]^N$. Then, with probability 1,*

$$\dim_{\mathscr{H}} X\big([0\,,1]^N\big) = \dim_{\mathscr{P}} X\big([0\,,1]^N\big) = \min\left\{d;\ \sum_{j=1}^{N}\frac{1}{H_j}\right\} \qquad (92)$$

and

$$\dim_{\mathscr{H}} \mathrm{Gr} X\big([0\,,1]^N\big) = \dim_{\mathscr{P}} \mathrm{Gr} X\big([0\,,1]^N\big)$$

$$= \min\left\{\sum_{j=1}^{k}\frac{H_k}{H_j} + N - k + (1-H_k)d,\ 1 \le k \le N;\ \sum_{j=1}^{N}\frac{1}{H_j}\right\} \qquad (93)$$

$$= \begin{cases} \sum_{j=1}^{N} H_j^{-1}, & \text{if } \sum_{j=1}^{N} H_j^{-1} \le d, \\ \sum_{j=1}^{k}(H_k/H_j) + N - k + (1-H_k)d, & \text{if } \sum_{j=1}^{k-1} H_j^{-1} \le d < \sum_{j=1}^{k} H_j^{-1}, \end{cases}$$

where $\sum_{j=1}^{0} H_j^{-1} := 0$.

The last equality in (93) is verified by the following lemma, whose proof is elementary and is omitted. Denote

$$\kappa := \min\left\{\sum_{j=1}^{k}\frac{H_k}{H_j} + N - k + (1-H_k)d,\, 1 \le k \le N\ ;\, \sum_{j=1}^{N}\frac{1}{H_j}\right\}. \qquad (94)$$

Lemma 6.2. *Assume (91) holds. We have*

(i) *If* $d \ge \sum_{j=1}^{N} H_j^{-1}$, *then* $\kappa = \sum_{j=1}^{N} H_j^{-1}$.
(ii) *If* $\sum_{j=1}^{\ell-1} H_j^{-1} \le d < \sum_{j=1}^{\ell} H_j^{-1}$ *for some* $1 \le \ell \le N$, *then*

$$\kappa = \sum_{j=1}^{\ell}\frac{H_\ell}{H_j} + N - \ell + (1-H_\ell)d \qquad (95)$$

and $\kappa \in (N - \ell + d\,, N - \ell + d + 1]$.

Because of (90) we can divide the proof of Theorem 6.1 into proving the upper bounds for the upper box dimensions and the lower bounds for the Hausdorff dimensions separately. The proofs are similar to those in [7] for fractional Brownian sheets. In the following, we first show that the upper bounds for $\overline{\dim}_{\mathscr{B}} X([0\,,1]^N)$ and $\overline{\dim}_{\mathscr{B}} \mathrm{Gr} X([0\,,1]^N)$ follow from Theorem 4.2 and a covering argument.

Proof of the upper bounds in Theorem 6.1. In order to prove the upper bound in (92), we note that clearly $\overline{\dim}_{\mathscr{B}} X([0\,,1]^N) \le d$ a.s., so it suffices to prove the following inequality:

$$\overline{\dim}_{\mathscr{B}} X\big([0\,,1]^N\big) \le \sum_{j=1}^{N}\frac{1}{H_j}, \qquad \text{a.s.} \qquad (96)$$

For any constants $0 < \gamma_j < H_j$ $(1 \le j \le N)$, it follows from Theorem 4.2 that there is a random variable A of finite moments of all orders such that for almost all $\omega \in \Omega$,

$$\sup_{s,t\in[0,1]^N} \frac{|X(s,\omega) - X(t,\omega)|}{\sum_{j=1}^N |s_j - t_j|^{\gamma_j}} \le A(\omega). \tag{97}$$

We fix an ω such that (97) holds and then suppress it. For any integer $n \ge 2$, we divide $[0,1]^N$ into m_n sub-rectangles $\{R_{n,i}\}$ with sides parallel to the axes and side-lengths n^{-1/H_j} $(j = 1,\dots,N)$, respectively. Then

$$m_n \le c_{6,1}\, n^{\sum_{j=1}^N (1/H_j)} \tag{98}$$

and $X([0,1]^N)$ can be covered by $X(R_{n,i})$ $(1 \le i \le m_n)$. By (97), we see that the diameter of the image $X(R_{n,i})$ satisfies

$$\operatorname{diam} X(R_{n,i}) \le c_{6,2}\, n^{-1+\delta}, \tag{99}$$

where $\delta = \max\{(H_j - \gamma_j)/H_j, 1 \le j \le N\}$. Consequently, for $\varepsilon_n = c_{6,2}\, n^{-1+\delta}$, $X([0,1]^N)$ can be covered by at most m_n balls in $\mathbf{R}^d$ of radius ε_n. That is,

$$N\big(X([0,1]^N),\, \varepsilon_n\big) \le c_{6,1}\, n^{\sum_{j=1}^N (1/H_j)}. \tag{100}$$

This implies

$$\overline{\dim}_{\mathscr{B}} X([0,1]^N) \le \frac{1}{1-\delta} \sum_{j=1}^N \frac{1}{H_j}, \quad \text{a.s.} \tag{101}$$

By letting $\gamma_j \uparrow H_j$ along rational numbers, we have $\delta \downarrow 0$ and (96) follows from (101).

Now we turn to the proof of the upper bound in (93). We will show that there are several different ways to cover $\operatorname{Gr} X([0,1]^N)$ by balls in $\mathbf{R}^{N+d}$ of the same radius, each of which leads to an upper bound for $\overline{\dim}_{\mathscr{B}} \operatorname{Gr} X([0,1]^N)$.

For each fixed integer $n \ge 2$, we have

$$\operatorname{Gr} X([0,1]^N) \subseteq \bigcup_{i=1}^{m_n} R_{n,i} \times X(R_{n,i}). \tag{102}$$

It follows from (99) and (102) that $\operatorname{Gr} X([0,1]^N)$ can be covered by m_n balls in $\mathbf{R}^{N+d}$ with radius $c_{6,2}\, n^{-1+\delta}$ and the same argument as the above yields

$$\overline{\dim}_{\mathscr{B}} \operatorname{Gr} X([0,1]^N) \le \sum_{j=1}^N \frac{1}{H_j}, \quad \text{a.s.} \tag{103}$$

We fix an integer $1 \le k \le N$. Observe that each $R_{n,i} \times X(R_{n,i})$ can be covered by $\ell_{n,k}$ balls (or cubes) in $\mathbf{R}^{N+d}$ of radius (or side-length) $n^{-(1/H_k)}$, where by (97) we have

$$\ell_{n,k} \le c\, n^{\sum_{j=k}^{N}(H_k^{-1}-H_j^{-1})} \times n^{(H_k^{-1}-1+\delta)d}, \qquad \text{a.s.} \tag{104}$$

Hence $\mathrm{Gr}X([0,1]^N)$ can be covered by $m_n \times \ell_{n,k}$ balls in $\mathbf{R}^{N+d}$ with radius $n^{-(1/H_k)}$. Consequently,

$$\overline{\dim}_{\mathscr{B}} \mathrm{Gr}X([0,1]^N) \le \sum_{j=1}^{k} \frac{H_k}{H_j} + N - k + (1 - H_k + \delta H_k)d, \qquad \text{a.s.} \tag{105}$$

Letting $\gamma_j \uparrow H_j$ along rational numbers, we derive that for every $k = 1, \ldots, N$,

$$\overline{\dim}_{\mathscr{B}} \mathrm{Gr}X([0,1]^N) \le \sum_{j=1}^{k} \frac{H_k}{H_j} + N - k + (1 - H_k)d. \tag{106}$$

Combining (103) and (106) yields the upper bound in (93).

For proving the lower bounds in Theorem 6.1, we will make use of the following elementary Lemmas 6.3 and 6.4. The former is proved in [110, p. 212] which will be used to derive a lower bound for $\dim_{\mathscr{H}} X([0,1]^N)$; the latter is proved in [7] which will be needed for determining a lower bound for $\dim_{\mathscr{H}} \mathrm{Gr}X([0,1]^N)$. Both lemmas will be useful in the proof of Theorem 7.1 in Section 7.

Lemma 6.3. *Let $0 < \alpha < 1$ and $\varepsilon > 0$ be given constants. Then for any constants $\delta > 2\alpha$, $M > 0$ and $p > 0$, there exists a positive and finite constant $c_{6,3}$, depending on ε, δ, p and M only, such that for all $0 < A \le M$,*

$$\int_{\varepsilon}^{1} ds \int_{\varepsilon}^{1} \frac{dt}{\left(A + |s - t|^{2\alpha}\right)^p} \le c_{6,3}\left(A^{-(p-\frac{1}{\delta})} + 1\right). \tag{107}$$

Lemma 6.4. *Let α, β and η be positive constants. For $A > 0$ and $B > 0$, let*

$$J := J(A, B) = \int_{0}^{1} \frac{dt}{(A + t^\alpha)^\beta (B + t)^\eta}. \tag{108}$$

Then there exist finite constants $c_{6,4}$ and $c_{6,5}$, depending on α, β, η only, such that the following hold for all real numbers A, $B > 0$ satisfying $A^{1/\alpha} \le c_{6,3} B$:

(i)   If $\alpha\beta > 1$, then

$$J \le c_{6,5} \frac{1}{A^{\beta - \alpha^{-1}} B^\eta}; \tag{109}$$

(ii)   If $\alpha\beta = 1$, then

$$J \le c_{6,5} \frac{1}{B^\eta} \log\left(1 + BA^{-1/\alpha}\right); \tag{110}$$

(iii) If $0 < \alpha\beta < 1$ and $\alpha\beta + \eta \neq 1$, then

$$J \leq c_{6,5} \left(\frac{1}{B^{\alpha\beta+\eta-1}} + 1 \right). \tag{111}$$

Proof of the lower bounds in Theorem 6.1. First we prove the lower bound in (92). Note that for any $\varepsilon \in (0,1)$, $\dim_{\mathscr{H}} X([0,1]^N) \geq \dim_{\mathscr{H}} X([\varepsilon,1]^N)$. It is sufficient to show that $\dim_{\mathscr{H}} X([\varepsilon,1]^N) \geq \gamma$ a.s. for every $0 < \gamma < \min\{d, \sum_{j=1}^{N} \frac{1}{H_j}\}$.

Let μ_X be the image measure of the Lebesgue measure on $[\varepsilon,1]^N$ under the mapping $t \mapsto X(t)$. Then the energy of μ_X of order γ can be written as

$$\int_{\mathbf{R}^d} \int_{\mathbf{R}^d} \frac{\mu_X(dx)\,\mu_X(dy)}{|x-y|^\gamma} = \int_{[\varepsilon,1]^N} \int_{[\varepsilon,1]^N} \frac{ds\,dt}{|X(s)-X(t)|^\gamma}.$$

Hence by Frostman's theorem [46, Chapter 10], it is sufficient to show that for every $0 < \gamma < \min\{d, \sum_{j=1}^{N} \frac{1}{H_j}\}$,

$$\mathscr{E}_\gamma = \int_{[\varepsilon,1]^N} \int_{[\varepsilon,1]^N} \mathrm{E}\left(\frac{1}{|X(s)-X(t)|^\gamma} \right) ds\,dt < \infty. \tag{112}$$

Since $0 < \gamma < d$, we have $0 < \mathrm{E}(|\Xi|^{-\gamma}) < \infty$, where Ξ is a standard d-dimensional normal vector. Combining this fact with Condition (C1), we have

$$\mathscr{E}_\gamma \leq c \int_\varepsilon^1 ds_1 \int_\varepsilon^1 dt_1 \cdots \int_\varepsilon^1 ds_N \int_\varepsilon^1 \frac{dt_N}{\left(\sum_{j=1}^N |s_j - t_j|^{2H_j} \right)^{\gamma/2}}. \tag{113}$$

We choose positive constants $\delta_2, \ldots, \delta_N$ such that $\delta_j > 2H_j$ for each $2 \leq j \leq N$ and

$$\frac{1}{\delta_2} + \cdots + \frac{1}{\delta_N} < \frac{\gamma}{2} < \frac{1}{2H_1} + \frac{1}{\delta_2} + \cdots + \frac{1}{\delta_N}. \tag{114}$$

This is possible since $\gamma < \sum_{j=1}^{N}(1/H_j)$. By applying Lemma 6.3 to (113) with

$$A = \sum_{j=1}^{N-1} |s_j - t_j|^{2H_j} \quad \text{and} \quad p = \gamma/2, \tag{115}$$

we find that $\mathscr{E}_\gamma$ is at most $c_{6,6}$ plus $c_{6,6}$ times

$$\int_\varepsilon^1 ds_1 \cdots \int_\varepsilon^1 ds_{N-1} \int_\varepsilon^1 \frac{dt_{N-1}}{\left(\sum_{j=1}^{N-1} |s_j - t_j|^{2H_j} \right)^{(\gamma/2)-(1/\delta_N)}}. \tag{116}$$

By repeatedly using Lemma 6.3 to the integral in (116) for $N-2$ steps, we deduce that

$$\mathcal{E}_\gamma \le c_{6,7} + c_{6,7} \int_\varepsilon^1 ds_1 \int_\varepsilon^1 \frac{dt_1}{\left(|s_1 - t_1|^{2H_1}\right)^{(\gamma/2)-((1/\delta_2)+\cdots+(1/\delta_N))}}. \tag{117}$$

Since the δ_j's satisfy (114), we have $2H_1[\gamma/2 - (\delta_2^{-1} + \cdots + \delta_N^{-1})] < 1$. Thus the integral in the right hand side of (117) is finite. This proves (112), and (92) follows.

Now we prove the lower bound in (93). Since $\dim_{\mathcal{H}} \mathrm{Gr} X([0,1]^N) \ge \dim_{\mathcal{H}} X([0,1]^N)$ always holds, we only need to consider the case when

$$\sum_{j=1}^{k-1} \frac{1}{H_j} \le d < \sum_{j=1}^{k} \frac{1}{H_j} \quad \text{for some } 1 \le k \le N. \tag{118}$$

Here and in the sequel, $\sum_{j=1}^0 (1/H_j) = 0$.

Let $0 < \varepsilon < 1$ and $0 < \gamma < \sum_{j=1}^k (H_k/H_j) + N - k + (1 - H_k)d$ be fixed, but arbitrary, constants. By Lemma 6.2, we may and will assume $\gamma \in (N - k + d, N - k + d + 1)$. In order to prove $\dim_{\mathcal{H}} \mathrm{Gr} X([\varepsilon, 1]^N) \ge \gamma$ a.s., again by Frostman's theorem, it is sufficient to show

$$\mathcal{G}_\gamma = \int_{[\varepsilon,1]^N} \int_{[\varepsilon,1]^N} \mathrm{E}\left[\frac{1}{\left(|s - t|^2 + |X(s) - X(t)|^2\right)^{\gamma/2}}\right] ds\, dt < \infty. \tag{119}$$

Since $\gamma > d$, we note that for a standard normal vector Ξ in $\mathbf{R}^d$ and any number $a \in \mathbf{R}$,

$$\mathrm{E}\left[\frac{1}{\left(a^2 + |\Xi|^2\right)^{\gamma/2}}\right] \le c_{6,8}\, a^{-(\gamma-d)}, \tag{120}$$

see [46, p. 279]. Consequently, we derive that

$$\mathcal{G}_\gamma \le c_{6,8} \int_{[\varepsilon,1]^N} \int_{[\varepsilon,1]^N} \frac{ds\, dt}{\sigma(s,t)^d\, |s - t|^{\gamma-d}}, \tag{121}$$

where $\sigma^2(s,t) = \mathrm{E}\left[(X_1(s) - X_1(t))^2\right]$. By Condition (C1) and a change of variables, we have

$$\mathcal{G}_\gamma \le c_{6,9} \int_0^1 dt_N \cdots \int_0^1 \frac{dt_1}{\left(\sum_{j=1}^N t_j^{H_j}\right)^d \left(\sum_{j=1}^N t_j\right)^{\gamma-d}}. \tag{122}$$

In order to show the integral in (122) is finite, we will integrate $[dt_1], \ldots, [dt_k]$ iteratively. Furthermore, we will assume $k > 1$ in (118) [If $k = 1$, we can use (111) to obtain (126) directly].

We integrate $[dt_1]$ first. Since $H_1 d > 1$, we can use (109) of Lemma 6.4 with $A = \sum_{j=2}^N t_j^{H_j}$ and $B = \sum_{j=2}^N t_j$ to get

$$\mathcal{G}_\gamma \le c_{6,10} \int_0^1 dt_N \cdots \int_0^1 \frac{dt_2}{\left(\sum_{j=2}^N t_j^{H_j}\right)^{d-(1/H_1)} \left(\sum_{j=2}^N t_j\right)^{\gamma-d}}. \tag{123}$$

We can repeat this procedure for integrating $dt_2, \ldots, dt_{k-1}$. Note that if $d = \sum_{j=1}^{k-1}(1/H_j)$, then we need to use (110) to integrate $[dt_{k-1}]$ and obtain

$$\mathscr{G}_\gamma \le c_{6,11} \int_0^1 dt_N \cdots \int_0^1 \frac{1}{\left(\sum_{j=k}^N t_j\right)^{\gamma-d}} \log\left(1 + \frac{1}{\sum_{j=k}^N t_j}\right) dt_k < \infty. \quad (124)$$

Note that the last integral is finite since $\gamma - d < N - k + 1$. On the other hand, if $d > \sum_{j=1}^{k-1}(1/H_j)$, then (109) gives

$$\mathscr{G}_\gamma \le c_{6,12} \int_0^1 dt_N \cdots \int_0^1 \frac{dt_k}{\left(\sum_{j=k}^N t_j^{H_j}\right)^{d-\sum_{j=1}^{k-1}(1/H_j)} \left(\sum_{j=k}^N t_j\right)^{\gamma-d}}. \quad (125)$$

We integrate $[dt_k]$ in (125) and by using (111), we see that $\mathscr{G}_\gamma$ is bounded above by

$$c_{6,13}\left[\int_0^1 dt_N \cdots \int_0^1 \frac{dt_{k+1}}{\left(\sum_{j=k+1}^N t_j\right)^{\gamma-d+H_k(d-\sum_{j=1}^{k-1}(1/H_j))-1}} + 1\right] < \infty, \quad (126)$$

since $\gamma - d + H_k(d - \sum_{j=1}^{k-1}(1/H_j)) - 1 < N - k$. Combining (124) and (126) yields (119). This completes the proof of Theorem 6.1.

There are several possible ways to strengthen and extend Theorem 6.1. For example, it would be interesting to determine the exact Hausdorff and packing measure functions for the range $X([0,1]^N)$ and graph $\mathrm{Gr}X([0,1]^N)$ for anisotropic Gaussian random fields. When X is the Brownian sheet or a fractional Brownian motion, the problems on exact Hausdorff measure functions have been considered by Ehm [34], Talagrand [85; 86], Xiao [99; 101; 102]. Here is a summary of the known results:

(i) Let $X^\alpha = \{X^\alpha(t), t \in \mathbf{R}^N\}$ be an (N, d)-fractional Brownian motion of index α. If $N < \alpha d$, then $\varphi_1(r) = r^{N/\alpha} \log\log 1/r$ ia an exact Hausdorff measure function for the range and graph of X^α. If $N > \alpha d$, then $X^\alpha([0,1]^N)$ a.s. has positive Lebesgue measure and interior points; and $\varphi_2(r) = r^{N+(1-\alpha)d}\left(\log\log 1/r\right)^{\frac{\alpha d}{N}}$ is an exact Hausdorff measure function for the graph of X^α. If $N = \alpha d$, then $\mathscr{H}^{\varphi_3}(X^\alpha([0,1]^N))$ is σ-finite almost surely, where $\varphi_3(r) = r^d \log(1/r)\log\log\log 1/r$. In the latter case the same is also true for the Hausdorff measure of the graph set of $X^\alpha(t)$. However, the lower bound problems for the Hausdorff measure of the range and graph remain open.

(ii) Let $W = \{W(t), t \in \mathbf{R}_+^N\}$ be the Brownian sheet in $\mathbf{R}^d$. If $2N < d$, then $\varphi_4(r) = r^{2N}\left(\log\log 1/r\right)^N$ ia an exact Hausdorff measure function for the range and graph of W. If $2N > d$, then $W([0,1]^N)$ a.s. has interior points and $\varphi_5(r) = r^{N+\frac{d}{2}}(\log\log 1/r)^{\frac{d}{2}}$ is an exact Hausdorff measure function for the graph of W. When $2N = d$, the problems for finding exact Hausdorff measure functions for the range and graph of W are completely open.

It is interesting to notice the subtle differences in the exact Hausdorff functions for the range and graph sets of fractional Brownian motion and the Brownian sheet, respectively. I believe the differences are a reflection of the two different types of local nondeterminism that they satisfy.

We remark that the methods in the aforementioned references rely respectively on specific properties of the Brownian sheet and fractional Brownian motion, and it is not clear whether these methods are applicable to Gaussian random fields satisfying (C3) or (C3′). It would be interesting to develop general methods that are applicable to larger classes of (Gaussian) random fields.

The problems on exact packing measure functions for $X([0,1]^N)$ and $\mathrm{Gr}X([0,1]^N)$ are related to the liminf properties of the occupation measures of X and are more difficult to study. When X is an (N,d)-fractional Brownian motion of index α and $N < \alpha d$, Xiao [100; 105] proved that $\varphi_6(r) = r^{N/\alpha}(\log\log 1/r)^{-N/(2\alpha)}$ is an exact packing measure function for $X([0,1]^N)$ and $\mathrm{Gr}X([0,1]^N)$. For all the other Gaussian fields including the Brownian sheet, the corresponding problems remain to be open.

On the other hand, it is a natural question to find $\dim_{\mathscr{H}} X(E)$ when $E \subseteq \mathbf{R}^N$ is an arbitrary Borel set, say a fractal set. It is not hard to see that, due to the anisotropy of X, the Hausdorff dimension of $X(E)$ can not be determined by $\dim_{\mathscr{H}} E$ and the index H alone, as shown by Example 6.6 below. This is in contrast with the cases of fractional Brownian motion or the Brownian sheet.

We start with the following Proposition 6.5 which determines $\dim_{\mathscr{H}} X(E)$ when E belongs to a special class of Borel sets in $\mathbf{R}^N$. Since the proof is almost the same as that of Proposition 3.1 in [97], we omit the proof.

Proposition 6.5. *Let* $X = \{X(t), t \in \mathbf{R}^N\}$ *be an* (N,d)-*Gaussian random field satisfying Condition (C1) on* $I = [0,1]^N$ *with parameters* $(H_1, \ldots, H_N)$. *Assume that* E_j $(j = 1, \ldots, N)$ *are Borel subsets of* $(0,1)$ *satisfying the following property:* $\exists \{j_1, \ldots, j_{N-1}\} \subseteq \{1, \ldots, N\}$ *such that* $\dim_{\mathscr{H}} E_{j_k} = \dim_{\mathscr{P}} E_{j_k}$ *for* $k = 1, \ldots, N-1$. *Let* $E = E_1 \times \cdots \times E_N \subseteq \mathbf{R}^N$, *then we have*

$$\dim_{\mathscr{H}} X(E) = \min\left\{d; \ \sum_{j=1}^{N} \frac{\dim_{\mathscr{H}} E_j}{H_j}\right\}, \quad \text{a.s.} \tag{127}$$

The following simple example illustrates that, in general, $\dim_{\mathscr{H}} E$ alone is not enough to determine the Hausdorff dimension of $X(E)$.

Example 6.6. Let $X = \{X(t), t \in \mathbf{R}^2\}$ be a $(2,d)$-Gaussian field with index $H = (H_1, H_2)$ and $H_1 < H_2$. Let $E = E_1 \times E_2$ and $F = E_2 \times E_1$, where $E_1 \subseteq (0,1)$ satisfies $\dim_{\mathscr{H}} E_1 = \dim_{\mathscr{P}} E_1$ and $E_2 \subseteq (0,1)$ is arbitrary. It is well known that

$$\dim_{\mathscr{H}} E = \dim_{\mathscr{H}} E_1 + \dim_{\mathscr{H}} E_2 = \dim_{\mathscr{H}} F. \tag{128}$$

See [36, p. 94]. However, by Proposition 6.5 we have

$$\dim_{\mathscr{H}} X(E) = \min\left\{d\,;\, \frac{\dim_{\mathscr{H}} E_1}{H_1} + \frac{\dim_{\mathscr{H}} E_2}{H_2}\right\}, \qquad (129)$$

and

$$\dim_{\mathscr{H}} X(F) = \min\left\{d\,;\, \frac{\dim_{\mathscr{H}} E_2}{H_1} + \frac{\dim_{\mathscr{H}} E_1}{H_2}\right\}. \qquad (130)$$

We see that $\dim_{\mathscr{H}} X(E) \neq \dim_{\mathscr{H}} X(F)$ in general unless $\dim_{\mathscr{H}} E_1 = \dim_{\mathscr{H}} E_2$.

Example 6.6 shows that for determining $\dim_{\mathscr{H}} X(E)$, we need to have more information about the geometry of E than its Hausdorff dimension.

In order to solve the problem for finding the Hausdorff dimension of the image $B^H(E)$ of fractional Brownian sheet B^H, Wu and Xiao [97] applied a measure-theoretic approach and introduced a notion of *Hausdorff dimension contour* for finite Borel measures and Borel sets.

Recall that the Hausdorff dimension of a Borel measure μ on $\mathbf{R}^N$ (or lower Hausdorff dimension as it is sometimes called) is defined by

$$\dim_{\mathscr{H}} \mu = \inf\left\{\dim_{\mathscr{H}} F\,:\, \mu(F) > 0 \text{ and } F \subseteq \mathbf{R}^N \text{ is a Borel set}\right\}. \qquad (131)$$

Hu and Taylor [44] proved the following characterization of $\dim_{\mathscr{H}} \mu$: If μ is a finite Borel measure on $\mathbf{R}^N$, then

$$\dim_{\mathscr{H}} \mu = \sup\left\{\gamma \geq 0\,:\, \limsup_{r\to 0+} \frac{\mu\big(B(t,r)\big)}{r^\gamma} = 0 \ \text{ for } \mu\text{-a.e. } t \in \mathbf{R}^N\right\}, \qquad (132)$$

where $B(t,r) = \{s \in \mathbf{R}^N : |s - t| \leq r\}$. It can be verified that for every Borel set $E \subseteq \mathbf{R}^N$, we have

$$\dim_{\mathscr{H}} E = \sup\left\{\dim_{\mathscr{H}} \mu\,:\, \mu \in \mathscr{M}_c^+(E)\right\}, \qquad (133)$$

where $\mathscr{M}_c^+(E)$ denotes the family of finite Borel measures on E with compact support in E.

From (132), we note that $\dim_{\mathscr{H}} \mu$ only describes the local behavior of μ in an isotropic way and is not quite informative if μ is highly anisotropic. To overcome this difficulty, Wu and Xiao [97] introduce the following notion of "dimension" for $E \subseteq (0,\infty)^N$ that is natural for studying $X(E)$.

Definition 6.7. *Given a Borel probability measure μ on $\mathbf{R}^N$, we define the set $\Lambda_\mu \subseteq \mathbf{R}_+^N$ by*

$$\Lambda_\mu = \left\{\lambda = (\lambda_1, \ldots, \lambda_N) \in \mathbf{R}_+^N\,:\, \limsup_{r\to 0+} \frac{\mu\left(R(t,r)\right)}{r^{\langle \lambda, H^{-1}\rangle}} = 0 \text{ for } \mu\text{-a.e.} \, t \in \mathbf{R}^N\right\},$$

where $R(t,r) = \prod_{j=1}^N [t_j - r^{1/H_j}, t_j + r^{1/H_j}]$ and $H^{-1} = (H_1^{-1}, \ldots, H_N^{-1})$.

The following lemma is proved in [97], which summarizes some basic properties of Λ_μ. Recall that $H_1 = \min\{H_j : 1 \leq j \leq N\}$.

Lemma 6.8. Λ_μ *has the following properties:*

(i) *The set Λ_μ is bounded:*

$$\Lambda_\mu \subseteq \left\{\lambda = (\lambda_1, \ldots, \lambda_N) \in \mathbf{R}_+^N : \langle \lambda, H^{-1} \rangle \leq \frac{N}{H_1} \right\}. \tag{134}$$

(ii) *For all $\beta \in (0, 1]^N$ and $\lambda \in \Lambda_\mu$, the Hadamard product of β and λ, $\beta \circ \lambda = (\beta_1 \lambda_1, \ldots, \beta_N \lambda_N) \in \Lambda_\mu$.*
(iii) *Λ_μ is convex; i.e., $b\lambda + (1 - b)\eta \in \Lambda_\mu$ for all $\lambda, \eta \in \Lambda_\mu$ and $0 < b < 1$.*
(iv) *For every $a \in (0, \infty)^N$, $\sup_{\lambda \in \Lambda_\mu} \langle \lambda, a \rangle$ is achieved on the boundary of Λ_μ.*

We call the boundary of Λ_μ, denoted by $\partial \Lambda_\mu$, the *Hausdorff dimension contour* of μ. See [97] for some examples for determining $\partial \Lambda_\mu$.

For any Borel set $E \subseteq \mathbf{R}^N$, we define

$$\Lambda(E) = \bigcup_{\mu \in \mathscr{M}_c^+(E)} \Lambda_\mu. \tag{135}$$

Similar to the case for measures, we call the set $\cup_{\mu \in \mathscr{M}_c^+(E)} \partial \Lambda_\mu$ the *Hausdorff dimension contour of E*. It follows from Lemma 6.8 that, for every $a \in (0, \infty)^N$, the supremum $\sup_{\lambda \in \Lambda(E)} \langle \lambda, a \rangle$ is determined by the Hausdorff dimension contour of E.

The same proof of Theorem 3.10 in [97] yields the following result.

Theorem 6.9. *Let $X = \{X(t), t \in \mathbf{R}^N\}$ be an (N, d)-Gaussian random field satisfying Condition (C1) on $I = [0, 1]^N$. Then for every Borel set $E \subseteq [0, 1]^N$,*

$$\dim_{\mathscr{H}} X(E) = \min\{d, s(H, E)\} \qquad \text{a.s.,} \tag{136}$$

where $s(H, E) = \sup_{\lambda \in \Lambda(E)} \langle \lambda, H^{-1} \rangle = \sup_{\mu \in \mathscr{M}_c^+(E)} s_\mu(E)$.

In the following, we give a more direct approach. Our results yield more geometric information about the quantity $s(H, E)$ as well.

For an arbitrary vector $(H_1, \ldots, H_N) \in (0, 1)^N$, we consider the metric space $(\mathbf{R}^N, \rho)$, where ρ is defined by (37). For any $\beta > 0$ and $E \subseteq \mathbf{R}^N$, define the β-dimensional Hausdorff measure [in the metric ρ] of E by

$$\mathscr{H}_\rho^\beta(E) = \lim_{\delta \to 0} \inf \left\{ \sum_{n=1}^\infty (2r_n)^\beta : E \subseteq \bigcup_{n=1}^\infty B_\rho(r_n), \, r_n \leq \delta \right\}. \tag{137}$$

This is a metric outer measure and all Borel sets are $\mathscr{H}_\rho^\beta$-measurable. The corresponding Hausdorff dimension of E is defined by

$$\dim_{\mathscr{H}}^\rho E = \inf \left\{ \beta > 0 : \mathscr{H}_\rho^\beta(E) = 0 \right\}. \tag{138}$$

In some special cases, Hausdorff measure and dimension of this type have been applied by Kaufman [48], Hawkes [42], Taylor and Watson [90], and Testard [91] to study the hitting probability of space-time processes of Brownian motion and other processes.

Note that the metric space $(\mathbf{R}^N, \rho)$ is complete and separable. Hence the following generalized Frostman's lemma is a consequence of Theorem 8.17 in [67] and a remark on page 117 of the same reference. It can also be proved by using a 1977 result of Assouad [46, p. 137] on the quasi-helix and the classical Frostman's lemma; see [91, p. 4] for a special case.

Lemma 6.10. *For any Borel set* $E \subseteq \mathbf{R}^N$, $\mathscr{H}_\rho^\beta(E) > 0$ *if and only if there exist a Borel probability measure on E and a positive constant c such that* $\mu(B_\rho(x, r)) \leq c\, r^\beta$ *for all* $x \in \mathbf{R}^N$ *and* $r > 0$.

We can now prove an equivalent result to Theorem 6.9, which extends the Hausdorff dimension result for the range of X in Theorem 6.1.

Theorem 6.11. *Let* $X = \{X(t), t \in \mathbf{R}^N\}$ *be an (N, d)-Gaussian random field satisfying Condition (C1) on $I = [0, 1]^N$. Then for every Borel set $E \subseteq [0, 1]^N$,*

$$\dim_{\mathscr{H}} X(E) = \min\{d\,;\dim_{\mathscr{H}}^\rho E\} \qquad \text{a.s.} \tag{139}$$

Proof. Since the idea for proving (139) is quite standard, we only give a sketch of it. For any $\gamma > \dim_{\mathscr{H}}^\rho E$, there is a covering $\{B_\rho(r_n), n \geq 1\}$ of E such that $\sum_{n=1}^\infty (2r_n)^\gamma \leq 1$. Note that $X(E) \subseteq \cup_{n=1}^\infty X(B_\rho(r_n))$ and the uniform modulus of continuity of X implies that the diameter of $X(B_\rho(r_n))$ is at most $cr_n^{1-\delta}$, where $\delta \in (0, 1)$ is a constant. We can show that $\dim_{\mathscr{H}} X(E) \leq \gamma/(1-\delta)$ almost surely. The desired upper bound follows from the arbitrariness of γ and δ.

To prove the lower bound, let $\gamma \in (0, \min\{d\,;\dim_{\mathscr{H}}^\rho E\})$ be fixed. Then by using the generalized Frostman's lemma [Lemma 6.10] one can show that there exists a probability measure μ on E such that

$$\int \int \frac{1}{\rho(s, t)^\gamma}\, \mu(ds)\, \mu(dt) < \infty. \tag{140}$$

This and Condition (C1) immediately imply

$$\mathrm{E} \int \int \frac{\mu(ds)\, \mu(dt)}{|X(s) - X(t)|^\gamma} < \infty. \tag{141}$$

Hence $\dim_{\mathscr{H}} X(E) \geq \min\{d\,;\dim_{\mathscr{H}}^\rho E\}$ almost surely. □

Combining Theorems 6.9 and 6.11, the invariance properties of $\dim_{\mathscr{H}}^\rho E$ and $s(H, E)$, we can derive the following alternative expression for $s(H, E)$. Of course, this can also be proved directly by using measure-theoretic methods.

Corollary 6.12. *For every Borel set $E \subseteq \mathbf{R}^N$, we have $\dim_{\mathcal{H}}^\rho E = s(H, E)$.*

As in the case of fractional Brownian sheets considered by Wu and Xiao [97], the image $X(E)$ has rich Fourier analytic and topological properties. For example, by modifying the proofs in [97], one can prove that if X is a Gaussian random field with stationary increments and spectral density satisfying (43) then $X(E)$ is a Salem set [46; 67] whenever $\dim_{\mathcal{H}}^\rho E \leq d$, and $X(E)$ has interior points whenever $\dim_{\mathcal{H}}^\rho E > d$ [It is an exercise to work out the details].

Finally, we consider the special case when $H = \langle \alpha \rangle$. Theorem 6.11 implies that for every Borel set $E \subseteq [0, 1]^N$,

$$\dim_{\mathcal{H}} X(E) = \min \left\{ d, \frac{1}{\alpha} \dim_{\mathcal{H}} E \right\} \qquad \text{a.s.} \tag{142}$$

The following theorem gives us a uniform version of (142).

Theorem 6.13. *Let $X = \{X(t), t \in \mathbf{R}^N\}$ be as in Theorem 6.11 with $H = \langle \alpha \rangle$. If $N \leq \alpha d$ and X satisfies either Condition (C3) or (C3'), then with probability 1*

$$\dim_{\mathcal{H}} X(E) = \frac{1}{\alpha} \dim_{\mathcal{H}} E \text{ for all Borel sets } E \subseteq I, \tag{143}$$

and

$$\dim_{\mathcal{P}} X(E) = \frac{1}{\alpha} \dim_{\mathcal{P}} E \text{ for all Borel sets } E \subseteq I. \tag{144}$$

The proof of Theorem 6.13 is reminiscent to those in [51; 70; 97]. The key step is to apply Condition (C3) or (C3') to prove the following lemma. For simplicity, assume $I = [0, 1]^N$.

Lemma 6.14. *Suppose the assumptions of Theorem 6.13 hold, and let $\delta > 0$ and $0 < 2\alpha - \delta < \beta < 2\alpha$ be given constants. Then with probability 1, for all integers n large enough, there do not exist more than $2^{n\delta d}$ distinct points of the form $t^j = 4^{-n} k^j$, where $k^j \in \{1, 2, \ldots, 4^n\}^N$, such that*

$$\left| X(t^i) - X(t^j) \right| < 3 \cdot 2^{-n\beta} \quad \text{for } i \neq j. \tag{145}$$

Proof. A proof of Lemma 6.14 under Condition (C3) is given in [97]; see also [51]. The proof under (C3') is similar and is left to the reader as an exercise. □

Both (142) and Theorem 6.13 imply that sometimes one can determine the packing dimension of the image $X(E)$ by the packing dimension of E. However, it is known that the conclusion is false if $N > \alpha d$ [87]. The method in [103] shows that if $X = \{X(t), t \in \mathbf{R}^N\}$ is a Gaussian random field satisfying (C1) with $H = \langle \alpha \rangle$ then for every Borel set $E \subseteq I$,

$$\dim_{\mathcal{P}} X(E) = \frac{1}{\alpha} \text{Dim}_{\alpha d} E \qquad \text{a.s.,} \tag{146}$$

where $\text{Dim}_s E$ is the packing dimension profile of E defined by Falconer and Howroyd [37]. However, the analogous problem for general anisotropic Gaussian random fields has not been settled.

7 Hausdorff Dimension of the Level Sets and Hitting Probabilities

Under Conditions (C1) and (C2), we can study fractal properties of the level set $L_x = \{t \in I : X(t) = x\}$ ($x \in \mathbf{R}^d$) and the hitting probabilities of Gaussian random field X.

The following result determines the Hausdorff and packing dimensions of the level set.

Theorem 7.1. *Let $X = \{X(t), t \in \mathbf{R}^N\}$ be an (N, d)-Gaussian random field satisfying Conditions (C1) and (C2) on $I = [\varepsilon, 1]^N$.*

(i) If $\sum_{j=1}^{N} H_j^{-1} < d$, then for every $x \in \mathbf{R}^d$, $L_x = \varnothing$ a.s.
(ii) If $\sum_{j=1}^{N} H_j^{-1} > d$, then for every $x \in \mathbf{R}^d$,

$$\dim_{\mathscr{H}} L_x = \dim_{\mathscr{P}} L_x$$

$$= \min \left\{ \sum_{j=1}^{k} \frac{H_k}{H_j} + N - k - H_k d, \ 1 \le k \le N \right\} \tag{147}$$

$$= \sum_{j=1}^{k} \frac{H_k}{H_j} + N - k - H_k d \quad \text{if} \quad \sum_{j=1}^{k-1} \frac{1}{H_j} \le d < \sum_{j=1}^{k} \frac{1}{H_j},$$

with positive probability.

Remark 7.2. In the critical case when $\sum_{j=1}^{N} H_j^{-1} = d$, it is believed that $L_x = \varnothing$ a.s. In the Brownian sheet case, this was proved by Orey and Pruitt [76, Theorem 3.4]. It also follows from a potential theoretic result of [50]. If X is a fractional Brownian motion of index $\alpha \in (0, 1)$, then an argument of [86] can be modified to show $L_x = \varnothing$ a.s. However, the problem whether $L_x = \varnothing$ a.s. for more general Gaussian random fields remains open. A proof would require Condition (C3) or (C3') and some extra conditions on the function $E(|X_1(t) - X_1(s)|^2)$.

Proof of Theorem 7.1. Similar to the proof of Theorem 5 in [7], we divide the proof of Theorem 7.1 into two steps. In Step one, we prove (i) and the upper bound for $\dim_{\mathscr{P}} L_x$ in (147); and in Step two we prove the lower bound for $\dim_{\mathscr{H}} L_x$ by constructing a random measure on L_x and using a capacity argument. Moreover, the last equality in (147) follows from Lemma 6.2.

First we prove

$$\overline{\dim}_{\mathscr{B}} L_x \le \min \left\{ \sum_{j=1}^{k} \frac{H_k}{H_j} + N - k - H_k d, 1 \le k \le N \right\} \quad \text{a.s.} \tag{148}$$

and $L_x = \varnothing$ a.s. whenever the right hand side of (148) is negative. It can be verified that the latter is equivalent to $\sum_{j=1}^{N} H_j^{-1} < d$.

For an integer $n \geq 1$, divide the interval $[\varepsilon, 1]^N$ into m_n sub-rectangles $R_{n,\ell}$ of side lengths n^{-1/H_j} $(j = 1, \cdots, N)$. Then $m_n \leq c \, n^{\sum_{j=1}^N (1/H_j)}$. Let $0 < \delta < 1$ be fixed and let $\tau_{n,\ell}$ be the lower-left vertex of $R_{n,\ell}$. Then

$$
\begin{aligned}
\mathrm{P}\{x \in X(R_{n,\ell})\} &\leq \mathrm{P}\left\{ \max_{s,t \in R_{n,\ell}} |X(s) - X(t)| \leq n^{-(1-\delta)}; \, x \in X(R_{n,\ell}) \right\} \\
&\quad + \mathrm{P}\left\{ \max_{s,t \in R_{n,\ell}} |X(s) - X(t)| > n^{-(1-\delta)} \right\} \\
&\leq \mathrm{P}\left\{ |X(\tau_{n,\ell}) - x| \leq n^{-(1-\delta)} \right\} + e^{-c \, n^{2\delta}} \\
&\leq c \, n^{-(1-\delta)d}.
\end{aligned}
\tag{149}
$$

In the above we have applied Lemma 2.1 in [85] to get the second inequality. If $\sum_{j=1}^N H_j^{-1} < d$, we choose $\delta > 0$ such that $(1-\delta)d > \sum_{j=1}^N H_j^{-1}$. Let N_n be the number of rectangles $R_{n,\ell}$ such that $x \in X(R_{n,\ell})$. It follows from (149) that

$$
\mathrm{E}(N_n) \leq c \, n^{\sum_{j=1}^N (1/H_j)} n^{-(1-\delta)d} \to 0 \quad \text{as } n \to \infty.
\tag{150}
$$

Since the random variables N_n are integer-valued, (150) and Fatou's lemma imply that a.s. $N_n = 0$ for infinitely many integers $n \geq 1$. Therefore $L_x = \emptyset$ almost surely.

Now we assume $\sum_{j=1}^N H_j^{-1} > d$ and define a covering $\{R'_{n,\ell}\}$ of L_x by $R'_{n,\ell} = R_{n,\ell}$ if $x \in X(R_{n,\ell})$ and $R'_{n,\ell} = \emptyset$ otherwise. We will show that there are N different ways to cover L_x by using cubes of the same side-lengths and each of these ways leads to an upper bound for $\overline{\dim}_{\mathscr{B}} L_x$.

For every $1 \leq k \leq N$, $R'_{n,\ell}$ can be covered by $n^{\sum_{j=k+1}^N (H_k^{-1} - H_j^{-1})}$ cubes of side-length n^{-1/H_k}. Thus we can cover the level set L_x by a sequence of cubes of side-length n^{-1/H_k}. Denote the number of such cubes by $M_{n,k}$. Using (149) again, we have

$$
\begin{aligned}
\mathrm{E}(M_{n,k}) &\leq c_{7,1} \, n^{\sum_{j=1}^N H_j^{-1}} n^{-(1-\delta)d} \cdot n^{\sum_{j=k+1}^N (H_k^{-1} - H_j^{-1})} \\
&= c_{7,1} \, n^{(N-k)H_k^{-1} + \sum_{j=1}^k H_j^{-1} - (1-\delta)d}.
\end{aligned}
\tag{151}
$$

Now let η be the constant defined by

$$
\eta = (N-k)H_k^{-1} + \sum_{j=1}^k H_j^{-1} - (1 - 2\delta)d.
\tag{152}
$$

We consider the sequence of integers $n_i = 2^i$ $(i \geq 1)$. Then by (151), the Markov inequality and the Borel-Cantelli lemma we see that almost surely $M_{n_i,k} \leq c \, n_i^\eta$ for all i large enough. This implies that $\overline{\dim}_{\mathscr{B}} L_x \leq H_k \eta$ almost surely. Letting $\delta \downarrow 0$ along rational numbers, we have

$$
\overline{\dim}_{\mathscr{B}} L_x \leq \sum_{j=1}^k \frac{H_k}{H_j} + N - k - H_k d \quad \text{a.s.}
\tag{153}
$$

Optimizing (153) over $k = 1, \ldots, N$ yields (148).

In order to prove the lower bound for $\dim_{\mathscr{H}} L_x$ in (147), we assume that $\sum_{j=1}^{k-1} H_j^{-1} \le d < \sum_{j=1}^{k} H_j^{-1}$ for some $1 \le k \le N$. Let $\delta > 0$ be a small constant such that

$$\gamma := \sum_{j=1}^{k} \frac{H_k}{H_j} + N - k - H_k(1+\delta)d > N - k. \tag{154}$$

This is possible by Lemma 6.2. Note that if we can prove that there is a constant $c_{7,2} > 0$, independent of δ, such that

$$P\{\dim_{\mathscr{H}} L_x \ge \gamma\} \ge c_{7,2}, \tag{155}$$

then the lower bound in (147) will follow by letting $\delta \downarrow 0$.

Our proof of (155) is based on the capacity argument due to Kahane [46]. Similar methods have been used by Adler [1], Testard [91], and Xiao [98] to certain Gaussian and stable random fields.

Let $\mathscr{M}_\gamma^+$ be the space of all non-negative measures on $\mathbf{R}^N$ with finite γ-energy [recall (84)]. It is known [1] that $\mathscr{M}_\gamma^+$ is a complete metric space under the metric

$$\|\mu\|_\gamma^2 = \iint \frac{\mu(dt)\,\mu(ds)}{|t-s|^\gamma}. \tag{156}$$

We define a sequence of random positive measures $\mu_n := \mu_n(x, \bullet)$ on the Borel sets C of $[\varepsilon, 1]^N$ by

$$\begin{aligned}
\mu_n(C) &= \int_C (2\pi n)^{d/2} \exp\left(-\frac{n\,|X(t) - x|^2}{2}\right) dt \\
&= \int_C \int_{\mathbf{R}^d} \exp\left(-\frac{|\xi|^2}{2n} + i\langle \xi, X(t) - x\rangle\right) d\xi\, dt.
\end{aligned} \tag{157}$$

It follows from [46, p. 206] or [91, p. 17] that if there exist positive and finite constants $c_{7,3}$, $c_{7,4}$ and $c_{7,5}$ such that

$$\mathrm{E}\big(\|\mu_n\|\big) \ge c_{7,3}, \qquad \mathrm{E}\big(\|\mu_n\|^2\big) \le c_{7,4}, \tag{158}$$

$$\mathrm{E}\big(\|\mu_n\|_\gamma\big) \le c_{7,5}, \tag{159}$$

where $\|\mu_n\| = \mu_n([\varepsilon, 1]^N)$ denotes the total mass of μ_n, then there is a subsequence of $\{\mu_n\}$, say $\{\mu_{n_k}\}$, such that $\mu_{n_k} \to \mu$ in $\mathscr{M}_\gamma^+$ and μ is strictly positive with probability $\ge c_{7,3}^2/(2c_{7,4})$. In this case, it follows from (157) that μ has its support in L_x almost surely. Moreover, (159) and the monotone convergence theorem together imply that the γ-energy of μ is finite. Hence Frostman's theorem yields (155) with $c_{7,2} = c_{7,3}^2/(2c_{7,4})$.

It remains to verify (158) and (159). By Fubini's theorem we have

$$
\begin{aligned}
\mathrm{E}\big(\|\mu_n\|\big) &= \int_{[\varepsilon,1]^N} \int_{\mathbf{R}^d} e^{-i\langle\xi,x\rangle} \exp\left(-\frac{|\xi|^2}{2n}\right) \mathrm{E}\exp\big(i\langle\xi,X(t)\rangle\big) d\xi\,dt \\
&= \int_{[\varepsilon,1]^N} \int_{\mathbf{R}^d} e^{-i\langle\xi,x\rangle} \exp\left(-\frac{1}{2}(n^{-1}+\sigma^2(t))|\xi|^2\right) d\xi\,dt \\
&= \int_{[\varepsilon,1]^N} \left(\frac{2\pi}{n^{-1}+\sigma^2(t)}\right)^{d/2} \exp\left(-\frac{|x|^2}{2(n^{-1}+\sigma^2(t))}\right) dt \\
&\geq \int_{[\varepsilon,1]^N} \left(\frac{2\pi}{1+\sigma^2(t)}\right)^{d/2} \exp\left(-\frac{|x|^2}{2\sigma^2(t)}\right) dt := c_{7,3}.
\end{aligned}
\tag{160}
$$

Denote by I_{2d} the identity matrix of order $2d$ and $\mathrm{Cov}(X(s),X(t))$ the covariance matrix of the random vector $(X(s),X(t))$. Let $\Gamma = n^{-1}I_{2d} + \mathrm{Cov}(X(s),X(t))$ and $(\xi,\eta)'$ be the transpose of the row vector (ξ,η). Then $\mathrm{E}(\|\mu_n\|^2)$ is equal to

$$
\begin{aligned}
&\int_{[\varepsilon,1]^N} \int_{[\varepsilon,1]^N} \int_{\mathbf{R}^d} \int_{\mathbf{R}^d} e^{-i\langle\xi+\eta,\,x\rangle} \exp\left(-\frac{1}{2}(\xi,\eta)\,\Gamma\,(\xi,\eta)'\right) d\xi\,d\eta\,ds\,dt \\
&= \int_{[\varepsilon,1]^N} \int_{[\varepsilon,1]^N} \frac{(2\pi)^d}{\sqrt{\det\Gamma}} \exp\left(-\frac{1}{2}(x,x)\,\Gamma^{-1}\,(x,x)'\right) ds\,dt \\
&\leq \int_{[\varepsilon,1]^N} \int_{[\varepsilon,1]^N} \frac{(2\pi)^d}{\big[\det\mathrm{Cov}(X_0(s),X_0(t))\big]^{d/2}} ds\,dt.
\end{aligned}
\tag{161}
$$

It follows from Conditions (C1), (C2) and (53) that for all $s,t \in [\varepsilon,1]^N$,

$$
\det\mathrm{Cov}\big(X_0(s),X_0(t)\big) \geq c_{7,6} \sum_{j=1}^N |s_j - t_j|^{2H_j}.
\tag{162}
$$

We combine (161), (162) and then apply Lemma 6.3, repeatedly, to obtain

$$
\mathrm{E}(\|\mu_n\|^2) \leq c_{7,7} \int_{[\varepsilon,1]^N} \int_{[\varepsilon,1]^N} \frac{ds\,dt}{\big[\sum_{j=1}^N |s_j - t_j|^{2H_j}\big]^{d/2}} := c_{7,4} < \infty.
\tag{163}
$$

This is similar to (113)–(117) in the proof of Theorem 6.1. Thus we have shown (158) holds.

Similar to (161), we find that $\mathrm{E}(\|\mu_n\|_\gamma)$ is equal to

$$
\begin{aligned}
&\int_{[\varepsilon,1]^N} \int_{[\varepsilon,1]^N} \frac{ds\,dt}{|s-t|^\gamma} \\
&\qquad \times \int_{\mathbf{R}^d} \int_{\mathbf{R}^d} e^{-i\langle\xi+\eta,\,x\rangle} \exp\left(-\frac{1}{2}(\xi,\eta)\,\Gamma\,(\xi,\eta)'\right) d\xi\,d\eta \\
&\leq c_{7,8} \int_{[\varepsilon,1]^N} \int_{[\varepsilon,1]^N} \frac{ds\,dt}{\big(\sum_{j=1}^N |s_j - t_j|\big)^\gamma \big(\sum_{j=1}^N |s_j - t_j|^{2H_j}\big)^{d/2}} \\
&\leq c_{7,9} \int_0^1 dt_N \cdots \int_0^1 \frac{dt_1}{\big(\sum_{j=1}^N t_j^{H_j}\big)^d \big(\sum_{j=1}^N t_j\big)^\gamma},
\end{aligned}
\tag{164}
$$

where the two inequalities follow from (162) and a change of variables. Note that the last integral in (164) is similar to (122) and can be estimated by using Lemma 6.4 in the same way as in the proof of (123)–(126). Moreover, we can take δ small enough, say $\delta < \delta_0$, so that the γ defined in (154) is bounded away from $N - k$ and $N - k + 1$. This implies that $E(\|\mu_n\|_\gamma) \leq c_{7,9}$ which is independent of δ. This proves (159) and hence Theorem 7.1.

In light of Theorem 7.1, it is of interest to further study the following question about fractal properties of the level sets.

Question 7.3. Determine the exact Hausdorff and packing measure functions for the level set L_x.

Questions 7.3 is closely related to the regularity properties such as the laws of the iterated logarithm of the local times of X. The latter will be considered in Section 8. When X is an (N, d)-fractional Brownian motion with index α, Xiao [102] proved that $\varphi_7(r) = r^{N-\alpha d}(\log \log 1/r)^{\alpha d/N}$ is an exact Hausdorff measure function for L_x. In Theorem 8.11 we give a partial result [i.e., lower bound] for the exact Hausdorff measure of the level set L_x. It seems that the method in [102] may be modified to determine an exact Hausdorff measure function for the level sets of Gaussian random fields satisfying (C3) or (C3').

So far no exact packing measure results have been proved for the level sets of fractional Brownian motion or the Brownian sheet. These problems are related to the liminf behavior of the local times of X which are more difficult to study.

More general than level sets, one can consider the following questions:

Question 7.4. Given a Borel set $F \subseteq \mathbf{R}^d$, when is $P\{X(I) \cap F \neq \varnothing\}$ positive?

Question 7.5. If $P\{X(I) \cap F \neq \varnothing\} > 0$, what are the Hausdorff and packing dimensions of the inverse image $X^{-1}(F) \cap I$?

Question 7.4 is an important question in potential theory of random fields. Complete answer has only been known for a few types of random fields with certain Markov structures. We mention that Khoshnevisan and Shi [50] proved that if X is an (N, d)-Brownian sheet, then for every Borel set $F \subseteq \mathbf{R}^d$,

$$P\{X(I) \cap F \neq \varnothing\} > 0 \Longleftrightarrow \mathscr{C}_{d-2N}(F) > 0. \tag{165}$$

Recall that $\mathscr{C}_\alpha$ denotes the Bessel-Riesz capacity of order α. Dalang and Nualart [26] have recently extended the methods of [50] and proved similar results for the solution of a system of d nonlinear hyperbolic stochastic partial differential equations with two variables. In this context, we also mention that Khoshnevisan and Xiao [52; 53; 54; 56] and Khoshnevisan, Xiao, and Zhong [57] have established systematic potential theoretical results for additive Lévy processes in $\mathbf{R}^d$. The arguments in the aforementioned work rely on the multiparameter martingale theory [49].

For random fields with general dependence structures, it is more difficult to solve Question 7.4 completely. Instead, one can look for sufficient conditions and necessary conditions on F so that $P\{X(I) \cap F \neq \varnothing\} > 0$. For example, when X is an (N, d)-fractional Brownian motion, Testard [91] and Xiao [104] proved the following results:

$$\mathscr{C}_{d-N/\alpha}(F) > 0 \Rightarrow P\{X(I) \cap F \neq \varnothing\} > 0 \Rightarrow \mathscr{H}^{d-N/\alpha}(F) > 0. \qquad (166)$$

Similar results for the solution to a non-linear stochastic heat equation with multiplicative noise have been proved recently by Dalang, Khoshnevisan and Nualart [23].

The following theorem is an analogue of (166) for all Gaussian random fields X satisfying Conditions (C1) and (C2).

Theorem 7.6. *Assume that an (N, d)-Gaussian random field $X = \{X(t), t \in \mathbf{R}^N\}$ satisfies Conditions (C1) and (C2) on I and $d > Q$. Then for every Borel set $F \subseteq \mathbf{R}^d$,*

$$c_{7,10}\, \mathscr{C}_{d-Q}(F) \leq P\{X(I) \cap F \neq \varnothing\} \leq c_{7,11}\, \mathscr{H}_{d-Q}(F), \qquad (167)$$

where $Q = \sum_{j=1}^{N} H_j^{-1}$, $c_{7,10}$ and $c_{7,11}$ are positive constants depending on I, F and H only.

Remark 7.7. When $d < Q$, Theorem 7.1 tells us that X hits points, hence (167) holds automatically. When $d = Q$, our proof below shows that the lower bound in (167) remains to be true provided $\mathscr{C}_0$ means the logarithmic capacity [see (86)]. This can be seen by estimating the integral in (173) when $d = Q$. However, if $\mathscr{C}_0(F) > 0$, then the upper bound in (167) becomes ∞, thus not informative.

Proof of Theorem 7.6. The lower bound in (167) can be proved by using a second moment argument. In fact one can follow the method in [26; 50; 91] to prove the lower bound in (167).

In the following, we provide an alternative proof for the lower bound which is similar to that of Theorem 7.1. For any Borel probability measure κ on F with $\mathscr{E}_{d-Q}(\kappa) < \infty$ and for all integers $n \geq 1$, we consider a family of random measures ν_n on I defined by

$$
\begin{aligned}
\int_I f(t)\, \nu_n(dt) &= \int_I \int_{\mathbf{R}^d} (2\pi n)^{d/2} \exp\big(-n\,|X(t) - x|^2\big)\, f(t)\, \kappa(dx)\, dt \\
&= \int_I \int_{\mathbf{R}^d} \int_{\mathbf{R}^d} \exp\Big(-\frac{|\xi|^2}{2n} + i\langle \xi, X(t) - x\rangle\Big)\, f(t)\, d\xi\, \kappa(dx)\, dt,
\end{aligned} \qquad (168)
$$

where f is an arbitrary measurable function on I. We claim that the following two inequalities hold:

$$E\big(\|\nu_n\|\big) \geq c_{7,12}, \qquad E\big(\|\nu_n\|^2\big) \leq c_{7,13}\, \mathscr{E}_{d-Q}(\kappa), \qquad (169)$$

where the constants $c_{7,12}$ and $c_{7,13}$ are independent of κ and n.

Since the proof of the first inequality in (169) is very similar to (160) in the proof of Theorem 7.1, we only prove the second inequality in (169).

Denote by I_{2d} the identity matrix of order $2d$ and $\mathrm{Cov}(X(s),X(t))$ the covariance matrix of the random vector $(X(s),X(t))$. Let $\Gamma_n = n^{-1}I_{2d} + \mathrm{Cov}(X(s),X(t))$ and $(\xi,\eta)'$ be the transpose of the row vector (ξ,η). Since Γ_n is positive definite, we have

$$\mathrm{E}\big(\|\nu_n\|^2\big) = \int_I \int_I \int_{\mathbf{R}^{2d}} \int_{\mathbf{R}^{2d}} e^{-i(\langle \xi,x\rangle+\langle \eta,y\rangle)}$$
$$\times \exp\Big(-\frac{1}{2}(\xi,\eta)\,\Gamma_n\,(\xi,\eta)'\Big)\,d\xi d\eta\,\kappa(dx)\kappa(dy)\,ds\,dt \qquad (170)$$
$$= \int_I \int_I \int_{\mathbf{R}^{2d}} \frac{(2\pi)^d}{\sqrt{\det\Gamma_n}}\,\exp\Big(-\frac{1}{2}(x,y)\,\Gamma_n^{-1}\,(x,y)'\Big)\,\kappa(dx)\kappa(dy)\,ds\,dt.$$

By modifying an argument from [91], we can prove that, under conditions (C1) and (C2), we have

$$\frac{(2\pi)^d}{\sqrt{\det\Gamma_n}}\,\exp\Big(-\frac{1}{2}(x,y)\,\Gamma_n^{-1}\,(x,y)'\Big) \leq \frac{c_{7,14}}{\max\{\rho^d(s,t),|x-y|^d\}} \qquad (171)$$

for all $s,t \in I$ and $x,y \in \mathbf{R}^d$; see [16] for details. Hence, it follows from (170) and (171) that

$$\mathrm{E}\big(\|\nu_n\|^2\big) \leq c_{7,14} \int_I \int_I \int_{\mathbf{R}^{2d}} \frac{1}{\max\{\rho^d(s,t),|x-y|^d\}}\,\kappa(dx)\kappa(dy)\,ds\,dt. \qquad (172)$$

We can verify that for all $x,y \in \mathbf{R}^d$,

$$\int_I \int_I \frac{ds\,dt}{\max\{\rho^d(s,t),|x-y|^d\}} \leq c_{7,15}\,|x-y|^{-(d-Q)}, \qquad (173)$$

where $c_{7,15} > 0$ is a constant. This can be done by breaking the integral in (173) over the regions $\{(s,t) \in I \times I : \rho(s,t) \leq |x-y|\}$ and $\{(s,t) \in I \times I : \rho(s,t) > |x-y|\}$, and estimate them separately. It is clear that (170), (173) and Fubini's theorem imply the second inequality in (169).

By using (169) and the Paley-Zygmund inequality [46], one can verify that there is a subsequence of $\{\nu_n, n \geq 1\}$ that converges weakly to a finite measure ν which is positive with positive probability [depending on $c_{7,12}$ and $c_{7,13}$ only] and ν also satisfies (169). Since ν is supported on $X^{-1}(F) \cap I$, we use the Paley-Zygmund inequality again to derive

$$\mathrm{P}\big\{X(I) \cap F \neq \varnothing\big\} \geq \mathrm{P}\{\|\nu\| > 0\} \geq \frac{[\mathrm{E}(\|\nu\|)]^2}{\mathrm{E}[\|\nu\|^2]} \geq \frac{c_{7,16}}{\mathscr{E}_{d-Q}(\kappa)}, \qquad (174)$$

where $c_{7,16} = c_{7,12}^2/c_{7,13}$. This implies the lower bound in (167).

Our proof of the upper bound in (167) relies on the following lemma on the hitting probability of X, whose proof will be deferred to the end of this section.

Lemma 7.8. *Let* $X = \{X(t), t \in \mathbf{R}^N\}$ *be an* (N, d)-*Gaussian random field satisfying Conditions (C1) and (C2) on* I. *Then there exists a constant* $c_{7,17} > 0$ *such that for all* $x \in I$ *and* $y \in \mathbf{R}^d$,

$$\mathrm{P}\left\{\inf_{t \in B_\rho(x, r)} |X(t) - y| \leq r\right\} \leq c_{7,17}\, r^d. \tag{175}$$

Now we proceed to prove the upper bound in (167) by using a simple covering argument. Choose and fix an arbitrary constant $\gamma > \mathscr{H}_{d-Q}(F)$. By the definition of $\mathscr{H}_{d-Q}(F)$, there is a sequence of balls $\{B(y_j, r_j), j \geq 1\}$ in $\mathbf{R}^d$ such that

$$F \subseteq \bigcup_{j=1}^\infty B(y_j, r_j) \quad \text{and} \quad \sum_{j=1}^\infty (2r_j)^{d-Q} \leq \gamma. \tag{176}$$

Clearly we have

$$\{F \cap X(I) \neq \varnothing\} \subseteq \bigcup_{j=1}^\infty \{B(y_j, r_j) \cap X(I) \neq \varnothing\}. \tag{177}$$

For every $j \geq 1$, we divide the interval I into $c r_j^{-Q}$ intervals of side-lengths r_j^{-1/H_ℓ} $(\ell = 1, \ldots, N)$. Hence I can be covered by at most $c r_j^{-Q}$ many balls of radius r_j in the metric ρ. It follows from Lemma 7.8 that

$$\mathrm{P}\{B(y_j, r_j) \cap X(I) \neq \varnothing\} \leq c r_j^{d-Q}. \tag{178}$$

Combining (177) and (178) we derive that $\mathrm{P}\{F \cap X(I) \neq \varnothing\} \leq c\gamma$. Since $\gamma > \mathscr{H}_{d-Q}(F)$ is arbitrary, the upper bound in (167) follows.

The following are some further remarks and open questions related to Theorem 7.6.

Remark 7.9. For any Borel set $F \subseteq \mathbf{R}^d$, Theorem 7.6 provides a sufficient condition and a necessary condition for $\mathrm{P}\{X^{-1}(F) \cap I \neq \varnothing\} > 0$. It is interesting to determine the Hausdorff and packing dimensions of $X^{-1}(F)$ when it is not empty. Recently, Biermé, Lacaux and Xiao (2007) determined the Hausdorff dimension of $X^{-1}(F)$. Namely, they proved that if $\dim_{\mathscr{H}} F > d - \sum_{\ell=1}^N (1/H_\ell)$, then

$$\left\|\dim_{\mathscr{H}}\left(X^{-1}(F) \cap I\right)\right\|_{L^\infty(\mathrm{P})}$$
$$= \min_{1 \leq k \leq N}\left\{\sum_{j=1}^k \frac{H_k}{H_j} + N - k - H_k\big(d - \dim_{\mathscr{H}} F\big)\right\}, \tag{179}$$

where, for all nonnegative random variables Y,

$$\|Y\|_{L^\infty(\mathrm{P})} = \sup\{\theta : Y \geq \theta \text{ on an event } E \text{ with } \mathrm{P}(E) > 0\}. \tag{180}$$

However, except for the special case of $F = \{x\}$, there have been no results on the packing dimension of $X^{-1}(F)$ for a given Borel set $F \subseteq \mathbf{R}^d$.

Remark 7.10. Note that the event on which (179) holds depends on F. Motivated by the results in [70], we may ask the following natural question: When $\sum_{\ell=1}^{N} H_\ell^{-1} > d$, is it possible to have a single event $\Omega_1 \subseteq \Omega$ of positive probability such that on Ω_1 (179) holds for all Borel sets $F \subseteq \mathbf{R}^d$?

Here are some partial answers. If in addition to Conditions (C1) and (C2), we also assume Condition (C3) or (C3′) holds and $H_1 = H_2 = \cdots = H_N$, then one can modify the proofs in [70] to show that the answer to the above question is affirmative. In general, it can be proved that, when $\sum_{\ell=1}^{N} H_\ell^{-1} > d$, the upper bound in (179) holds almost surely for all Borel sets $F \subseteq \mathbf{R}^d$. But it is likely that the lower bound may not hold uniformly due to the anisotropy of X.

Remark 7.11. The method for proving Theorem 7.6 may be extended to provide necessary conditions and sufficient conditions for $P\{X(E) \cap F \neq \emptyset\} > 0$, where $E \subseteq (0, \infty)^N$ and $F \subseteq \mathbf{R}^d$ are arbitrary Borel sets. Some difficulties arise when both E and F are fractal sets. Testard [91] obtained some partial results for fractional Brownian motion and, for every fixed Borel set $E \subseteq (0, \infty)^N$ (or $F \subseteq \mathbf{R}^d$), Xiao [104] characterized the "smallest" set F (or E) such that $P\{X(E) \cap F \neq \emptyset\} > 0$. No such results on anisotropic Gaussian random fields have been proved.

Finally, let us prove Lemma 7.8. There are two ways to prove (175). One is to use the argument in the proof of Proposition 4.4 of [22] and the other is reminiscent to the proof of Lemma 3.1 in [100]. While the former method is slightly simpler, the latter can be adapted to establish hitting probability estimates of the form (194) below for anisotropic Gaussian random fields. Hence we will use an argument similar to that in [100].

Proof of Lemma 7.8. For every integer $n \geq 1$, let $\varepsilon_n = r \exp(-2^{n+1})$ and denote by $N_n = N_\rho(B_\rho(x, r), \varepsilon_n)$ the minimum number of ρ-balls of radius ε_n that are needed to cover $B_\rho(x, r)$. Note that $N_n \leq c \exp(Q2^{n+1})$ [recall that $Q = \sum_{\ell=1}^{N}(1/H_\ell)$].

Let $\{t_i^{(n)}, 1 \leq i \leq N_n\}$ be a set of the centers of open balls with radius ε_n that cover $B_\rho(x, r)$. Denote

$$r_n = \beta \varepsilon_n \, 2^{(n+1)/2}, \qquad (181)$$

where $\beta \geq c_{4,10}$ is a constant to be determined later. Here $c_{4,10}$ is the constant in (69).

For all integers $n \geq 1, 1 \leq j \leq n$ and $1 \leq i \leq N_n$, we consider the following events

$$A_i^{(j)} = \left\{ |X(t_i^{(j)}) - y| \leq r + \sum_{k=j}^{\infty} r_k \right\},$$

$$A^{(n)} = \bigcup_{j=1}^{n} \bigcup_{i=1}^{N_j} A_i^{(j)} = A^{(n-1)} \cup \bigcup_{i=1}^{N_n} A_i^{(n)}. \qquad (182)$$

Then by a chaining argument, the triangle inequality and (69), we have

$$P\left\{\inf_{t\in B_\rho(x,r)}|X(t)-y|\le r\right\}\le\lim_{n\to\infty}P\big(A^{(n)}\big). \tag{183}$$

By (182), we have

$$P\big(A^{(n)}\big)\le P\big(A^{(n-1)}\big)+P\big(A^{(n)}\backslash A^{(n-1)}\big) \tag{184}$$

and

$$P\big(A^{(n)}\backslash A^{(n-1)}\big)\le\sum_{i=1}^{N_n}P\big(A_i^{(n)}\backslash A_{i'}^{(n-1)}\big), \tag{185}$$

where i' is chosen so that $\rho(t_i^{(n)},t_{i'}^{(n-1)})<\varepsilon_{n-1}$. Note that

$$\begin{aligned}
&P\big(A_i^{(n)}\backslash A_{i'}^{(n-1)}\big)\\
&=P\left\{|X(t_i^{(n)})-y|<r+\sum_{k=n}^\infty r_k\ ,\ |X(t_{i'}^{(n-1)})-y|>r+\sum_{k=n-1}^\infty r_k\right\}\\
&\le P\left\{|X(t_i^{(n)})-y|<r+\sum_{k=n}^\infty r_k\ ,\ |X(t_i^{(n)})-X(t_{i'}^{(n-1)})|\ge r_{n-1}\right\}.
\end{aligned} \tag{186}$$

By the elementary properties of Gaussian random variables, we can write

$$\frac{X(t_i^{(n)})-X(t_{i'}^{(n-1)})}{\sigma(t_i^{(n)},t_{i'}^{(n-1)})}=\eta\,\frac{X(t_i^{(n)})}{\sigma(t_i^{(n)})}+\sqrt{1-\eta^2}\,\Xi\ , \tag{187}$$

where

$$\eta=\frac{E\left[\left(X_1(t_i^{(n)})-X_1(t_{i'}^{(n-1)})\right)X_1(t_i^{(n)})\right]}{\sigma(t_i^{(n)},t_{i'}^{(n-1)})\sigma(t_i^{(n)})} \tag{188}$$

and where Ξ is a centered Gaussian random vector in $\mathbf{R}^d$ with the identity matrix as its covariance matrix and, moreover, Ξ is independent of $X(t_i^{(n)})$.

We observe that

$$r+\sum_{k=n}^\infty r_k\le r+\sum_{k=0}^\infty r_k\le\left(1+c\int_0^\infty\exp(-\alpha x^2)\,dx\right)r:=c_{7,18}\,r. \tag{189}$$

It follows from Condition (C1) that (186) is bounded by

$$\begin{aligned}
&P\left\{|X(t_i^{(n)})-y|\le c_{7,18}\,r\ ,\ |\Xi|\ge\frac{1}{\sqrt{1-\eta^2}}\left[\frac{r_{n-1}}{\rho(t_i^{(n)},t_{i'}^{(n-1)})}-\eta\,\frac{X(t_i^{(n)})}{\sigma(t_i^{(n)})}\right]\right\}\\
&\le P\left\{|X(t_i^{(n)})-y|\le c_{7,18}r\ ,\ |\Xi|\ge\frac{\beta d}{2}\,2^{n/2}\right\}\\
&\quad+P\left\{|X(t_i^{(n)})-y|\le c_{7,18}r\ ,\ \eta\,\frac{|X(t_i^{(n)})|}{\sigma(t_i^{(n)})}\ge\frac{\beta d}{2}\,2^{n/2}\right\}\\
&:=I_1+I_2.
\end{aligned} \tag{190}$$

By the independence of Ξ and $X(t_i^{(n)})$, we have

$$
\begin{aligned}
I_1 &= P\left\{\left|X(t_i^{(n)}) - y\right| \le c_{7,18}\, r\right\} \cdot P\left\{|\Xi| \ge \beta d 2^{-1+(n/2)}\right\} \\
&\le c_{7,19}\, r^d \exp\left(-\frac{(\beta d)^2}{16} 2^n\right).
\end{aligned}
\tag{191}
$$

On the other hand,

$$
\begin{aligned}
I_2 &\le \int_{\{|u-y|\le c_{7,18}\, r,\ |u|\ge \beta d 2^{-1+(n/2)}\sigma(t_i^{(n)})\}} \left(\frac{1}{2\pi}\right)^{d/2} \frac{1}{\sigma^d(t_i^{(n)})} \exp\left(-\frac{|u|^2}{2\sigma^2(t_i^{(n)})}\right) du \\
&\le c_{7,20} \int_{\{|u-y|\le c_{7,18}\, r\}} \frac{1}{\sigma^d(t_i^{(n)})} \exp\left(-\frac{|u|^2}{4\sigma^2(t_i^{(n)})}\right) du \cdot \exp\left(-\frac{(\beta d)^2}{16} 2^n\right) \\
&\le c_{7,21}\, r^d \exp\left(-\frac{(\beta d)^2}{16} 2^n\right).
\end{aligned}
\tag{192}
$$

Combining (184) through (192) and choosing $\beta \ge c_{4,10}$ satisfying $(\beta d)^2 > 32$, we obtain

$$
\begin{aligned}
P\left(A^{(n)}\right) &\le P\left(A^{(n-1)}\right) + c_{7,22}\, N_n\, r^d \exp\left(-\frac{(\beta d)^2}{16} 2^n\right) \\
&\le c_{7,23}\left[\sum_{k=0}^{\infty} N_k \exp\left(-\frac{(\beta d)^2}{16} 2^k\right)\right] r^d \\
&\le c_{7,24}\, r^d.
\end{aligned}
\tag{193}
$$

Therefore, (175) follows from (183) and (193).

When X is an (N,d)-fractional Brownian motion of index α, Xiao [104] proved the following hitting probability result: If $N < \alpha d$, then there exist positive and finite constants $c_{7,25}$ and $c_{7,26}$, depending only on N, d and α, such that for any $r > 0$ small enough and any $y \in \mathbf{R}^d$ with $|y| \ge r$, we have

$$
\begin{aligned}
c_{7,25}\left(\frac{r}{|y|}\right)^{d-\frac{N}{\alpha}} &\le P\left\{\exists\, t \in \mathbf{R}^N \text{ such that } |X(t) - y| < r\right\} \\
&\le c_{7,26}\left(\frac{r}{|y|}\right)^{d-\frac{N}{\alpha}}.
\end{aligned}
\tag{194}
$$

It would be interesting and useful to establish analogous results for all Gaussian random fields satisfying Conditions (C1) and (C2). Such an estimate will be useful in studying the escape rate and exact packing measure of the sample paths of Gaussian random fields; see [105] for the special case of fractional Brownian motion.

8 Local Times and Their Joint Continuity

We start by briefly recalling some aspects of the theory of local times. For excellent surveys on local times of random and deterministic vector fields, we refer to [31; 41].

Let $X(t)$ be a Borel vector field on $\mathbf{R}^N$ with values in $\mathbf{R}^d$. For any Borel set $T \subseteq \mathbf{R}^N$, the occupation measure of X on T is defined as the following Borel measure on $\mathbf{R}^d$:

$$\mu_T(\bullet) = \lambda_N \{t \in T : X(t) \in \bullet\}. \tag{195}$$

If μ_T is absolutely continuous with respect to the Lebesgue measure λ_d, we say that $X(t)$ has a *local time* on T, and define its local time, $L(\bullet, T)$, as the Radon–Nikodým derivative of μ_T with respect to λ_d, i.e.,

$$L(x, T) = \frac{d\mu_T}{d\lambda_d}(x) \qquad \forall x \in \mathbf{R}^d. \tag{196}$$

In the above, x is the so-called *space variable*, and T is the *time variable*. Sometimes, we write $L(x, t)$ in place of $L(x, [0, t])$. It is clear that if X has local times on T, then for every Borel set $S \subseteq T$, $L(x, S)$ also exists.

By standard martingale and monotone class arguments, one can deduce that the local times of X have a version, still denoted by $L(x, T)$, such that it is a kernel in the following sense:

(i) For each fixed $S \subseteq T$, the function $x \mapsto L(x, S)$ is Borel measurable in $x \in \mathbf{R}^d$.

(ii) For every $x \in \mathbf{R}^d$, $L(x, \cdot)$ is Borel measure on $\mathscr{B}(T)$, the family of Borel subsets of T.

Moreover, $L(x, T)$ satisfies the following *occupation density formula*: For every Borel set $T \subseteq \mathbf{R}^N$ and for every measurable function $f : \mathbf{R}^d \to \mathbf{R}_+$,

$$\int_T f(X(t)) \, dt = \int_{\mathbf{R}^d} f(x) L(x, T) \, dx. \tag{197}$$

See [41, Theorems 6.3 and 6.4].

Suppose we fix a rectangle $T = \prod_{i=1}^N [a_i, a_i + h_i]$ in $\mathscr{A}$. Then, whenever we can choose a version of the local time, still denoted by $L(x, \prod_{i=1}^N [a_i, a_i + t_i])$, such that it is a continuous function of $(x, t_1, \cdots, t_N) \in \mathbf{R}^d \times \prod_{i=1}^N [0, h_i]$, X is said to have a *jointly continuous local time* on T. When a local time is jointly continuous, $L(x, \bullet)$ can be extended to be a finite Borel measure supported on the level set

$$X_T^{-1}(x) = \{t \in T : X(t) = x\}; \tag{198}$$

see [1] for details. In other words, local times often act as a natural measure on the level sets of X for applying the capacity argument. As such, they are useful in studying the various fractal properties of level sets and inverse images of the vector field X [13; 34; 70; 81; 102].

First we consider the existence of the local times of Gaussian random fields.

Theorem 8.1. *Let $X = \{X(t), t \in \mathbf{R}^N\}$ be an (N, d)-Gaussian field defined by (36) and suppose Condition (C1) is satisfied on I. Then X has a local time $L(x, I) \in L^2(\mathrm{P} \times \lambda_d)$ if and only if $d < \sum_{j=1}^{N}(1/H_j)$. In the latter case, $L(x, I)$ admits the following L^2 representation:*

$$L(x, I) = (2\pi)^{-d} \int_{\mathbf{R}^d} e^{-i\langle y, x\rangle} \int_I e^{i\langle y, X(s)\rangle} ds\, dy, \qquad \forall x \in \mathbf{R}^d. \tag{199}$$

Proof. The Fourier transform of the occupation measure μ_I is

$$\widehat{\mu}_I(\xi) = \int_I e^{i\langle \xi, X(t)\rangle} dt. \tag{200}$$

By applying Fubini's theorem twice, we have

$$\mathrm{E} \int_{\mathbf{R}^d} |\widehat{\mu}(\xi)|^2 d\xi = \int_I ds \int_I dt \int_{\mathbf{R}^d} \mathrm{E} \exp\left(i\langle \xi, X(s) - X(t)\rangle\right) d\xi. \tag{201}$$

We denote the right hand side of (201) by $\mathscr{J}(I)$. It follows from the Plancherel theorem that X has a local time $L(\cdot, I) \in L^2(\mathrm{P} \times \lambda_d)$ if and only if $\mathscr{J}(I) < \infty$; see [41, Theorem 21.9] or [46]. Hence it is sufficient to prove

$$\mathscr{J}(I) < \infty \quad \text{if and only if} \quad d < \sum_{j=1}^{N} \frac{1}{H_j}. \tag{202}$$

For this purpose, we use the independence of the coordinate processes of X and Condition (C1) to deduce

$$\begin{aligned}
\mathscr{J}(I) &= \int_I \int_I \frac{ds\, dt}{\left[\mathrm{E}(X_0(s) - X_0(t))^2\right]^{d/2}} \\
&\asymp \int_I \int_I \frac{ds\, dt}{\left(\sum_{j=1}^{N} |s_j - t_j|^{2H_j}\right)^{d/2}}.
\end{aligned} \tag{203}$$

By using Lemma 8.6 below, it is elementary to verify that the last integral in (203) is finite if and only if $d < \sum_{j=1}^{N}(1/H_j)$. This proves (202), and hence Theorem 8.1. $\qquad\square$

The following result on the joint continuity of the local times is similar to those proved by Xiao and Zhang [110], Ayache, Wu and Xiao [6] for fractional Brownian sheets.

Theorem 8.2. *Let $X = \{X(t), t \in \mathbf{R}^N\}$ be an (N, d)-Gaussian field defined by (36) and we assume Conditions (C1) and (C3') are satisfied on I. If $d < \sum_{j=1}^{N}(1/H_j)$, then X has a jointly continuous local time on I.*

Remark 8.3. The conclusion of Theorem 8.2 can also be proved to hold for all Gaussian random fields satisfying Conditions (C1) and (C3). The proof follows a similar line, but some modifications are needed in order to prove analogous estimates in Lemmas 8.4 and 8.8. This is left to the reader as an exercise.

To prove Theorem 8.2 we will, similar to [6; 34; 102], first use the Fourier analytic arguments to derive estimates on the moments of the local times [see Lemmas 8.4 and 8.8 below] and then apply a multiparameter version of Kolmogorov continuity theorem [49]. It will be clear that Condition (C3′) plays an essential role in the proofs of Lemmas 8.4 and 8.8.

Our starting points is the following identities about the moments of the local time and its increments. It follows from [41, (25.5) and (25.7)] or [77] that for all $x, y \in \mathbf{R}^d$, $T \in \mathscr{A}$ and all integers $n \geq 1$,

$$
\mathrm{E}\left[L(x,T)^n\right] = (2\pi)^{-nd} \int_{T^n} \int_{\mathbf{R}^{nd}} e^{-i\sum_{j=1}^n \langle u^j, x \rangle}
$$
$$
\times \mathrm{E}\left[e^{i\sum_{j=1}^n \langle u^j, X(t^j) \rangle}\right] d\bar{u}\, d\bar{t} \tag{204}
$$

and for all even integers $n \geq 2$,

$$
\mathrm{E}\left[(L(x,T) - L(y,T))^n\right] \tag{205}
$$
$$
= (2\pi)^{-nd} \int_{T^n} \int_{\mathbf{R}^{nd}} \prod_{j=1}^n \left[e^{-i\langle u^j, x \rangle} - e^{-i\langle u^j, y \rangle}\right] \mathrm{E}\left[e^{i\sum_{j=1}^n \langle u^j, X(t^j) \rangle}\right] d\bar{u}\, d\bar{t},
$$

where $\bar{u} = (u^1, \dots, u^n)$, $\bar{t} = (t^1, \dots, t^n)$, and each $u^j \in \mathbf{R}^d$, $t^j \in T \subseteq (0, \infty)^N$. In the coordinate notation we then write $u^j = (u_1^j, \dots, u_d^j)$.

Lemma 8.4. *Suppose the assumptions of Theorem 8.2 hold. Let $\tau \in \{1, \dots, N\}$ be the integer such that*

$$
\sum_{\ell=1}^{\tau-1} \frac{1}{H_\ell} \leq d < \sum_{\ell=1}^{\tau} \frac{1}{H_\ell}, \tag{206}
$$

then there exists a positive and finite constant $c_{8,1}$, depending on N, d, H and I only, such that for all hypercubes $T = [a, a + \langle r \rangle] \subseteq I$ with side-length $r \in (0, 1)$, all $x \in \mathbf{R}^d$ and all integers $n \geq 1$,

$$
\mathrm{E}\left[L(x,T)^n\right] \leq c_{8,1}^n\, n!\, r^{n\,\beta_\tau}, \tag{207}
$$

where $\beta_\tau = \sum_{\ell=1}^{\tau}(H_\tau/H_\ell) + N - \tau - H_\tau d$.

Remark 8.5. (i) It is important to note that, when (206) holds, $\beta_\tau = \sum_{\ell=1}^{\tau}(H_\tau/H_\ell) + N - \tau - H_\tau d$ is the Hausdorff dimension of the level set L_x; see Theorem 7.1. Combining (207) with the upper density theorem of [80], one can obtain some information on the exact Hausdorff measure of L_x [see Corollary 8.11 below].

(ii) We point out that the upper bound in (207) is not sharp, and one may be able to prove the following inequality:

$$\mathbb{E}\big[L(x,T)^n\big] \le c_{8,2}^n \,(n!)^{N-\beta_\tau}\, r^{n\,\beta_\tau}. \tag{208}$$

If this is indeed true, then one can conjecture that the function $\varphi_8(r) = r^{\beta_\tau}\big(\log\log 1/r\big)^{N-\beta_\tau}$ is an exact Hausdorff measure function for L_x.

For proving Lemma 8.4, we will make use of the following elementary lemma [which is stronger than Lemma 6.3].

Lemma 8.6. *Let α, β and A be positive constants. Then*

$$\int_0^1 \frac{1}{(A+t^\alpha)^\beta}\, dt \asymp \begin{cases} A^{-(\beta-\frac{1}{\alpha})} & \text{if } \alpha\beta > 1, \\ \log\big(1+A^{-1/\alpha}\big) & \text{if } \alpha\beta = 1, \\ 1 & \text{if } \alpha\beta < 1. \end{cases} \tag{209}$$

Proof of Lemma 8.4. Since $X_1,\cdots,X_d$ are independent copies of X_0, it follows from (204) that for all integers $n \ge 1$,

$$\mathbb{E}\big[L(x,T)^n\big] \le (2\pi)^{-nd} \int_{T^n} \prod_{k=1}^d \left\{ \int_{\mathbf{R}^n} e^{-\frac{1}{2}\mathrm{Var}(\sum_{j=1}^n u_j^k X_0(t^j))}\, dU_k \right\} d\bar{t}, \tag{210}$$

where $U_k = (u_k^1,\cdots,u_k^n) \in \mathbf{R}^n$. It is clear that in order to bound the integral in $d\bar{t}$ it is sufficient to consider only the integral over $T_{\neq}^n = \{\bar{t} \in T^n : t^1,\dots,t^n \text{ are distinct}\}$ [the set of $\bar{t} \in \mathbf{R}^{Nn}$ having $t^i = t^j$ for some $i \neq j$ has (Nn)-dimensional Lebesgue measure 0]. It follows from Lemma 3.4 that for every $\bar{t} \in T_{\neq}^n$, the covariance matrix of $X_0(t^1),\cdots,X_0(t^n)$ is invertible. We denote the inverse matrix by $R(t^1,\cdots,t^n)$, and let $(Z_1,\cdots,Z_n)$ be the Gaussian vector with mean zero and the covariance matrix $R(t^1,\cdots,t^n)$. Then the density function of $(Z_1,\cdots,Z_n)$ is given by

$$\frac{\big[\det \mathrm{Cov}\big(X_0(t^1),\dots,X_0(t^n)\big)\big]^{1/2}}{(2\pi)^{n/2}} e^{-\frac{1}{2}U\mathrm{Cov}(X_0(t^1),\dots,X_0(t^n))U'}, \tag{211}$$

where $U = (u^1,\cdots,u^n) \in \mathbf{R}^n$ and U' is the transpose of U. Hence for each $1 \le k \le d$,

$$\int_{\mathbf{R}^n} e^{-\frac{1}{2}\mathrm{Var}(\sum_{j=1}^n u_k^j X_0(t^j))}\, dU_k = \frac{(2\pi)^{n/2}}{\big[\det \mathrm{Cov}\big(X_0(t^1),\dots,X_0(t^n)\big)\big]^{1/2}}. \tag{212}$$

Combining (210) and (212), we derive

$$\mathbb{E}\big[L(x,T)^n\big] \le (2\pi)^{-nd/2} \int_{T^n} \frac{1}{\big[\det \mathrm{Cov}(X_0(t^1),\dots,X_0(t^n))\big]^{d/2}}\, d\bar{t}. \tag{213}$$

It follows from Condition (C3′) and (53) that

$$\det \mathrm{Cov}\big(X_0(t^1), \ldots, X_0(t^n)\big) = \prod_{j=1}^{n} \mathrm{Var}\big(X_0(t^j) \big| X_0(t^i), j < i \leq n\big) \tag{214}$$

$$\geq c_{8,3}^n \prod_{j=1}^{n} \min_{j < i \leq n+1} \rho(t^j, t^i)^2,$$

where $t^{n+1} = 0$. This and (213) together imply that

$$\mathrm{E}\big[L(x,T)^n\big] \leq c_{8,4}^n \int_{T^n} \prod_{j=1}^{n} \frac{1}{\big[\min_{j<i\leq n+1} \rho(t^j, t^i)\big]^d} \, d\bar{t}. \tag{215}$$

We will estimate the integral in (215) by integrating in the order dt^1, $dt^2, \ldots, dt^n$. Considering first the integral in dt^1, we have

$$\int_T \frac{1}{\big[\min_{1<i\leq n+1} \rho(t^j, t^i)\big]^d} \, dt^1 \leq \sum_{i=2}^{n+1} \int_T \frac{1}{\big[\rho(t^j, t^i)\big]^d} \, dt^1 \tag{216}$$

$$\leq c\,n \int_{[0,r]^N} \frac{ds_1 \cdots ds_N}{\big[\sum_{k=1}^{N} s_k^{H_k}\big]^d},$$

where the last inequality follows from a change of variables. Integrating the last integral in the order $ds_1, \cdots, ds_N$ and applying (209) in Lemma 6.3, we can show that, because of (206), the last integrand in (216) only affects the integral in $ds_1, \ldots, ds_\tau$ which contributes [up to a constant] the factor $r^{\sum_{\ell=1}^{\tau}(H_\tau/H_\ell) - H_\tau d}$; and the integral in $ds_{\tau+1}, \ldots, ds_N$ contributes the factor $r^{N-\tau}$. In other words, we have

$$\int_{[0,r]^N} \frac{ds_1 \cdots ds_N}{\big[\sum_{k=1}^{N} s_k^{H_k}\big]^d} \leq c_{8,5} \, r^{\sum_{\ell=1}^{\tau}(H_\tau/H_\ell) + N - \tau - H_\tau d}. \tag{217}$$

This and (216) imply

$$\int_T \frac{dt^1}{\big[\min_{1<i\leq n+1} \rho(t^j, t^i)\big]^d} \leq c_{8,6} \, n \, r^{\sum_{\ell=1}^{\tau}(H_\tau/H_\ell) + N - \tau - H_\tau d}. \tag{218}$$

Repeating the same procedure for integrating in $dt^2, \ldots, dt^n$ in (215) yields (207). This proves Lemma 8.4.

Remark 8.7. In the proof of Lemma 8.4, we have assumed T is a hypercube $T = [a, a + \langle r \rangle]$. This is only for convenience and one can consider arbitrary closed intervals $T = \prod_{\ell=1}^{N}[a_\ell, a_\ell + r_\ell] \subseteq I$. The argument is the same as above, but (216) becomes

$$\int_T \frac{dt^1}{\left[\min\limits_{1 < i \leq n+1} \rho(t^j, t^i)\right]^d} \leq cn \int_{\prod_{k=1}^{N}[0, r_k]} \frac{ds_1 \cdots ds_N}{\left[\sum_{k=1}^{N} s_k^{H_k}\right]^d}. \tag{219}$$

Choose N positive numbers $p_1, \ldots, p_N \in (0, 1)$ defined by

$$p_k = \frac{H_k^{-1}}{\sum_{i=1}^{N} H_i^{-1}} \qquad (k = 1, \ldots, N). \tag{220}$$

Then $\sum_{k=1}^{N} p_k = 1$. By using the following inequality

$$\sum_{k=1}^{N} s_k^{H_k} \geq \sum_{k=1}^{N} p_k\, s_k^{H_k} \geq \prod_{k=1}^{N} s_k^{p_k H_k} \qquad \forall s \in (0, \infty)^N, \tag{221}$$

one can verify that

$$\int_{\prod_{k=1}^{N}[0, r_k]} \frac{ds_1 \cdots ds_N}{\left[\sum_{k=1}^{N} s_k^{H_k}\right]^d} \leq c\, \lambda_N(T)^{1-\nu}, \tag{222}$$

where $\nu = d/(\sum_{i=1}^{N} H_i^{-1}) \in (0, 1)$. Combining (215), (219) and (222) we derive that

$$E\big[L(x, T)^n\big] \leq c_{8,7}^n\, n!\, \lambda_N(T)^{n(1-\nu)} \tag{223}$$

holds for every interval $T \subseteq I$. We will apply this inequality in the proof of Theorem 8.2 below.

Lemma 8.4 implies that for all $n \geq 1$, $L(x, T) \in L^n(\mathbf{R}^d)$ a.s. [41, p. 42)]. Our next lemma estimates the moments of the increments of $L(x, T)$ in the space variable x.

Lemma 8.8. *Assume (206) holds for some $\tau \in \{1, \ldots, N\}$. Then there exists a positive and finite constant $c_{8,8}$, depending on N, d, H and I only, such that for all hypercubes $T = [a, a + \langle r \rangle] \subseteq I$, $x, y \in \mathbf{R}^d$ with $|x - y| \leq 1$, all even integers $n \geq 1$ and all $\gamma \in (0, 1)$ small enough,*

$$E\Big[\big(L(x, T) - L(y, T)\big)^n\Big] \leq c_{8,8}^n\, (n!)^{(1+\gamma)}\, |x - y|^{n\gamma}\, r^{n(\beta_\tau - 2H_\tau \gamma)}. \tag{224}$$

In order to prove Lemma 8.8, we will make use of the following lemma essentially due to Cuzick and DuPreez [19]; see also [55].

Lemma 8.9. *Let $Z_1, \ldots, Z_n$ be mean zero Gaussian variables which are linearly independent, then for any nonnegative function $g : \mathbf{R} \to \mathbf{R}_+$,*

$$\int_{\mathbf{R}^n} g(v_1) e^{-\frac{1}{2}\mathrm{Var}\left(\sum_{j=1}^n v_j Z_j\right)} \, dv_1 \cdots dv_n$$

$$= \frac{(2\pi)^{(n-1)/2}}{\left[\det \mathrm{Cov}\left(Z_1, \cdots, Z_n\right)\right]^{1/2}} \int_{-\infty}^{\infty} g\left(\frac{v}{\sigma_1}\right) e^{-v^2/2} \, dv, \tag{225}$$

where $\sigma_1^2 = \mathrm{Var}(Z_1 \mid Z_2, \ldots, Z_n)$ denotes the conditional variance of Z_1 given $Z_2, \ldots, Z_n$.

Proof of Lemma 8.8. Let $\gamma \in (0,1)$ be a constant whose value will be determined later. Note that by the elementary inequalities

$$|e^{iu} - 1| \leq 2^{1-\gamma} |u|^\gamma \qquad \text{for all } u \in \mathbf{R} \tag{226}$$

and $|u + v|^\gamma \leq |u|^\gamma + |v|^\gamma$, we see that for all $u^1, \ldots, u^n$, $x, y \in \mathbf{R}^d$,

$$\prod_{j=1}^n \left| e^{-i\langle u^j, x\rangle} - e^{-i\langle u^j, y\rangle} \right| \leq 2^{(1-\gamma)n} |x - y|^{n\gamma} \sum{}' \prod_{j=1}^n |u_{k_j}^j|^\gamma, \tag{227}$$

where the summation $\sum'$ is taken over all the sequences $(k_1, \cdots, k_n) \in \{1, \cdots, d\}^n$.

It follows from (205) and (227) that for every even integer $n \geq 2$,

$$\mathrm{E}\left[\left(L(x, T) - L(y, T)\right)^n\right] \leq (2\pi)^{-nd} 2^{(1-\gamma)n} |x - y|^{n\gamma}$$

$$\times \sum{}' \int_{T^n} \int_{\mathbf{R}^{nd}} \prod_{m=1}^n |u_{k_m}^m|^\gamma \, \mathrm{E}\left[e^{-i\sum_{j=1}^n \langle u^j, X(t^j)\rangle}\right] \, d\bar{u} \, d\bar{t}$$

$$\leq c_{8,9}^n |x - y|^{n\gamma} \sum{}' \int_{T^n} d\bar{t} \tag{228}$$

$$\times \prod_{m=1}^n \left\{ \int_{\mathbf{R}^{nd}} |u_{k_m}^m|^{n\gamma} e^{-\frac{1}{2}\mathrm{Var}\left(\sum_{j=1}^n \langle u^j, X(t^j)\rangle\right)} \, d\bar{u} \right\}^{1/n},$$

where the last inequality follows from the generalized Hölder inequality.

Now we fix a vector $\bar{k} = (k_1, \ldots, k_n) \in \{1, \cdots, d\}^n$ and n distinct points $t^1, \ldots, t^n \in T$ [the set of such points has full (nN)-dimensional Lebesgue measure]. Let $\mathscr{M} = \mathscr{M}(\bar{k}, \bar{t}, \gamma)$ be defined by

$$\mathscr{M} = \prod_{m=1}^n \left\{ \int_{\mathbf{R}^{nd}} |u_{k_m}^m|^{n\gamma} e^{-\frac{1}{2}\mathrm{Var}\left(\sum_{j=1}^n \langle u^j, X(t^j)\rangle\right)} \, d\bar{u} \right\}^{1/n}. \tag{229}$$

Note that X_ℓ $(1 \leq \ell \leq N)$ are independent copies of X_0. By Lemma 3.4, the random variables $X_\ell(t^j)$ $(1 \leq \ell \leq N, 1 \leq j \leq n)$ are linearly independent. Hence Lemma 8.9 gives

$$\int_{\mathbf{R}^{nd}} |u_{k_m}^m|^{n\gamma} e^{-\frac{1}{2}\text{Var}\left(\sum_{j=1}^n \langle u^j, X(t^j)\rangle\right)} d\bar{u}$$

$$= \frac{(2\pi)^{(nd-1)/2}}{\left[\det \text{Cov}\left(X_0(t^1), \ldots, X_0(t^n)\right)\right]^{d/2}} \int_{\mathbf{R}} \left(\frac{v}{\sigma_m}\right)^{n\gamma} e^{-v^2/2} dv \qquad (230)$$

$$\leq \frac{c_{8,10}^n \, (n!)^\gamma}{\left[\det \text{Cov}\left(X_0(t^1), \ldots, X_0(t^n)\right)\right]^{d/2}} \frac{1}{\sigma_m^{n\gamma}},$$

where σ_m^2 is the conditional variance of $X_{k_m}(t^m)$ given $X_i(t^j)$ ($i \neq k_m$ or $i = k_m$ but $j \neq m$), and the last inequality follows from Stirling's formula.

Combining (229) and (230) we obtain

$$\mathscr{M} \leq \frac{c_{8,11}^n \, (n!)^\gamma}{\left[\det \text{Cov}\left(X_0(t^1), \ldots, X_0(t^n)\right)\right]^{d/2}} \prod_{m=1}^n \frac{1}{\sigma_m^\gamma}. \qquad (231)$$

The second product in (231) will be treated as a "perturbation" factor and will be shown to be small when integrated. For this purpose, we use again the independence of the coordinate processes of X and Condition (C3$'$) to derive

$$\sigma_m^2 = \text{Var}\left(X_{k_m}(t^m) \mid X_{k_m}(t^j), \mathbf{j} \neq m\right)$$
$$\geq c_{8,12}^2 \, \min\left\{\rho(t^m, t^j)^2 : j = 0 \text{ or } j \neq m\right\}. \qquad (232)$$

Now we define a permutation π of $\{1, \cdots, n\}$ such that

$$\rho(t^{\pi(1)}, 0) = \min_{1 \leq j \leq n} \rho(t^j, 0). \qquad (233)$$

and once $t^{\pi(m-1)}$ has been defined, we choose $t^{\pi(m)}$ such that

$$\rho(t^{\pi(m)}, t^{\pi(m-1)})$$
$$= \min\left\{\rho(t^j, t^{\pi(m-1)}), \, j \in \{1, \cdots, n\} \setminus \{\pi(1), \cdots, \pi(m-1)\}\right\}. \qquad (234)$$

By this definition, we see that for every $m = 1, \cdots, n$,

$$\min\left\{\rho(t^{\pi(m)}, t^j) : j = 0 \text{ or } j \neq \pi(m)\right\}$$
$$= \min\left\{\rho(t^{\pi(m)}, t^{\pi(m-1)}), \, \rho(t^{\pi(m+1)}, t^{\pi(m)})\right\}. \qquad (235)$$

It follows from (232), (235) and (53) that

$$\prod_{m=1}^{n} \frac{1}{\sigma_m^{\gamma}} \leq c_{8,12}^{-n\gamma} \prod_{m=1}^{n} \frac{1}{\min\{\rho(t^{\pi(m)}, t^j)^{\gamma} : j = 0 \text{ or } j \neq \pi(m)\}}$$

$$\leq c_{8,13}^{n} \prod_{m=1}^{n} \frac{1}{\min\{\rho(t^{\pi(m)}, t^{\pi(m-1)})^{\gamma}, \rho(t^{\pi(m+1)}, t^{\pi(m)})^{\gamma}\}}$$

$$\leq c_{8,13}^{n} \prod_{m=1}^{n} \frac{1}{\rho(t^{\pi(m)}, t^{\pi(m-1)})^{2\gamma}} \tag{236}$$

$$\leq c_{8,14}^{n} \prod_{m=1}^{n} \frac{1}{\left[\text{Var}(X_0(t^{\pi(m)})|X_0(t^{\pi(i)}), i = 1, \cdots, m-1)\right]^{\gamma}}$$

$$= \frac{c_{8,15}^{n}}{\left[\det \text{Cov}(X_0(t^1), \cdots, X_0(t^n))\right]^{\gamma}}.$$

Combining (231) and (236), we obtain

$$\mathcal{M} \leq \frac{c_{8,16}^{n} (n!)^{\gamma}}{\left[\det \text{Cov}(X_0(t^1), \cdots, X_0(t^n))\right]^{\frac{d}{2}+\gamma}}$$

$$\leq \frac{c_{8,17}^{n} (n!)^{\gamma}}{\prod_{j=1}^{n} \left[\min_{j < i \leq n+1} \rho(t^j, t^i)\right]^{d+2\gamma}}, \tag{237}$$

where the last step follows from Condition (C3′) and (53).

It follows from (228), (229), (231) and (237) that

$$\text{E}\left[\left(L(x+y, T) - L(x, T)\right)^n\right]$$

$$\leq c_{8,18}^{n} |y|^{n\gamma} (n!)^{\gamma} \int_{T^n} \prod_{j=1}^{n} \frac{1}{\left[\min_{j < i \leq n+1} \rho(t^j, t^i)\right]^{d+2\gamma}} d\bar{t}. \tag{238}$$

Note that the last integral in (238) is similar to that in (215) and can be estimated by integrating in the order $dt^1, dt^2, \ldots, dt^n$. To this end, we take $\gamma \in (0, 1)$ small such that

$$\sum_{\ell=1}^{\tau-1} \frac{1}{H_\ell} \leq d + 2\gamma < \sum_{\ell=1}^{\tau} \frac{1}{H_\ell}. \tag{239}$$

Then, similar to (216)–(218), we derive

$$\int_{T^n} \prod_{j=1}^{n} \frac{1}{\left[\min_{j < i \leq n+1} \rho(t^j, t^i)\right]^{d+2\gamma}} d\bar{t}$$

$$\leq c_{8,19}^{n} (n!)^{1+\gamma} r^{n\left(\sum_{\ell=1}^{\tau}(H_\tau/H_\ell)+N-\tau-H_\tau(d+2\gamma)\right)}. \tag{240}$$

It is now clear that (224) follows from (238) and (240). This proves Lemma 8.8.

Now we are ready to prove Theorem 8.2.

Proof of Theorem 8.2. It follows from Lemma 8.8 and the multiparameter version of Kolmogorov's continuity theorem [49] that, for every fixed interval $T \in \mathscr{A}$ such that $T \subseteq I$, X has almost surely a local time $L(x, T)$ that is continuous for all $x \in \mathbf{R}^d$.

To prove the joint continuity, observe that for all $x, y \in \mathbf{R}^d$, $s, t \in I$ and all even integers $n \geq 1$, we have

$$\mathrm{E}\left[(L(x, [\varepsilon, s]) - L(y, [\varepsilon, t]))^n\right] \tag{241}$$
$$\leq 2^{n-1} \left\{ \mathrm{E}\left[(L(x, [\varepsilon, s]) - L(x, [\varepsilon, t]))^n\right] + \mathrm{E}\left[(L(x, [\varepsilon, t]) - L(y, [\varepsilon, t]))^n\right]\right\}.$$

Since the difference $L(x, [\varepsilon, s]) - L(x, [\varepsilon, t])$ can be written as a sum of finite number (only depends on N) of terms of the form $L(x, T_j)$, where each $T_j \in \mathscr{A}$ is a closed subinterval of I with at least one edge length $\leq |s - t|$, we can use Lemma 8.4 and Remark 7.7, to bound the first term in (241). On the other hand, the second term in (241) can be dealt with using Lemma 8.8 as above. Consequently, for some $\gamma \in (0, 1)$ small, the right hand side of (241) is bounded by $c_{8,20}^n (|x - y| + |s - t|)^{n\gamma}$, where $n \geq 2$ is an arbitrary even integer. Therefore the joint continuity of the local times $L(x, t)$ follows again from the multiparameter version of Kolmogorov's continuity theorem. This finishes the proof of Theorem 8.2.

The proof of Theorem 8.2 also provides some information about the modulus of continuity of $L(x, t)$ as a function of x and t. It would be interesting to establish sharp local and uniform Hölder conditions for the local time, because such results bear rich information about the irregular properties of the sample functions of X [1; 13; 41; 102].

By applying Lemma 8.4 and the Borel-Cantelli lemma, one can easily derive the following law of the iterated logarithm for the local time $L(x, \cdot)$: There exists a positive constant $c_{8,21}$ such that for every $x \in \mathbf{R}^d$ and $t \in (0, 1)^N$,

$$\limsup_{r \to 0} \frac{L(x, U(t, r))}{\varphi_9(r)} \leq c_{8,21}, \tag{242}$$

where $U(t, r)$ denotes the open or closed ball [in the Euclidean metric] centered at t with radius r and $\varphi_9(r) = r^{\beta_\tau} \log\log(1/r)$. It follows from Fubini's theorem that, with probability one, (242) holds for λ_N-almost all $t \in I$. Now we prove a stronger version of this result, which is useful in determining the exact Hausdorff measure of the level set L_x.

Theorem 8.10. *Let X be an (N, d)-Gaussian random field defined by (36). We assume Conditions (C1) and (C3') are satisfied on I and $d < \sum_{j=1}^{N}(1/H_j)$. Let $\tau \in \{1, \ldots, N\}$ be the integer so that (206) holds and let L be the jointly* *continuous local time of X. Then, there is a finite constant $c_{8,22}$ such that with probability one,*

$$\limsup_{r\to 0} \frac{L(x,U(t,r))}{\varphi_9(r)} \leq c_{8,22} \tag{243}$$

holds for $L(x,\cdot)$-almost all $t \in I$.

Proof. The proof is similar to that of Theorem 4.1 in [6]. See also [102].

For every integer $k \geq 1$, we consider the random measure $\mu_k := \mu_k(x, \bullet)$ on the Borel subsets C of I defined by (157) [where the integer n is replaced by k]. Then, by the occupation density formula (197) and the continuity of the function $y \mapsto L(y,C)$, one can verify that almost surely $\mu_k(C) \to L(x,C)$ as $k \to \infty$ for every Borel set $C \subseteq I$.

For every integer $m \geq 1$, denote $f_m(t) = L(x,U(t,2^{-m}))$. From the proof of Theorem 8.2 we can see that almost surely the functions $f_m(t)$ are continuous and bounded. Hence we have almost surely, for all integers $m, n \geq 1$,

$$\int_I [f_m(t)]^n \, L(x,dt) = \lim_{k\to\infty} \int_I [f_m(t)]^n \, \mu_k(dt). \tag{244}$$

It follows from (244), (157) and the proof of Proposition 3.1 of [77] that for every positive integer $n \geq 1$,

$$\begin{aligned}
\mathrm{E} &\int_I [f_m(t)]^n \, L(x,dt) \\
&= \left(\frac{1}{2\pi}\right)^{(n+1)d} \int_I \int_{U(t,2^{-m})^n} \int_{\mathbf{R}^{(n+1)d}} e^{-i\sum_{j=1}^{n+1} \langle x,u^j \rangle} \\
&\qquad\qquad\qquad \times \mathrm{E}\exp\left(i\sum_{j=1}^{n+1} \langle u^j, X(s^j) \rangle\right) \, d\bar{u}d\bar{s},
\end{aligned} \tag{245}$$

where $\bar{u} = (u^1,\ldots,u^{n+1}) \in \mathbf{R}^{(n+1)d}$ and $\bar{s} = (t,s^1,\ldots,s^n)$. Similar to the proof of (207) we have that the right hand side of Eq. (245) is at most

$$\begin{aligned}
c_{8,23}^n &\int_I \int_{U(t,2^{-m})^n} \frac{d\bar{s}}{[\det \mathrm{Cov}(X_0(t),X_0(s^1),\ldots,X_0(s^n))]^{d/2}} \\
&\leq c_{8,24}^n \, n! \, 2^{-mn\beta_\tau},
\end{aligned} \tag{246}$$

where $c_{8,24}$ is a positive and finite constant depending on N, d, H, and I only.

Let $\gamma > 0$ be a constant whose value will be determined later. We consider the random set

$$I_m(\omega) = \{t \in I : f_m(t) \geq \gamma\varphi_9(2^{-m})\}.$$

Denote by ν_ω the restriction of the random measure $L(x,\cdot)$ on I, that is, $\nu_\omega(E) = L(x,E \cap I)$ for every Borel set $E \subseteq \mathbf{R}_+^N$. Now we take $n = \lfloor \log m \rfloor$, where $\lfloor y \rfloor$ denotes the integer part of y. Then by applying (246) and Stirling's formula, we have

$$
\begin{aligned}
\mathrm{E}\big[\nu_\omega(I_m)\big] &\leq \frac{\mathrm{E} \int_I [f_m(t)]^n \, L(x, dt)}{[\gamma \varphi_9(2^{-m})]^n} \\
&\leq \frac{c_{8,25}^n \, n! \, 2^{-mn\beta_\tau}}{\gamma^n 2^{-mn\beta_\tau} (\log m)^n} \leq m^{-2},
\end{aligned}
\tag{247}
$$

provided $\gamma > 0$ is chosen large enough, say, $\gamma \geq c_{8,25} \, e^2 := c_{8,26}$. This implies that

$$
\mathrm{E}\left(\sum_{m=1}^{\infty} \nu_\omega(I_m)\right) < \infty.
\tag{248}
$$

Therefore, with probability 1 for ν_ω almost all $t \in I$, we have

$$
\limsup_{m \to \infty} \frac{L(x, U(t, 2^{-m}))}{\varphi_9(2^{-m})} \leq c_{8,26}.
\tag{249}
$$

Finally, for any $r > 0$ small enough, there exists an integer m such that $2^{-m} \leq r < 2^{-m+1}$ and (249) is applicable. Since the function $\varphi_9(r)$ is increasing near $r = 0$, (243) follows from (249) and a monotonicity argument.   □

Since $L(x, \cdot)$ is a random Borel measure supported on the level set $L_x = \{t \in I : X(t) = x\}$, Theorem 8.10 and the upper density theorem of [80] imply the following partial result on the exact Hausdorff measure of L_x.

Corollary 8.11. *Assume the conditions of Theorem 8.10 are satisfied. Then there exists a positive constant $c_{8,27}$ such that for every $x \in \mathbf{R}^d$, we have*

$$
\mathcal{H}^{\varphi_9}(L_x) \geq c_{8,27} \, L(x, I), \qquad a.s.
\tag{250}
$$

Proof. The proof is left to the reader as an exercise.   □

We should mention that the estimates in Lemmas 8.4 and 8.8 are not sharp and it would be interesting to improve them. In the rest of this section, we consider the special case when $H = \langle \alpha \rangle$ and $\alpha \in (0, 1)$. Many sample path properties of such Gaussian random fields have been studied in [1; 49; 83; 102; 108]. By applying Lemma 2.3 in [102] in place of (216), we prove the following sharp estimates.

Lemma 8.12. *Let X be an (N, d)-Gaussian random field satisfying the conditions (C1) and (C3') with $H = \langle \alpha \rangle$. We assume that $N > \alpha d$. Then there exists a positive and finite constant $c_{8,28}$, depending on N, d, α and I only, such that for all intervals $T = [a, a + \langle r \rangle] \subseteq I$ with edge length $r \in (0, 1)$, all $x \in \mathbf{R}^d$ and all integers $n \geq 1$,*

$$E[L(x,T)^n] \le c_{8,28}^n (n!)^{\alpha d/N} r^{n(N-\alpha d)} \tag{251}$$

and for any $0 < \gamma < \min\{1, (N/\alpha - d)/2\}$, there exists a positive and finite constant $c_{8,29}$ such that

$$E[(L(x,T) - L(y,T))^n]$$
$$\le c_{8,29}^n (n!)^{2\gamma + \alpha(d+\gamma)/N} |x - y|^{n\gamma} r^{n(N-\alpha(d+\gamma))}. \tag{252}$$

Note that, for a Gaussian random field X satisfying the assumptions of Lemma 8.12, Eq. (251) allows us to improve the results in Theorem 8.10 and Corollary 8.11 by replacing the Hausdorff measure function $\varphi_9(r)$ by $\varphi_7(r) = r^{N-\alpha d}(\log\log 1/r)^{\alpha d/N}$. Moreover, by combining Lemma 8.12 and the argument in [102], one can establish the following sharp local and uniform Hölder conditions for the maximum local time $L^*(\bullet)$ of X defined by

$$L^*(T) = \sup_{x \in \mathbf{R}^d} L(x,T) \qquad \forall\, T \subseteq I. \tag{253}$$

Theorem 8.13. *Let X be an (N,d)-Gaussian random field satisfying the conditions (C1) and (C3') with $H = \langle \alpha \rangle$ and $N > \alpha d$. The following statements hold:*

(i) *There exists a positive and finite constant $c_{8,30}$ such that for every $t \in I$,*

$$\limsup_{r \to 0} \frac{L^*(U(t,r))}{\varphi_7(r)} \le c_{8,30} \qquad a.s., \tag{254}$$

where $U(t,r) = \{s \in I : |s - t| \le r\}$.
(ii) *There exists a positive and finite constant $c_{8,31}$ such that*

$$\limsup_{r \to 0} \sup_{t \in I} \frac{L^*(U(t,r))}{\varphi_{10}(r)} \le c_{8,31} \qquad a.s., \tag{255}$$

where $\varphi_{10}(r) = r^{N-\alpha d}(\log 1/r)^{\alpha d/N}$.

Proof. The proofs of (254) and (255) are based on Lemma 8.12 and a chaining argument, which is the same as those of Theorems 1.1 and 1.2 in [102]; see also [34]. We leave the details to the reader.   □

Similar to [102; 108], one can apply Lemma 8.12 and Theorem 8.13 to derive further results, such as the Chung-type laws of the iterated logarithm, modulus of nowhere differentiability, tail probability of the local times, for (N,d)-Gaussian random fields satisfying the conditions (C1) and (C3') with $H = \langle \alpha \rangle$. These are left to the reader as exercises.

The following is our final remark.

Remark 8.14. Both Conditions (C3) and (C3′) are useful in studying the existence and regularity of the self-intersection local times of X which, in turn, provide information on the fractal dimensions of the sets of multiple points and multiple times of X. When X is an (N, d)-fractional Brownian sheet, these problems have been studied in [69]. It is expected that analogous results also hold for Gaussian random fields satisfying Conditions (C1) and (C3′).

References

[1] R. J. Adler (1981), *The Geometry of Random Fields.* John Wiley & Sons Ltd., New York.

[2] L. Arnold and P. Imkeller (1996), Stratonovich calculus with spatial parameters and anticipative problems in multiplicative ergodic theory. *Stoch. Process. Appl.* **62**, 19–54.

[3] A. Ayache (2004), Hausdorff dimension of the graph of the fractional Brownian sheet. *Rev. Mat. Iberoamericana,* **20**, 395–412.

[4] A. Ayache, S. Leger and M. Pontier (2002), Drap Brownien fractionnaire. *Potential Theory* **17**, 31–43.

[5] A. Ayache and M. S. Taqqu (2003), Rate optimality of wavelet series approximations of fractional Brownian sheet. *J. Fourier Anal. Appl.* **9**, 451–471.

[6] A. Ayache, D. Wu and Y. Xiao (2006), Joint continuity of the local times of fractional Brownian sheets. *Ann. Inst. H. Poincaré Probab. Statist.*, (to appear).

[7] A. Ayache and Y. Xiao (2005), Asymptotic properties and Hausdorff dimensions of fractional Brownian sheets. *J. Fourier Anal. Appl.* **11**, 407–439.

[8] R. Balan and D. Kim (2006), The stochastic heat equation driven by a Gaussian noise: Markov property. *Preprint.*

[9] E. Belinski and W. Linde (2002), Small ball probabilities of fractional Brownian sheets via fractional integration operators. *J. Theoret. Probab.* **15**, 589–612.

[10] A. Benassi, S. Jaffard and D. Roux (1997), Elliptic Gaussian random processes. *Rev. Mat. Iberoamericana* **13**, 19–90.

[11] D. A. Benson, M. M. Meerschaert, B. Baeumer and H. P. Scheffler (2006), Aquifer operator-scaling and the effect on solute mixing and dispersion. *Water Resources Research* **42**, no. 1, W01415 (18 pp.).

[12] C. Berg and G. Forst (1975), *Potential Theory on Locally Compact Abelian Groups.* Springer-Verlag, New York-Heidelberg.

[13] S. M. Berman (1972), Gaussian sample function: uniform dimension and Hölder conditions nowhere. *Nagoya Math. J.* **46**, 63–86.

[14] S. M. Berman (1973), Local nondeterminism and local times of Gaussian processes. *Indiana Univ. Math. J.* **23**, 69–94.

[15] H. Biermé, A. Estrade, M. M. Meerschaert and Y. Xiao (2008), Sample path properties of operator scaling Gaussian random fields. *In Preparation.*

[16] H. Biermé, C. Lacaux and Y. Xiao (2007), Hitting probabilities and the Hausdorff dimension of the inverse images of anisotropic Gaussian random fields. *Preprint.*

[17] H. Biermé, M. M. Meerschaert and H.-P. Scheffler (2007), Operator scaling stable random fields. *Stoch. Process. Appl.* **117**, 312–332.

[18] A. Bonami and A. Estrade (2003), Anisotropic analysis of some Gaussian models. *J. Fourier Anal. Appl.* **9**, 215–236.

[19] J, Cuzick and J. DuPreez (1982), Joint continuity of Gaussian local times. *Ann. Probab.* **10**, 810–817.

[20] R. C. Dalang (1999), Extending martingale measure stochastic integral with applications to spatially homogeneous s.p.d.e.'s. *Electron. J. Probab.* **4**, no. 6, 1–29. Erratum in *Electron. J. Probab.* **6** (2001), no. 6, 1–5.

[21] R. C. Dalang and N. E. Frangos (1998), The stochastic wave equation in two spatial dimensions. *Ann. Probab.* **26**, 187–212.

[22] R. C. Dalang, D. Khoshnevisan and E. Nualart (2007), Hitting probabilities for systems of non-linear stochastic heat equations with additive noise. *Latin Amer. J. Probab. Statist. (Alea)* **3**, 231–271.

[23] R. C. Dalang, D. Khoshnevisan and E. Nualart (2007), Hitting probabilities for the non-linear stochastic heat equation with multiplicative noise. *Submitted.*

[24] R. C. Dalang and C. Mueller (2003), Some non-linear s.p.e.e.'s that are second order in time. *Electron. J. Probab.* **8**, No. 1, 1–21.

[25] R. C. Dalang, C. Mueller and L. Zambotti, (2006), Hitting properties of parabolic s.p.d.e.'s with reflection. *Ann. Probab.* **34**, 1423–1450.

[26] R. C. Dalang and E. Nualart (2004), Potential theory for hyperbolic SPDEs. *Ann. Probab.* **32**, 2099–2148.

[27] R. C. Dalang and M. Sanz-Solé (2005), Regularity of the sample paths of a class of second-order spde's. *J. Funct. Anal.* **227**, 304–337.

[28] R. C. Dalang and M. Sanz-Solé (2007), Hölder regularity of the solution to the stochastic wave equation in dimension three. *Memoirs Amer. Math. Soc.*, (to appear).

[29] S. Davies and P. Hall (1999), Fractal analysis of surface roughness by using spatial data (with discussion). *J. Roy. Statist. Soc. Ser. B* **61**, 3–37.

[30] P. Doukhan, G. Oppenheim and M. S. Taqqu (2003), *Theory and Applications of Long-Range Dependence.* Birkhaüser, Boston.

[31] M. Dozzi (2002), Occupation density and sample path properties of N-parameter processes. *Topics in Spatial Stochastic Processes (Martina Franca, 2001)*, pp. 127–169, *Lecture Notes in Math.* **1802**, Springer, Berlin.

[32] T. Dunker (2000), Estimates for the small ball probabilities of the fractional Brownian sheet. *J. Theoret. Probab.* **13**, 357–382.

[33] K. Dzhaparidze and H. van Zanten (2005), Optimality of an explicit series of the fractional Browinan sheet. *Statist. Probab. Lett.* **71**, 295–301.

[34] W. Ehm (1981), Sample function properties of multi-parameter stable processes. *Z. Wahrsch. verw Gebiete* **56**, 195–228.

[35] N. Eisenbaum and D. Khoshnevisan (2002), On the most visited sites of symmetric Markov processes. *Stoch. Process. Appl.* **101**, 241–256.

[36] K. J. Falconer (1990), *Fractal Geometry – Mathematical Foundations And Applications.* John Wiley & Sons Ltd., Chichester.

[37] K. J. Falconer and J. D. Howroyd (1997), Packing dimensions for projections and dimension profiles. *Math. Proc. Camb. Philo. Soc.* **121**, 269–286.

[38] T. Funaki (1983), Random motion of strings and related stochastic evolution equations. *Nagoya Math. J.* **89**, 129–193.

[39] T. Funaki, M. Kikuchi and J. Potthoff (2006), Direction-dependent modulus of continuity for random fields. *Preprint.*

[40] A. M. Garsia, E. Rodemich and H. Jr. Rumsey (1970), A real variable lemma and the continuity of paths of some Gaussian processes. *Indiana Univ. Math. J.* **20**, 565–578.

[41] D. Geman and J. Horowitz (1980), Occupation densities. *Ann. Probab.* **8**, 1–67.

[42] J. Hawkes (1978), Measures of Hausdorff type and stable processes. *Mathematika* **25**, 202–210.

[43] E. Herbin (2006), From N parameter fractional Brownian motions to N parameter multifractional Brownian motions. *Rocky Mount. J. Math.* **36**, 1249–1284.

[44] X. Hu and S. J. Taylor (1994), Fractal properties of products and projections of measures in $\mathbf{R}^d$. *Math. Proc. Camb. Philos. Soc.*, **115**, 527–544.

[45] Y. Hu, B. Øksendal and T. Zhang (2000), Stochastic partial differential equations driven by multiparameter fractional white noise. *Stochastic Processes, Physics and Geometry: new interplays, II (Leipzig, 1999)*, 327–337, Amer. Math. Soc., Providence, RI.

[46] J.-P. Kahane (1985), *Some Random Series of Functions.* 2nd edition, Cambridge University Press, Cambridge.

[47] A. Kamont (1996), On the fractional anisotropic Wiener field. *Probab. Math. Statist.* **16**, 85–98.

[48] R. Kaufman (1972), Measures of Hausdorff-type, and Brownian motion. *Mathematika* **19**, 115–119.

[49] D. Khoshnevisan (2002), *Multiparameter Processes: An Introduction to Random Fields.* Springer, New York.

[50] D. Khoshnevisan and Z. Shi (1999), Brownian sheet and capacity. *Ann. Probab.* **27**, 1135–1159.

[51] D. Khoshnevisan, D. Wu and Y. Xiao (2006), Sectorial local nondeterminism and the geometry of the Brownian sheet. *Electron. J. Probab.* **11**, 817–843.

[52] D. Khoshnevisan and Y. Xiao (2002), Level sets of additive Lévy processes. *Ann. Probab.* **30**, 62–100.

[53] D. Khoshnevisan and Y. Xiao (2003), Weak unimodality of finite measures, and an application to potential theory of additive Lévy processes. *Proc. Amer. Math. Soc.* **131**, 2611–2616.

[54] D. Khoshnevisan and Y. Xiao (2005), Lévy processes: capacity and Hausdorff dimension. *Ann. Probab.* **33**, 841–878.

[55] D. Khoshnevisan and Y. Xiao (2007a), Images of the Brownian sheet. *Trans. Amer. Math. Soc.* **359**, 3125–3151.

[56] D. Khoshnevisan and Y. Xiao (2007b), Harmonic analysis of additive Lévy processes. *Submitted.*

[57] D. Khoshnevisan, Y. Xiao and Y. Zhong (2003), Measuring the range of an additive Lévy process. *Ann. Probab.* **31**, 1097–1141.

[58] N. Kôno (1975), Asymptotoc behavior of sample functions of Gaussian random fields. *J. Math. Kyoto Univ.* **15**, 671–707.

[59] T. Kühn and W. Linde (2002), Optimal series representation of fractional Brownian sheet. *Bernoulli* **8**, 669–696.

[60] S. Kwapień and J. Rosiński (2004), Sample Hölder continuity of stochastic processes and majorizing measures. In: *Seminar on Stochastic Analysis, Random Fields and Applications IV*, pp. 155–163, Progr. Probab., **58**, Birkhäuser, Basel.

[61] M. Ledoux (1996), Isoperimetry and Gaussian analysis. *Lecture Notes in Math.* **1648**, 165–294, Springer-Verlag, Berlin.

[62] W. V. Li and Q.-M. Shao (2001), Gaussian processes: inequalities, small ball probabilities and applications. In *Stochastic Processes: Theory and Methods.* Handbook of Statistics, **19**, (C. R. Rao and D. Shanbhag, editors), pp. 533–597, North-Holland.

[63] M. A. Lifshits (1999), Asymptotic behavior of small ball probabilities. In: *Probab. Theory and Math. Statist., Proc. VII International Vilnius Conference (1998).* Vilnius, VSP/TEV, pp. 453–468.

[64] W. Linde (2007), Non-determinism of linear operators and lower entropy estimates. *Preprint.*

[65] D. J. Mason and Y. Xiao (2001), Sample path properties of operator self-similar Gaussian random fields. *Teor. Veroyatnost. i Primenen.* **46**, 94–116.

[66] D. M. Mason and Z. Shi (2001), Small deviations for some multi-parameter Gaussian processes. *J. Theoret. Probab.* **14**, 213–239.

[67] P. Mattila (1995), *Geometry of Sets and Measures in Euclidean Spaces.* Cambridge University Press, Cambridge.

[68] M. M. Meerschaert, W. Wang and Y. Xiao (2007), Fernique-type inequalities and moduli of continuity for anisotropic Gaussian random fields. *Preprint.*

[69] M. M. Meerschaert, D. Wu and Y. Xiao (2008), Self-intersection local times of fractional Brownian sheets. *In Preparation.*

[70] D. Monrad and L. D. Pitt (1987), Local nondeterminism and Hausdorff dimension. *Progress in Probability and Statistics. Seminar on Stochastic Processes 1986*, (E, Cinlar, K. L. Chung, R. K. Getoor, Editors), pp. 163–189, Birkhäuser, Boston.

[71] D. Monrad and H. Rootzén (1995), Small values of Gaussian processes and functional laws of the iterated logarithm. *Probab. Thory Relat. Fields* **101**, 173–192.

[72] T. S. Mountford (1989), Uniform dimension results for the Brownian sheet. *Ann. Probab.* **17**, 1454–1462.

[73] C. Mueller and R. Tribe (2002), Hitting probabilities of a random string. *Electron. J. Probab.* **7**, Paper No. 10, 1–29.

[74] D. Nualart (2006), Stochastic heat equation driven by fractional noise. *Preprint.*

[75] B. Øksendal and T. Zhang (2000), Multiparameter fractional Brownian motion and quasi-linear stochastic partial differential equations. *Stoch. Stoch. Rep.* **71**, 141–163.

[76] S. Orey and W. E. Pruitt (1973), Sample functions of the N-parameter Wiener process. *Ann. Probab.* **1**, 138–163.

[77] L. D. Pitt (1978), Local times for Gaussian vector fields. *Indiana Univ. Math. J.* **27**, 309–330.

[78] L. D. Pitt and R. S. Robeva (2003), On the sharp Markov property for Gaussian random fields and spectral synthesis in spaces of Bessel potentials. *Ann. Probab.* **31**, 1338–1376.

[79] R. S. Robeva and L. D. Pitt (2004), On the equality of sharp and germ σ-fields for Gaussian processes and fields. *Pliska Stud. Math. Bulgar.* **16**, 183–205.

[80] C. A. Rogers and S. J. Taylor (1961), Functions continuous and singular with respect to a Hausdorff measure. *Mathematika* **8**, 1–31.

[81] J. Rosen (1984), Self-intersections of random fields. *Ann. Probab.* **12**, 108–119.

[82] G. Samorodnitsky and M. S. Taqqu (1994), *Stable non-Gaussian Random Processes: Stochastic models with infinite variance.* Chapman & Hall, New York.

[83] N.-R. Shieh and Y. Xiao (2006), Images of Gaussian random fields: Salem sets and interior points. *Studia Math.* **176**, 37–60.

[84] M. Talagrand (1993), New Gaussian estimates for enlarged balls. *Geometric Funct. Anal.* **3**, 502–526.

[85] M. Talagrand (1995), Hausdorff measure of trajectories of multiparameter fractional Brownian motion. *Ann. Probab.* **23**, 767–775.

[86] M. Talagrand (1998), Multiple points of trajectories of multiparameter fractional Brownian motion. *Probab. Theory Relat. Fields* **112**, 545–563.

[87] M. Talagrand and Y. Xiao (1996), Fractional Brownian motion and packing dimension. *J. Theoret. Probab.* **9**, 579–593.

[88] S. J. Taylor (1986), The measure theory of random fractals. *Math. Proc. Camb. Philos. Soc.* **100**, 383–406.

[89] S. J. Taylor and C. Tricot (1985), Packing measure and its evaluation for a Brownian path. *Trans. Amer. Math. Soc.* **288**, 679–699.

[90] S. J. Taylor and A. A. Watson (1985), A Hausdorff measure classification of polar sets for the heat equation. *Math. Proc. Camb. Philos. Soc.* **97**, 325–344.

[91] F. Testard (1986), Polarité, points multiples et géométrie de certain processus gaussiens. *Publ. du Laboratoire de Statistique et Probabilités de l'U.P.S.* Toulouse, 01–86.

[92] C. Tricot (1982), Two definitions of fractional dimension. *Math. Proc. Camb. Philos. Soc.* **91**, 57–74.

[93] C. A. Tudor and Y. Xiao (2007), Sample path properties of bifractional Brownian motion. *Bernoulli* **14**, 1023–1052.

[94] J. B. Walsh (1986), An introduction to stochastic partial differential equations. *École d'été de probabilités de Saint-Flour, XIV—1984,* 265–439, Lecture Notes in Math., **1180**, Springer, Berlin.

[95] W. Wang (2007), Almost-sure path properties of fractional Brownian sheet. *Ann. Inst. H. Poincaré Probab. Statist.* **43**, 619–631.

[96] D. Wu and Y. Xiao (2006), Fractal properties of random string processes. *IMS Lecture Notes-Monograph Series–High Dimensional Probability.* **51**, 128–147.

[97] D. Wu and Y. Xiao (2007), Geometric properties of fractional Brownian sheets. *J. Fourier Anal. Appl.* **13**, 1–37.

[98] Y. Xiao (1995), Dimension results for Gaussian vector fields and index-α stable fields. *Ann. Probab.* **23**, 273–291.

[99] Y. Xiao (1996a), Hausdorff measure of the sample paths of Gaussian random fields. *Osaka J. Math.* **33**, 895–913.

[100] Y. Xiao (1996b), Packing measure of the sample paths of fractional Brownian motion. *Trans. Amer. Math. Soc.* **348**, 3193–3213.

[101] Y. Xiao (1997a), Hausdorff measure of the graph of fractional Brownian motion. *Math. Proc. Camb. Philos. Soc.* **122**, 565–576.

[102] Y. Xiao (1997b), Hölder conditions for the local times and the Hausdorff measure of the level sets of Gaussian random fields. *Probab. Theory Relat. Fields* **109**, 129–157.

[103] Y. Xiao (1997c), Packing dimension of the image of fractional Brownian motion. *Statist. Probab. Lett.* **33**, 379–387.

[104] Y. Xiao (1999), Hitting probabilities and polar sets for fractional Brownian motion. *Stoch. Stoch. Rep.* **66**, 121–151.

[105] Y. Xiao (2003), The packing measure of the trajectories of multiparameter fractional Brownian motion. *Math. Proc. Camb. Philos. Soc.* **135**, 349–375.

[106] Y. Xiao (2004), Random fractals and Markov processes. In: *Fractal Geometry and Applications: A Jubilee of Benoit Mandelbrot*, (M. L. Lapidus and M. van Frankenhuijsen, editors), pp. 261–338, American Mathematical Society.

[107] Y. Xiao (2006), Properties of local nondeterminism of Gaussian and stable random fields and their applications. *Ann. Fac. Sci. Toulouse Math.* **XV**, 157–193.

[108] Y. Xiao (2007), Strong local nondeterminism and sample path properties of Gaussian random fields. In: *Asymptotic Theory in Probability and Statistics with Applications* (T.-L. Lai, Q.-M. Shao and L. Qian, eds), pp. 136–176, Higher Education Press, Beijing.

[109] Y. Xiao (2008), Spectral conditions for strong local nondeterminism of Gaussian random fields. *In Preparation*.

[110] Y. Xiao and T. Zhang (2002), Local times of fractional Brownian sheets. *Probab. Theory Relat. Fields* **124**, 204–226.

[111] A. M. Yaglom (1957), Some classes of random fields in n-dimensional space, related to stationary random processes. *Th. Probab. Appl.* **2**, 273–320.

List of Participants

Tom Alberts
New York University
New York, USA
alberts@cims.nyu.edu

Hakima Bessaih
University of Wyoming
Laramie, USA
bessaih@uwyo.edu

Ryan Card
University of Washington
Seattle, USA
card@math.washington.edu

Robert C. Dalang
Ecole Polytechnique Fédérale
Lausanne, Switzerland
robert.dalang@epfl.ch

Sabrina Guettes
University of Wisconsin
Madison, USA
guettes@math.wisc.edu

Gunnar Gunnarsson
University of California
Santa Barbara, USA
gungun@math.ucsb.edu

Samuel Isaacson
University of Utah
Salt Lake City, USA
isaacson@math.utah.edu

Karim Khader
University of Utah
Salt Lake City, USA
khader@math.utah.edu

Davar Khoshnevisan
University of Utah
Salt Lake City, USA
davar@math.utah.edu

An Le
Utah State University
Logan, USA
anle@math.utah.edu

Scott McKinley
Ohio State University
Columbus, USA
mckinley@math.ohio-state.edu

Carl Mueller
University of Rochester
Rochester, USA
cmlr+new@math.rochester.edu

David Nualart
University of Kansas
Lawrence, USA
nualart@math.ku.edu

Chris Orum
Los Alamos National Laboratory
New Mexico, USA
orumc@lanl.gov

Michael Purcell
University of Utah
Salt Lake City, USA
purcell@math.utah.edu

Firas Rassoul-Agha
University of Utah
Salt Lake City, USA
firas@math.utah.edu

Jean-François Renaud
Université de Montreal
Montréal, Canada
renaud@dms.umontreal.ca

Boris Rozovsky
University of Southern California
Los Angeles, USA
rozovski@math.usc.edu

Shang-Yuan Shiu
University of Utah
Salt Lake City, USA
shiu@math.utah.edu

Blerta Shtylla
University of Utah
Salt Lake City, USA
shtyllab@math.utah.edu

Andrejs Treibergs
University of Utah
Salt Lake City, USA
treiberg@math.utah.edu

Dongsheng Wu
Michigan State University
East Lansing, USA
wudongsh@msu.edu

Yimin Xiao
Michigan State University
East Lansing, USA
xiao@stt.msu.edu

Index

Lecture Notes in Mathematics

For information about earlier volumes
please contact your bookseller or Springer
LNM Online archive: springerlink.com

Vol. 1822: S.-N. Chow, R. Conti, R. Johnson, J. Mallet-Paret, R. Nussbaum, Dynamical Systems. Cetraro, Italy 2000. Editors: J. W. Macki, P. Zecca (2003)

Vol. 1823: A. M. Anile, W. Allegretto, C. Ringhofer, Mathematical Problems in Semiconductor Physics. Cetraro, Italy 1998. Editor: A. M. Anile (2003)

Vol. 1824: J. A. Navarro González, J. B. Sancho de Salas, $\mathscr{C}^\infty$ – Differentiable Spaces (2003)

Vol. 1825: J. H. Bramble, A. Cohen, W. Dahmen, Multiscale Problems and Methods in Numerical Simulations, Martina Franca, Italy 2001. Editor: C. Canuto (2003)

Vol. 1826: K. Dohmen, Improved Bonferroni Inequalities via Abstract Tubes. Inequalities and Identities of Inclusion-Exclusion Type. VIII, 113 p, 2003.

Vol. 1827: K. M. Pilgrim, Combinations of Complex Dynamical Systems. IX, 118 p, 2003.

Vol. 1828: D. J. Green, Gröbner Bases and the Computation of Group Cohomology. XII, 138 p, 2003.

Vol. 1829: E. Altman, B. Gaujal, A. Hordijk, Discrete-Event Control of Stochastic Networks: Multimodularity and Regularity. XIV, 313 p, 2003.

Vol. 1830: M. I. Gil', Operator Functions and Localization of Spectra. XIV, 256 p, 2003.

Vol. 1831: A. Connes, J. Cuntz, E. Guentner, N. Higson, J. E. Kaminker, Noncommutative Geometry, Martina Franca, Italy 2002. Editors: S. Doplicher, L. Longo (2004)

Vol. 1832: J. Azéma, M. Émery, M. Ledoux, M. Yor (Eds.), Séminaire de Probabilités XXXVII (2003)

Vol. 1833: D.-Q. Jiang, M. Qian, M.-P. Qian, Mathematical Theory of Nonequilibrium Steady States. On the Frontier of Probability and Dynamical Systems. IX, 280 p, 2004.

Vol. 1834: Yo. Yomdin, G. Comte, Tame Geometry with Application in Smooth Analysis. VIII, 186 p, 2004.

Vol. 1835: O.T. Izhboldin, B. Kahn, N.A. Karpenko, A. Vishik, Geometric Methods in the Algebraic Theory of Quadratic Forms. Summer School, Lens, 2000. Editor: J.-P. Tignol (2004)

Vol. 1836: C. Năstăsescu, F. Van Oystaeyen, Methods of Graded Rings. XIII, 304 p, 2004.

Vol. 1837: S. Tavaré, O. Zeitouni, Lectures on Probability Theory and Statistics. Ecole d'Eté de Probabilités de Saint-Flour XXXI-2001. Editor: J. Picard (2004)

Vol. 1838: A.J. Ganesh, N.W. O'Connell, D.J. Wischik, Big Queues. XII, 254 p, 2004.

Vol. 1839: R. Gohm, Noncommutative Stationary Processes. VIII, 170 p, 2004.

Vol. 1840: B. Tsirelson, W. Werner, Lectures on Probability Theory and Statistics. Ecole d'Eté de Probabilités de Saint-Flour XXXII-2002. Editor: J. Picard (2004)

Vol. 1841: W. Reichel, Uniqueness Theorems for Variational Problems by the Method of Transformation Groups (2004)

Vol. 1842: T. Johnsen, A. L. Knutsen, K_3 Projective Models in Scrolls (2004)

Vol. 1843: B. Jefferies, Spectral Properties of Noncommuting Operators (2004)

Vol. 1844: K.F. Siburg, The Principle of Least Action in Geometry and Dynamics (2004)

Vol. 1845: Min Ho Lee, Mixed Automorphic Forms, Torus Bundles, and Jacobi Forms (2004)

Vol. 1846: H. Ammari, H. Kang, Reconstruction of Small Inhomogeneities from Boundary Measurements (2004)

Vol. 1847: T.R. Bielecki, T. Björk, M. Jeanblanc, M. Rutkowski, J.A. Scheinkman, W. Xiong, Paris-Princeton Lectures on Mathematical Finance 2003 (2004)

Vol. 1848: M. Abate, J. E. Fornaess, X. Huang, J. P. Rosay, A. Tumanov, Real Methods in Complex and CR Geometry, Martina Franca, Italy 2002. Editors: D. Zaitsev, G. Zampieri (2004)

Vol. 1849: Martin L. Brown, Heegner Modules and Elliptic Curves (2004)

Vol. 1850: V. D. Milman, G. Schechtman (Eds.), Geometric Aspects of Functional Analysis. Israel Seminar 2002-2003 (2004)

Vol. 1851: O. Catoni, Statistical Learning Theory and Stochastic Optimization (2004)

Vol. 1852: A.S. Kechris, B.D. Miller, Topics in Orbit Equivalence (2004)

Vol. 1853: Ch. Favre, M. Jonsson, The Valuative Tree (2004)

Vol. 1854: O. Saeki, Topology of Singular Fibers of Differential Maps (2004)

Vol. 1855: G. Da Prato, P.C. Kunstmann, I. Lasiecka, A. Lunardi, R. Schnaubelt, L. Weis, Functional Analytic Methods for Evolution Equations. Editors: M. Iannelli, R. Nagel, S. Piazzera (2004)

Vol. 1856: K. Back, T.R. Bielecki, C. Hipp, S. Peng, W. Schachermayer, Stochastic Methods in Finance, Bressanone/Brixen, Italy, 2003. Editors: M. Fritelli, W. Runggaldier (2004)

Vol. 1857: M. Émery, M. Ledoux, M. Yor (Eds.), Séminaire de Probabilités XXXVIII (2005)

Vol. 1858: A.S. Cherny, H.-J. Engelbert, Singular Stochastic Differential Equations (2005)

Vol. 1859: E. Letellier, Fourier Transforms of Invariant Functions on Finite Reductive Lie Algebras (2005)

Vol. 1860: A. Borisyuk, G.B. Ermentrout, A. Friedman, D. Terman, Tutorials in Mathematical Biosciences I. Mathematical Neurosciences (2005)

Vol. 1861: G. Benettin, J. Henrard, S. Kuksin, Hamiltonian Dynamics – Theory and Applications, Cetraro, Italy, 1999. Editor: A. Giorgilli (2005)

Vol. 1862: B. Helffer, F. Nier, Hypoelliptic Estimates and Spectral Theory for Fokker-Planck Operators and Witten Laplacians (2005)

Vol. 1863: H. Führ, Abstract Harmonic Analysis of Continuous Wavelet Transforms (2005)

Vol. 1864: K. Efstathiou, Metamorphoses of Hamiltonian Systems with Symmetries (2005)

Vol. 1865: D. Applebaum, B.V. R. Bhat, J. Kustermans, J. M. Lindsay, Quantum Independent Increment Processes I. From Classical Probability to Quantum Stochastic Calculus. Editors: M. Schürmann, U. Franz (2005)

Vol. 1866: O.E. Barndorff-Nielsen, U. Franz, R. Gohm, B. Kümmerer, S. Thorbjønsen, Quantum Independent Increment Processes II. Structure of Quantum Lévy Processes, Classical Probability, and Physics. Editors: M. Schürmann, U. Franz, (2005)

Vol. 1867: J. Sneyd (Ed.), Tutorials in Mathematical Biosciences II. Mathematical Modeling of Calcium Dynamics and Signal Transduction. (2005)

Vol. 1868: J. Jorgenson, S. Lang, $Pos_n(R)$ and Eisenstein Series. (2005)

Vol. 1869: A. Dembo, T. Funaki, Lectures on Probability Theory and Statistics. Ecole d'Eté de Probabilités de Saint-Flour XXXIII-2003. Editor: J. Picard (2005)

Vol. 1870: V.I. Gurariy, W. Lusky, Geometry of Müntz Spaces and Related Questions. (2005)

Vol. 1871: P. Constantin, G. Gallavotti, A.V. Kazhikhov, Y. Meyer, S. Ukai, Mathematical Foundation of Turbulent Viscous Flows, Martina Franca, Italy, 2003. Editors: M. Cannone, T. Miyakawa (2006)

Vol. 1925: M. du Sautoy, L. Woodward, Zeta Functions of Groups and Rings (2008)

Vol. 1926: L. Barreira, V. Claudia, Stability of Nonautonomous Differential Equations (2008)

Vol. 1927: L. Ambrosio, L. Caffarelli, M.G. Crandall, L.C. Evans, N. Fusco, Calculus of Variations and Non-Linear Partial Differential Equations. Cetraro, Italy 2005. Editors: B. Dacorogna, P. Marcellini (2008)

Vol. 1928: J. Jonsson, Simplicial Complexes of Graphs (2008)

Vol. 1929: Y. Mishura, Stochastic Calculus for Fractional Brownian Motion and Related Processes (2008)

Vol. 1930: J.M. Urbano, The Method of Intrinsic Scaling. A Systematic Approach to Regularity for Degenerate and Singular PDEs (2008)

Vol. 1931: M. Cowling, E. Frenkel, M. Kashiwara, A. Valette, D.A. Vogan, Jr., N.R. Wallach, Representation Theory and Complex Analysis. Venice, Italy 2004. Editors: E.C. Tarabusi, A. D'Agnolo, M. Picardello (2008)

Vol. 1932: A.A. Agrachev, A.S. Morse, E.D. Sontag, H.J. Sussmann, V.I. Utkin, Nonlinear and Optimal Control Theory. Cetraro, Italy 2004. Editors: P. Nistri, G. Stefani (2008)

Vol. 1933: M. Petkovic, Point Estimation of Root Finding Methods (2008)

Vol. 1934: C. Donati-Martin, M. Émery, A. Rouault, C. Stricker (Eds.), Séminaire de Probabilités XLI (2008)

Vol. 1935: A. Unterberger, Alternative Pseudodifferential Analysis (2008)

Vol. 1936: P. Magal, S. Ruan (Eds.), Structured Population Models in Biology and Epidemiology (2008)

Vol. 1937: G. Capriz, P. Giovine, P.M. Mariano (Eds.), Mathematical Models of Granular Matter (2008)

Vol. 1938: D. Auroux, F. Catanese, M. Manetti, P. Seidel, B. Siebert, I. Smith, G. Tian, Symplectic 4-Manifolds and Algebraic Surfaces. Cetraro, Italy 2003. Editors: F. Catanese, G. Tian (2008)

Vol. 1939: D. Boffi, F. Brezzi, L. Demkowicz, R.G. Durán, R.S. Falk, M. Fortin, Mixed Finite Elements, Compatibility Conditions, and Applications. Cetraro, Italy 2006. Editors: D. Boffi, L. Gastaldi (2008)

Vol. 1940: J. Banasiak, V. Capasso, M.A.J. Chaplain, M. Lachowicz, J. Miękisz, Multiscale Problems in the Life Sciences. From Microscopic to Macroscopic. Będlewo, Poland 2006. Editors: V. Capasso, M. Lachowicz (2008)

Vol. 1941: S.M.J. Haran, Arithmetical Investigations. Representation Theory, Orthogonal Polynomials, and Quantum Interpolations (2008)

Vol. 1942: S. Albeverio, F. Flandoli, Y.G. Sinai, SPDE in Hydrodynamic. Recent Progress and Prospects. Cetraro, Italy 2005. Editors: G. Da Prato, M. Röckner (2008)

Vol. 1943: L.L. Bonilla (Ed.), Inverse Problems and Imaging. Martina Franca, Italy 2002 (2008)

Vol. 1944: A. Di Bartolo, G. Falcone, P. Plaumann, K. Strambach, Algebraic Groups and Lie Groups with Few Factors (2008)

Vol. 1945: F. Brauer, P. van den Driessche, J. Wu (Eds.), Mathematical Epidemiology (2008)

Vol. 1946: G. Allaire, A. Arnold, P. Degond, T.Y. Hou, Quantum Transport. Modelling, Analysis and Asymptotics. Cetraro, Italy 2006. Editors: N.B. Abdallah, G. Frosali (2008)

Vol. 1947: D. Abramovich, M. Mariño, M. Thaddeus, R. Vakil, Enumerative Invariants in Algebraic Geometry and String Theory. Cetraro, Italy 2005. Editors: K. Behrend, M. Manetti (2008)

Vol. 1948: F. Cao, J-L. Lisani, J-M. Morel, P. Musé, F. Sur, A Theory of Shape Identification (2008)

Vol. 1949: H.G. Feichtinger, B. Helffer, M.P. Lamoureux, N. Lerner, J. Toft, Pseudo-Differential Operators. Quantization and Signals. Cetraro, Italy 2006. Editors: L. Rodino, M.W. Wong (2008)

Vol. 1950: M. Bramson, Stability of Queueing Networks, Ecole d'Eté de Probabilités de Saint-Flour XXXVI-2006 (2008)

Vol. 1951: A. Moltó, J. Orihuela, S. Troyanski, M. Valdivia, A Non Linear Transfer Technique for Renorming (2008)

Vol. 1952: R. Mikhailov, I.B.S. Passi, Lower Central and Dimension Series of Groups (2008)

Vol. 1953: K. Arwini, C.T.J. Dodson, Information Geometry (2008)

Vol. 1954: P. Biane, L. Bouten, F. Cipriani, N. Konno, N. Privault, Q. Xu, Quantum Potential Theory. Editors: U. Franz, M. Schuermann (2008)

Vol. 1955: M. Bernot, V. Caselles, J.-M. Morel, Optimal Transportation Networks (2008)

Vol. 1956: C.H. Chu, Matrix Convolution Operators on Groups (2008)

Vol. 1957: A. Guionnet, On Random Matrices: Macroscopic Asymptotics, Ecole d'Eté de Probabilités de Saint-Flour XXXVI-2006 (2008)

Vol. 1958: M.C. Olsson, Compactifying Moduli Spaces for Abelian Varieties (2008)

Vol. 1959: Y. Nakkajima, A. Shiho, Weight Filtrations on Log Crystalline Cohomologies of Families of Open Smooth Varieties (2008)

Vol. 1960: J. Lipman, M. Hashimoto, Foundations of Grothendieck Duality for Diagrams of Schemes (2009)

Vol. 1961: G. Buttazzo, A. Pratelli, E. Stepanov, S. Solimini, Optimal Urban Networks via Mass Transportation (2009)

Vol. 1962: R. Dalang, D. Khoshnevisan, C. Mueller, D. Nualart, Y. Xiao, A Minicourse on Stochastic Partial Differential Equations (2009)

Vol. 1963: W. Siegert, Local Lyapunov Exponents (2009)

Recent Reprints and New Editions

Vol. 1702: J. Ma, J. Yong, Forward-Backward Stochastic Differential Equations and their Applications. 1999 – Corr. 3rd printing (2007)

Vol. 830: J.A. Green, Polynomial Representations of GL_n, with an Appendix on Schensted Correspondence and Littelmann Paths by K. Erdmann, J.A. Green and M. Schoker 1980 – 2nd corr. and augmented edition (2007)

Vol. 1693: S. Simons, From Hahn-Banach to Monotonicity (Minimax and Monotonicity 1998) – 2nd exp. edition (2008)

Vol. 470: R.E. Bowen, Equilibrium States and the Ergodic Theory of Anosov Diffeomorphisms. With a preface by D. Ruelle. Edited by J.-R. Chazottes. 1975 – 2nd rev. edition (2008)

Vol. 523: S.A. Albeverio, R.J. Høegh-Krohn, S. Mazzucchi, Mathematical Theory of Feynman Path Integral. 1976 – 2nd corr. and enlarged edition (2008)

Vol. 1764: A. Cannas da Silva, Lectures on Symplectic Geometry 2001 – Corr. 2nd printing (2008)

LECTURE NOTES IN MATHEMATICS  Springer

Edited by J.-M. Morel, F. Takens, B. Teissier, P.K. Maini

Editorial Policy (for Multi-Author Publications: Summer Schools/Intensive Courses)

1. Lecture Notes aim to report new developments in all areas of mathematics and their applications - quickly, informally and at a high level. Mathematical texts analysing new developments in modelling and numerical simulation are welcome. Manuscripts should be reasonably self-contained and rounded off. Thus they may, and often will, present not only results of the author but also related work by other people. They should provide sufficient motivation, examples and applications. There should also be an introduction making the text comprehensible to a wider audience. This clearly distinguishes Lecture Notes from journal articles or technical reports which normally are very concise. Articles intended for a journal but too long to be accepted by most journals, usually do not have this "lecture notes" character.

2. In general SUMMER SCHOOLS and other similar INTENSIVE COURSES are held to present mathematical topics that are close to the frontiers of recent research to an audience at the beginning or intermediate graduate level, who may want to continue with this area of work, for a thesis or later. This makes demands on the didactic aspects of the presentation. Because the subjects of such schools are advanced, there often exists no textbook, and so ideally, the publication resulting from such a school could be a first approximation to such a textbook. Usually several authors are involved in the writing, so it is not always simple to obtain a unified approach to the presentation.

   For prospective publication in LNM, the resulting manuscript should not be just a collection of course notes, each of which has been developed by an individual author with little or no co-ordination with the others, and with little or no common concept. The subject matter should dictate the structure of the book, and the authorship of each part or chapter should take secondary importance. Of course the choice of authors is crucial to the quality of the material at the school and in the book, and the intention here is not to belittle their impact, but simply to say that the book should be planned to be written by these authors jointly, and not just assembled as a result of what these authors happen to submit.

   This represents considerable preparatory work (as it is imperative to ensure that the authors know these criteria before they invest work on a manuscript), and also considerable editing work afterwards, to get the book into final shape. Still it is the form that holds the most promise of a successful book that will be used by its intended audience, rather than yet another volume of proceedings for the library shelf.

3. Manuscripts should be submitted either to Springer's mathematics editorial in Heidelberg, or to one of the series editors. Volume editors are expected to arrange for the refereeing, to the usual scientific standards, of the individual contributions. If the resulting reports can be forwarded to us (series editors or Springer) this is very helpful. If no reports are forwarded or if other questions remain unclear in respect of homogeneity etc, the series editors may wish to consult external referees for an overall evaluation of the volume. A final decision to publish can be made only on the basis of the complete manuscript; however a preliminary decision can be based on a pre-final or incomplete manuscript. The strict minimum amount of material that will be considered should include a detailed outline describing the planned contents of each chapter.

   Volume editors and authors should be aware that incomplete or insufficiently close to final manuscripts almost always result in longer evaluation times. They should also be aware that parallel submission of their manuscript to another publisher while under consideration for LNM will in general lead to immediate rejection.

4. Manuscripts should in general be submitted in English. Final manuscripts should contain at least 100 pages of mathematical text and should always include

   – a general table of contents;
   – an informative introduction, with adequate motivation and perhaps some historical remarks: it should be accessible to a reader not intimately familiar with the topic treated;
   – a global subject index: as a rule this is genuinely helpful for the reader.

   Lecture Notes volumes are, as a rule, printed digitally from the authors' files. We strongly recommend that all contributions in a volume be written in LaTeX2e. To ensure best results, authors are asked to use the LaTeX2e style files available from Springer's web-server at

   ftp://ftp.springer.de/pub/tex/latex/svmultt1/ (for summer schools/tutorials).

   Additional technical instructions are available on request from: lnm@springer.com.

5. Careful preparation of the manuscripts will help keep production time short besides ensuring satisfactory appearance of the finished book in print and online. After acceptance of the manuscript authors will be asked to prepare the final LaTeX source files (and also the corresponding dvi-, pdf- or zipped ps-file) together with the final printout made from these files. The LaTeX source files are essential for producing the full-text online version of the book. For the existing online volumes of LNM see: www.springerlink.com/content/110312

   The actual production of a Lecture Notes volume takes approximately 12 weeks.

6. Volume editors receive a total of 50 free copies of their volume to be shared with the authors, but no royalties. They and the authors are entitled to a discount of 33.3% on the price of Springer books purchased for their personal use, if ordering directly from Springer.

7. Commitment to publish is made by letter of intent rather than by signing a formal contract. Springer-Verlag secures the copyright for each volume. Authors are free to reuse material contained in their LNM volumes in later publications: a brief written (or e-mail) request for formal permission is sufficient.

Addresses:

Professor J.-M. Morel, CMLA,
École Normale Supérieure de Cachan,
61 Avenue du Président Wilson,
94235 Cachan Cedex, France
E-mail: Jean-Michel.Morel@cmla.ens-cachan.fr

Professor F. Takens, Mathematisch Instituut,
Rijksuniversiteit Groningen, Postbus 800,
9700 AV Groningen, The Netherlands
E-mail: F.Takens@math.rug.nl

Professor B. Teissier,
Institut Mathématique de Jussieu,
UMR 7586 du CNRS,
Équipe "Géométrie et Dynamique",
175 rue du Chevaleret
75013 Paris, France
E-mail: teissier@math.jussieu.fr

For the "Mathematical Biosciences Subseries" of LNM:

Professor P.K. Maini, Center for Mathematical Biology
Mathematical Institute, 24-29 St Giles,
Oxford OX1 3LP, UK
E-mail: maini@maths.ox.ac.uk

Springer, Mathematics Editorial I, Tiergartenstr. 17,
69121 Heidelberg, Germany,
Tel.: +49 (6221) 487-8259
Fax: +49 (6221) 4876-8259
E-mail: lnm@springer.com